PROCESSING TECHNOLOGIES FOR MILK AND MILK PRODUCTS

Methods, Applications, and Energy Usage

Innovations in Agricultural and Biological Engineering

PROCESSING TECHNOLOGIES FOR MILK AND MILK PRODUCTS

Methods, Applications, and Energy Usage

Edited by
Ashok K. Agrawal, PhD
Megh R. Goyal, PhD, PE

Apple Academic Press Inc.	Apple Academic Press Inc.
3333 Mistwell Crescent	9 Spinnaker Way
Oakville, ON L6L 0A2 Canada	Waretown, NJ 08758 USA

© 2017 by Apple Academic Press, Inc.

First issued in paperback 2021

Exclusive worldwide distribution by CRC Press, a member of Taylor & Francis Group
No claim to original U.S. Government works

ISBN 13: 978-1-77-463663-3 (pbk)
ISBN 13: 978-1-77-188548-5 (hbk)

Library and Archives Canada Cataloguing in Publication

Processing technologies for milk and milk products : methods, applications, and energy usage / edited by Ashok K. Agrawal, PhD, Megh R. Goyal, PhD, PE.

(Innovations in agricultural and biological engineering)
Includes bibliographical references and index.
Issued in print and electronic formats.
ISBN 978-1-77188-548-5 (hardcover).--ISBN 978-1-315-20740-7 (PDF)

1. Dairy products. 2. Dairy products industry. I. Agrawal, Ashok K., 1942-, editor II. Goyal, Megh Raj, editor III. Series: Innovations in agricultural and biological engineering

| SF250.5.P76 2017 | 637 | C2017-902587-2 C2017-902588-0 |

Library of Congress Cataloging-in-Publication Data

Names: Agrawal, Ashok K., 1942- editor. | Goyal, Megh Raj, editor.
Title: Processing technologies for milk and milk products : methods, applications, and energy usage / editors: Ashok K. Agrawal (PhD), Megh R. Goyal, (PhD, PE).
Description: Waretown, NJ : Apple Academic Press, 2017. | Series: Innovations in agricultural and biological engineering | Includes bibliographical references and index.
Identifiers: LCCN 2017017447 (print) | LCCN 2017018586 (ebook) | ISBN 9781315207407 (ebook) | ISBN 9781771885485 (hardcover : alk. paper)
Subjects: LCSH: Dairy products. | Dairy products industry.
Classification: LCC SF250.5 (ebook) | LCC SF250.5 .P75 2017 (print) | DDC 636.2/142--dc23
LC record available at https://lccn.loc.gov/2017017447

Apple Academic Press also publishes its books in a variety of electronic formats. Some content that appears in print may not be available in electronic format. For information about Apple Academic Press products, visit our website at **www.appleacademicpress.com** and the CRC Press website at **www.crcpress.com**

CONTENTS

LIST OF CONTRIBUTORS

Ashok K. Agrawal, PhD
Professor and Head, Department of Dairy Engineering, College of Dairy Science and Food Technology, Chhattisgarh Kamdhenu Vishwavidyalaya (CGKV), Raipur – 492012, India; Mobile: +91-9827871020; E-mail: akagrawal.raipur@gmail.com; akagrwal_codt@yahoo.co.in

Shakeel Asgar, PhD
Associate Professor, Department of Dairy Technology, College of Dairy Science and Food Technology, Chhattisgarh Kamdhenu Vishwavidyalaya (CGKV), Raipur – 492012, India; Mobile: + 91-9826152155; E-mail: s.asgar@gmail.com

A. G. Bhadania, PhD
Professor and Head, Seth M.C. College of Dairy Science, Anand Agricultural University (AAU), Anand (Gujarat), India; Tel.: +91-02692–225838; Mobile: +91-9898611856; E-mail:bhadania@gmail.com; de@aau.in

P. Bisen, MTech
Former Student, Department of Dairy Engineering, College of Dairy Science and Food Technology, Chhattisgarh Kamdhenu Vishwa Vidyalaya (CGKV), Raipur – 492012, India; Mobile: +91-986890101; E-mail: pankaj.bisen007@gmail.com

Bhavesh B. Chavhan, MTech
Assistant Professor, Department of Dairy Engineering, College of Dairy Technology Maharashtra Animal and Fisheries Sciences University, Udgir – Nagpur, 413517, India; Mobile: +91-9168135153; E-mail: bhaveshchavhan@gmail.com

Santosh S. Chopde, MTech
Assistant Professor, Department of Dairy Engineering, College of Dairy Technology at Udgir, COVAS Campus, Kavalkhed Road, Udgi – 413517 (MS), India; Mobile: +91-9011799266; E-mail: santosh.der@gmail.com

Vandana Choubey, MTech
Junior Manager, C. G. Soya Products Pvt. Ltd. Kharora, Raipur (C.G.), 492001, India; Mobile: +91-9827928210; E-mail: choubey.vandana222@gmail.com

G. P. Deshmukh, MTech
PhD Research Scholar, Department of Dairy Engineering, National Dairy Research Institute, Bangalore, India; Mobile: +91-8087907871; E-mail: gajanannnn@gmail.com

Krishan Dewangan, BTech
Junior Research Fellow, Department of Dairy Engineering, College of Dairy Science and Food Technology, Chhattisgarh Kamdhenu Vishwavidyalaya (CGKV), Raipur – 492012, India; Mobile: +91-9039217758, E-mail: dewangankrishan@gmail.com

A. V. Dhotre, PhD
Assistant Professor, Department of Dairy Engineering, College of Dairy Technology, MAFSU-Nagpur, Warud (Pusad) – 445206, Dist. Yavatmal (M.S.), India; E-mail: anantdhotre@yahoo.com

Megh R. Goyal, PhD, PE
Retired Faculty in Agricultural and Biomedical Engineering from General Engineering Department, University of Puerto Rico – Mayaguez Campus; and Senior Technical Editor-in-Chief in Agriculture Sciences and Biomedical Engineering, Apple Academic Press, Inc., USA. E-mail: goyalmegh@gmail.com

Adarsh M. Kalla, MTech
Assistant Professor, Department of Dairy Engineering, Dairy Science College, Karnataka Veterinary Animal and Fisheries Sciences University, Mahagoan Cross, Kalaburagi – 585316, Karnataka, India; Mobile: +91-9035815171; E-mail: adarshkalla002@gmail.com

Anil K. Khare, PhD
Professor and Head, Department of Zoology, Dr. R.B. Government Naveen Girls College, Raipur – 492012, India; Mobile: +91-9826550084; E-mail: dr.anilganesh@yahoo.com

Archana Khare, PhD
Professor and Head, Department of Dairy Chemistry, College of Dairy Science and Food Technology, Chhattisgarh Kamdhenu Vishwa Vidyalaya, Raipur (C.G.) – 492006, India; Mobile: +91-8871625370; Tel.: +91-7712442339; E-mail: cgkvarchana@gmail.com

Rekha R. Menon, PhD
Senior Scientist, ICAR-National Dairy Research Institute, Bangalore – 560030, Karnataka, India; Mobile: +91-9916703069; E-mail: rekhamn@gmail.com

Bhushan D. Meshram, PhD (Pursuing)
Assistant Professor, College of Dairy Technology, Pusad – 445204 (M.S.) India; Mobile: +91-9423133120; E-mail: bdmeshram@gmail.com

Ruchi Patel, MTech
PhD Research Scholar, Dairy Engineering Department, Sheth M.C. College of Dairy Science, Anand Agricultural University, Anand – 388110, Gujarat, India; Mobile: +91-8347995739; E-mail: ruchi26patel@yahoo.com

Madhav R. Patil, PhD
Associate Dean, College of Dairy Technology, Udgir, COVAS Campus, Kavalkhed Road, Udgir – 413517 (M.S.), India; E-mail:cdtudgir@gmail.com

Dinesh Chandra Rai, PhD
Professor and Head, Department of Animal Husbandry and Dairying, Institute of Agricultural Sciences, Banaras Hindu University, Varanasi – 221005, Uttar Pradesh, India; Mobile: +91-9415256645, E-mail: dcrai.bhu@gmail.com

Devaraju Rajanna, MTech
Assistant Professor, Department of Dairy Engineering, Karnataka Veterinary Animal and Fisheries Sciences University, Dairy Science College, Mahagoan Cross, Kalaburagi – 585316, Karnataka, India; Mobile: +91-8088168268; E-mail: draj.raju20@gmail.com

C. Sahu, MTech
Assistant Professor, Department of Dairy Engineering, College of Dairy Science and Food Technology, Chhattisgarh Kamdhenu Vishwa Vidyalaya (CGKV), Raipur – 492012, India; Mobile: +91-8120343764; E-mail: ercsahu2003@yahoo.com

K. K. Sandey, PhD (Pursuing)
Assistant Professor, Department of Dairy Engineering, College of Dairy Science and Food Technology, Chhattisgarh Kamdhenu Vishwavidyalaya (CGKV), Raipur – 492012, India; Mobile: +91-8602006313; E-mail: kksandey@gmail.com

Geetesh Sinha, MTech
Senior Research Fellow, Department of Dairy Engineering, College of Dairy Science and Food Technology, Chhattisgarh Kamdhenu Vishwavidyalaya (CGKV), Raipur – 492012, India; Mobile: +91-8878165265; E-mail: geeteshsinha20@gmail.com

Janakkumar B. Upadhyay, PhD
Professor, Dairy Engineering Department, Sheth M.C. College of Dairy Science, Anand Agricultural University, Anand – 388110, Gujarat, India; Mobile: +91-7600530282; E-mail: drjbupadhyay.de.dsc@aau.in; jbupadhyay2004@yahoo.co.in

A. N. Vyahaware, MTech
Guest Lecturer, College of Dairy Technology, Warud – 445204 (M.S.) India; Mobile: +91-95459-23305; E-mail: aishwarya16.av@gmail.com

Prashant G. Wasnik, PhD
Associate Professor and Head, Department of Dairy Engineering, College of Dairy Technology, Pusad – 445204 (M.S.) India; Mobile: +91-9422866519; E-mail: pgwasnik@gmail.com

Ashok Kumar Yadav, MTech
PhD Research Scholar, Center of Food Science and Technology, Institute of Agricultural Sciences, Banaras Hindu University, Varanasi – 221005, Uttar Pradesh, India; Mobile: +91-9451716338; E-mail: ashokbhu99@gmail.com

LIST OF ABBREVIATIONS

ANN	artificial neural network
APP	atmospheric pressure plasma
ASD	adjustable speed drive
ASME	American Society of Mechanical Engineers
BHM	batch type halwasan making machine
C.O.P.	coefficient of performance
Ca	calcium
CBM	Continuous Basundi Making Machine
CFC	chlorofluorocarbon
CFL	compact fluorescent lamp
CHP	combined heat and power system
CIP	clean in place
CLA	conjugated linoleic acid
CP	cold plasma
CPV	conical process vat
CSLM	confocal scanning microscopy
CT	computed tomography
Cu	copper
DC	direct current
DM unit	demineralization unit
DMF	dual media filter
DOE	US Department of Energy
Dpi	dot per inch
DWB	dry weight basis
Eq.	equation
ET	electrical tomography
ETC	evacuated tube collectors
EU	European Union
FAO	Food and Agriculture Organization
FD	fractal dimension
FDA	Food and Drug Administration

FPC	flat plate collectors
FSSA	Food Safety and Standards Act
FSSAI	Food Safety and Standards Authority of India
FVR	fruit and vegetable residue
GAU	Gujarat Agricultural University
GB	Gigabyte
GC	gas chromatography
GHG	greenhouse gases
GM	region growing and merging
GMO	genetically modified organism
HACCP	Hazard Analysis and Critical Control Point
HAV	Hepatitis A virus
HHP	high hydrostatic pressure processing
HIPEF	high intensity pulsed electric field
HP	high pressure
HPHT	high pressure high temperature
HPLC	high pressure liquid chromatography
HPLT	high pressure low temperature
HPP	high pressure processing
HTST	high temperature short time
i.e.	exempli gratia
IBR	Indian Boiler Regulation
IEA	International Energy Agency
IIT	Indian Institute of Technology
ISSHE	Inclined Scraped Surface Heat Exchanger
LPG	liquefied petroleum gas
MB/Mb	mega bite
MCWC	microwave circulation water combination
MeV	mega electron volt
MFD	microwave freeze drying
MNRE	Ministry of New and Renewable Energy
MPa	mega pascal
MRI	magnetic resonance image
MS	monosonication
MTS	monothermosonication
MVD	microwave vacuum drying

MW	microwave
MW	megawatt
NaCAS	sodium caseinate
NDDB	National Dairy Development Board
NDRI	National Dairy Research Institute
NH_4	ammonia
NIR	near infra red
NRFI	normalized reference fruit image
OHC	oxygen holding capacity
ORAL	oxygen radical absorbance capacities
PA	particle analysis
PATP	pressure assisted thermal processing
PATS	pressure assisted thermal sterilization
PCA	principal component analysis
PDCAAS	protein digestibility corrected amino acid score
PE	polyethylene
PEF	pulsed electric field
PHE	plate heat exchanger
PI	pixel intensity
PL	pulsed light
PP	polypropylene
Ppi	pixels per inch
ppm	parts per million
PS	polystyrene
PUFA	poly-unsaturated fatty acid
PV	solar photovoltaic
RCT	rennet coagulation time
RF	radio frequency
RGB	Red Green Blue
RMRD	Raw Milk Reception Doc
RO	reverse osmosis
RO unit	reverse osmosis unit
ROI	region of interest
ROS	reactive oxygen species
rpm	revolutions per minute
RPM	rotations per minute

RTD	resistance temperature detector
RTE	ready to eat
Se	selenium
SLS	solar street-lighting system
SM	region splitting and merging
SMP	skim milk powder
SMUF	simulated milk ultra-filtrate
SNF	solids not fat (Milk)
Spi	squares per inch
SS	stainless steel
SSHE	scraped surface heat exchanger
SV	slab voxel
SWH	solar water heater
SWHS	solar water heater system
T.V.	texturized vegetable
TFSSHE	thin film scraped surface heat exchanger
TIDP	traditional Indian dairy products
TR	tone
TS	thermosonication
TS	total solid
U-value	overall heat transfer coefficient
UHP	ultra high pressure processing
UHT	ultra heat treatment
UHT	ultra high temperature
US	ultrasonication
USDA	United States Department of Agriculture
UV	ultraviolet
V	wind velocity (m/s)
VFD	variable frequency drive
Vit.	vitamin
VSD	variable speed drive
WHC	water holding capacity
WPI	whey protein isolate
Zn	zinc

LIST OF SYMBOLS

”	inches
@	at
%	percentage
°	degree
°C	degree celsius
A	rotor area (m^2)
C. jejuni	*Campylobactor jejuni*
cfu/g	colony-forming units per gram
Cs	cesium
g	gram
K	kelvin
K	potassium
kg	kilogram
kGy	kilogray
KJ/kg	kilo Joules per kilo gram
km/h	kilometer per hour
kPa	kilo Pascal
KV/cm	kilovolt per centimeter
kW	kilowatt
kWh	kilowatt-hour
L. monocytogenes	*Listeria monocytognes*
m	meter
m/s	meter per second
Mg	magnesium
MHz	mega hertz
min	minute
MJ	mega joule
mJ	milli joule
mm	millimeter
Na	sodium

P	phosphorus
P	power
W	watt
W/cm^2	watt per square centimeter
w/w	weight-to-weight ratio
Y. enterocolitica	*Yersiniaenterocolitica*
μ	Mu
μΦ	microfarad
μμ	micrometer
μs	microsecond
ρ	air density (kg/m^3)

FOREWORD BY UMESH K. MISHRA

The present scenario of dairying in the world is promising in many ways. The world's milk production has reached about 800 million tons per annum, which will continue to increase in the future also. The entire array of dairying, particularly milk processing and dairy product manufacturing, has achieved exceptional growth momentum throughout the world. Efficient production, processing, and marketing of milk is largely credited for this transformation. In developing countries, a substantial quantity of milk is now also processed in organized processing plants, thus assuring a quality milk supply to masses. The step up in milk production is essential to meet the increasing demand generated by increasing populations and income levels. There is an urgent need to take care of the health of children and other consumers by offering them safely processed milk and milk products.

In line with the modern industrial trends, new processes and corresponding new equipment are being introduced. The development of highly sensitive measuring and control devices have made it possible to incorporate automatic operation with high degree of mechanization to meet the huge demand of quality milk and milk products. In order to encourage dairying on economic terms and to stimulate awareness about emerging techniques, there is a need to disseminate the unseen knowledge among all stakeholders of dairy and food industry. The technical keenness in dairy technocrats will help in better utilization of existing equipment and better planning of future equipment, instruments, process controls and automation. It would enable them to conserve various processing utilities, thereby facilitating production with minimum environmental impact.

This book has the ready-made solution to quench the quest of qualified dairy personnel. It would fulfill a thirst for knowledge about recent developments in processing of milk and dairy products. This compendium deals with the salient aspects of the milk processing and dairy product manufacturing. The cutthroat competition of modern era necessitates the knowledge of fundamentals of emerging processes, acquaintance with innovative equipment, and self-assurance regarding instruments and automatic control systems for commercial processing of milk and dairy products.

The task of providing comprehensive information about pertinent topics, more ubiquitous in the future milk and dairy product industry, is nicely done. The authors and editors deserve accolades for this stupendous work. Various authors have distinguished themselves in their interest in the engineering and technology of milk and milk product processing.

I am confident that this book will be very useful for students, as a reference book for researchers, and as a source book for teachers. It would be equally useful for all professionals working in various capacities in the milk and dairy product manufacturing industry.

Umesh K. Mishra, PhD
Vice Chancellor
Chhattisgarh Kamdhenu University – Durg
Krishak Nagar, Raipur–492012,
Chhattisgarh, India
Tel. +91-7882623462
E-mail: vccgkv2012@gmail.com

PREFACE 1 BY ASHOK K. AGRAWAL

This volume, *Processing Technologies for Milk and Dairy Products: Methods, Applications, and Energy Usage,* is compiled with the view to fill the gap due to non-availability of relevant books having recent information on these subjects. The milk processing and dairy product preparation is the sunrise sector of food processing, and its scope is expanding with leaps and bounds due to rapid urbanization, improving life styles and higher paying capacity. The expected growth in milk consumption is based mostly on the number of growing consumers who are earning more than US$ 5 per day, and this is increasing. The speed of growth has to be higher than the sum of the inflation rate and percentage of population increase. The demand for quality milk products is increasing throughout the world. Food patterns are changing from eating plant protein to animal protein due to increasing incomes.

The so-called dairy basket now contains some non-traditional dairy products also in demand by the common consumer. This is particularly true for developing countries, which are witnessing popularity of Western products along with their own conventional dairy products. Coordinated efforts from all stakeholders are needed to sustain and develop this trend in the future also. This book is compiled with a view to meet the key requirement of fulfilling the technical competency in budding technocrats of dairy industry. This book contains particularly those topics that would inculcate inspiration among its readers by exposing them to recent advances in the field of the milk and dairy product processing.

The processing of milk and conversion of raw milk into some dairy products adds significant value to the output from the dairy processing plants. The broad objective of this book is to provide readers with an improved understanding about processing of milk and manufacturing of the dairy products. This book is divided into three sections: (i) Innovative Techniques in Dairy Engineering; (ii) Processing Methods and Their Applications in the Dairy Industry; and (iii) Energy Usage in Dairy Engineering: Sources, Conservation, and Requirements.

The role of processing in dairy industry involves the designing of the equipment/storage structure. During preparation of dairy products from raw milk, many different types of engineering processes are involved. There are various established and promising techniques for completion of these processes. In dairy and food processing, quality of food is most important because it is deciding factor for its on-going demand. Looking to this, a separate chapter on uses and potential of digital image analysis as a tool for food quality evaluation is included. The chapter on passivation (a method to ensure quality of dairy and food processing equipment) shows the necessity and logic behind the pre-treatment given to processing equipment. The chapter on technology of protein rich vegetable-based formulated foods deals with processing of those food products that are gaining importance as some wellness scientists have pointed out drawbacks associated with non-vegetarian foods and have advocated adoption of vegan foods.

There are some types of equipment that are often used during manufacturing of various dairy products. Hence, a critical review is made on utilization of scraped surface heat exchanger in manufacturing of various dairy products. Another emerging technique on high-pressure processing is discussed by two distinguished authors. The new discoveries in various field of science have their inbuilt applications in food processing. The emergence of the novel thermal technologies, for example, microwave heating, allows producing high-quality products with improvements in terms of heating efficiency and, consequently, in energy savings. In the near future, more and more thermal processes in the dairy product manufacturing field may be diverted toward the use of microwaves. Hence, a discussion on microwave is also included.

In today's world, quality has to be maintained for a huge quantity of food. Better and newer techniques are being continuously evolved that are particularly capable of tackling and storing millions of liters of milk. Hence, a chapter is incorporated to highlight this aspect it collects some information about mass milk storage structures called milk silos. Milk is required to be kept in a safe condition from its production up to consumption.

In this book, a chapter is devoted for utilization of alternate sources of energy for processing of milk and dairy products while the other two chapters deal with its conservation for achieving lower cost of milk processing. Upon identifying the various outlets of energy in milk processing and dairy product manufacturing, 'pumps' would come out at the top. An article is included to focus the reduction of power/energy consumption by pumps.

Renewable energy sources are continuously replenished by natural processes, for example, solar energy, wind energy, bio-energy, hydropower, etc. There are number of renewable energy sources that can be easily integrated into the dairy industry for energy conservation. Considering this as a prominent matter, a chapter on renewable energy is also included. Gone are the days when water was available in plenty and free of cost. For a layman, it may be unbelievable that in a dairy plant, quantity of water handled is much more than quantity of milk processed. Hence, its conservation is important from not only from a quantity-wise perspective but also energywise. In dairy processing beside equipment, utilities like steam, refrigerant, vacuum, compressed air, are topics that need attention, as a major cost of processing is attributed to this aspect. Conserving consumption of this would ultimately result in huge savings of energy, time and operation period, and labor. Cooling of milk is another facet of dairying that needs a sincere approach to maintain quality of milk.

I hope this book will fulfill the expectations of all dairy professionals who are constantly working for betterment of human life by ensuring the availability of milk as nature's nectar, which is time and again proved only next to elixir. The very purpose of the publication of this compilation would be suitably accomplished if it caters to the need of budding scientists, engineers and technologists working in the field of processing of milk and dairy products.

Completion of this book would have been impossible without significant input from many colleagues and fellow dairy professionals. The opportunity to collaborate with these scientists and engineers has made this a memorable experience. The continuous support of Dr. Megh R. Goyal deserves special mention and lots of appreciation throughout the compilation and editing of this book. Dr. Goyal has excelled as a devoted editor of Apple Academic Press, Inc. (AAP) to bring out quality book volumes under the book series *Innovations in Agricultural and Biological Engineering*.

Ours is a profession with a future, and it needs professional books in all focus areas. I hope that my colleagues will seriously consider publishing in this book series as I find AAP to be a world-renowned source.

—*Ashok K. Agrawal, PhD*

PREFACE 2 BY MEGH R. GOYAL

On March 30 of 1983, I was driving on Puerto Rico Highway 52 to visit some historic places in Cayey. While driving, I drank the chocolate milk and threw the empty bottle on the road. My five-year-old son (today, he is 38 years old) requested me to stop the car to pick up the empty bottle from the road and added that "*we should not contaminate our beautiful island.*" At that time, there were no recycling plants for solid waste in Puerto Rico. The attention by my child helped me to love my mother planet by conserving our natural resources. Therefore, in my introduction to this book volume, I want to emphasize the importance of treating solid/liquid/gas wastes that are generated in the operations of milk and milk products. One must follow all local regulations for waste disposal, taking into consideration human health. I invite my fellow colleagues to suggest book volumes on this focus area under my book series. Apple Academic Press, Inc., has identified this topic as one of the priority areas to help save our planet from degradation and perish. We, engineers, are meant to help all persons at all levels and at all places. Can you help us?

Agricultural waste treatment is the treatment of waste produced in the course of agricultural activities. Agriculture is a highly intensified industry in many parts of the world, producing a range of waste requiring a variety of treatment technologies and management practices. Farms with large livestock and poultry operations, such as factory farms, can be a major source of waste. In the United States, these facilities are called concentrated animal feeding operations or confined animal feeding operations (CAFOs) and are being subjected to increasing US government regulations. Agricultural waste also includes waste from dairy farms, dairy industries, and processing industries involved in milk and milk products.

https://en.wikipedia.org/wiki/Dairy#Waste_disposal mentions that "*in countries where cows are grazed outside year-round, there is little waste disposal to deal with. The most concentrated waste is at the milking shed, where the animal waste may be liquefied (during the water-washing process) or left in a more solid form, either to be returned to be used on farm ground as organic fertilizer. In the associated milk processing factories, most of the waste is washing water that is treated, usually by composting, and spread on farm fields in either liquid or solid form. This is much different from half a century ago, when the main products were butter, cheese and casein, and the rest of the milk had to be disposed of as waste (sometimes as animal feed). In dairy-intensive areas, various methods have been proposed for disposing of large quantities of milk. Large application rates of milk onto land, or disposing in a hole, is problematic as the residue from the decomposing milk will block the soil pores and thereby reduce the water infiltration rate through the soil profile. As recovery of this effect can take time, any land based application needs to be well managed and considered. Other waste milk disposal methods commonly employed include solidification and disposal at a solid waste landfill, disposal at a wastewater treatment plant, or discharge into a sanitary sewer."*

"The constituents of animal wastewater typically contain: Strong organic content — much stronger than human sewage; High concentration of solids; High nitrate and phosphorus content; Antibiotics; Synthetic hormones; Often high concentrations of parasites and their eggs; Spores of Cryptosporidium (a protozoan) resistant to drinking water treatment processes; Spores of Giardia; Human pathogenic bacteria such as Brucella and Salmonella. Milking parlor wastes are often treated in admixture with human sewage in a local sewage treatment plant. This ensures that disinfectants and cleaning agents are sufficiently diluted and amenable to treatment. Running milking wastewaters into a farm slurry lagoon is a possible option although this tends to consume lagoon capacity very quickly. Land spreading is also a treatment option," according to https://en.wikipedia.org/wiki/Agricultural_wastewater_treatment#Milking_parlor_.28dairy_farming.29_wastes.

In dairy industries, a large amount of waste is discarded in the form of diluted milk that may have whey liquid, detergents, sanitizers and other chemicals used for the purpose of sterilization. Clean-in-Place (CIP) is also a process of cleaning the place, where a huge amount of contaminated water is released. Along with these drippings, problem and accidental leakage from the packaging process and from the CIP respectively occur. This leakage will end up in the sewer system and create a lot of problems.

Whey is one of main product left after manufacturing of cheese. It is watery in nature and left after the separation of curd when the milk had been coagulated with enzyme or acid. The generation is quite high, approximately 9 liters of whey per one kg of cheese. Due to higher Biological Oxygen Demand (BOD: approx. 40 g/l), it is not advisable to dispose of this liquid without pretreatment. It will trigger diseases and health risks. The higher value of BOD is due to the presence of sugar, called lactose. Although the concentration of lactose is around 5%, it can cause damages. However, industries are opting for new techniques of utilizing the whey or discarding it after pretreatment. But some losses still occur in small-scale industries, where rules and regulations are not followed properly. In the current scenario, different methods have been developed to dry the whey to blend with edible food to prepare food at lesser cost. It can be blended with different categories of foods. The technique of reverse osmosis (RO) can also be utilized to manufacture a protein-based product.

Another area for whey utilization is the preparation of single cell protein (SCP). The SCP and other protein concentrates are gaining value in the market for their fortification in the value addition of different products. The application of whey utilization is quite large and includes a number of techniques for producing products: fermentation method to produce ethyl alcohol and lactic acid; formation of concentrate (whey protein, dried whey); and pasteurization technique to process whey cream and sweet whey. Membrane filtration has also been designed to separate compounds from whey. A filtration technique called ultra-filtration can be used for the segregation of proteins from the solution, which have mainly lactose. Similarly number of techniques like evaporation, spray drying, and crystallization are used for the purpose of separation of valuable nutrients like protein, minerals matter, and other chemical compounds. But these techniques are quite expensive. Therefore industries are looking for some innovative, affordable, and economical techniques. These days, ultra-filtration and RO are gaining importance for the purpose of separation. Concentrate of whey protein is known for its nutritive value throughout the world.

At the 49th annual meeting of the Indian Society of Agricultural Engineers at Punjab Agricultural University (PAU) during February 22–25 of 2015, a group of ABEs and FEs convinced me that there is a dire need to publish book volumes on the focus areas of agricultural and biological engineering (ABE). This is how the idea was born for the new book series titled *Innovations in Agricultural and Biological Engineering*. This book *Processing Technologies for Milk and Dairy Products: Methods, Applications, and Energy Usage,* is twelfth volume in this book series, and it contributes to the ocean of knowledge on dairy engineering.

The contribution by the cooperating authors to this book volume has been most valuable in the compilation. Their names are mentioned in each chapter and in the list of contributors. This book would not have been written without the valuable cooperation of these investigators, many of whom are renowned scientists who have worked in the field of food engineering throughout their professional careers. I am glad to introduce Dr. Ashok K Agrawal, who is Professor and Head of Department of Dairy Engineering in the College of Dairy Science and Food Technology at Chhattisgarh Kamdhenu Vishwavidyalaya (CGKV), Raipur- India. He is a professor/ researcher and is specialized in dairy and dairy products. Without his support, and leadership qualities as lead editor of this book and his extraordinary work on dairy engineering applications, readers would not have this quality publication.

I will like to thank editorial staff, Sandy Jones Sickels, Vice President, and Ashish Kumar, Publisher and President at Apple Academic Press, Inc., for making every effort to publish the book. Special thanks are due to the AAP production staff for the quality production of this book.

I request that readers offer their constructive suggestions that may help to improve the next edition.

I express my deep admiration to Subhadra D. Goyal for understanding and collaboration during the preparation of this book volume.

Can anyone live without food or milk? As an educator, there is a piece of advice to one and all in the world: *"Permit that our almighty God, our Creator, provider of all and excellent Teacher, feed our life with Healthy Milk and Milk Products and His Grace; and Get married to your profession."*

—Megh R. Goyal, PhD, PE
Senior Editor-in-Chief

WARNING/DISCLAIMER

PLEASE READ CAREFULLY

The goal of this book volume, *Processing Technologies for Milk and Dairy Products: Methods, Applications, and Energy Usage,* is to guide the world community on how to manage efficiently for technology available for different processes of milk and milk products.

The editors, the contributing authors, the publisher, and the printer have made every effort to make this book as complete and as accurate as possible. However, there still may be grammatical errors or mistakes in the content or typography. Therefore, the contents in this book should be considered as a general guide and not a complete solution to address any specific situation in food engineering. For example, one type of dairy technology does not fit all cases in dairy engineering/science/technology.

The editors, the contributing authors, the publisher, and the printer shall have neither liability nor responsibility to any person, any organization or entity with respect to any loss or damage caused, or alleged to have caused, directly or indirectly, by information or advice contained in this book. Therefore, the purchaser/reader must assume full responsibility for the use of the book or the information therein.

The mention of commercial brands and trade names are only for technical purposes. It does not mean that a particular product is endorsed over another product or equipment not mentioned. The editors, cooperating authors, educational institutions, and the publisher, Apple Academic Press, Inc., do not have any preference for a particular product.

All web-links that are mentioned in this book were active on December 31, 2016. The editors, the contributing authors, the publisher, and the printing company shall have neither liability nor responsibility, if any of the web-links is inactive at the time of reading of this book.

ABOUT THE LEAD EDITOR

Ashok Kumar Agrawal, PhD, is a distinguished engineer, researcher, and professor of dairy engineering at the College of Dairy Science and Food Technology at Chhattisgarh Kamdhenu University–Durg, Raipur, in India, where he heads the Dairy Engineering Department. He also served as Dean of the college. Dr. Agrawal is also acting as Principal Investigator of the National Agricultural Development Project on the operation of small milk processing plants with the utilization of solar energy. He is a life member of the Indian Dairy Engineers Association, the Indian Society of Agricultural Engineers, and the Indian Dairy Association.

His major areas of research are parboiling and storage studies of granular and horticultural crops. He is engaged in the teaching of various dairy and food engineering courses to undergraduate, postgraduate, and doctoral students. He has supervised about 20 students for their research work leading to advanced degrees in dairy engineering and agricultural process and food engineering.

He has edited the proceedings of the National Conference on Dairy Engineers for the Cause of Rural India and the 5th Convention of the Indian Dairy Engineers Association. He has published about 60 research papers in national and international journals; 25 research articles in national and international conferences/symposiums as well as some popular articles. Dr. Agrawal is also a reviewer for some several journal, including the Indian Journal of Dairy Science and the Journal of Food Processing and Preservation. He was conferred a Best Teacher Award by his own university and has received awards for his presentations at many national and international conferences.

He obtained his BTech (Agricultural Engineering) degree from the College of Agricultural Engineering, J. N. Agricultural University, Jabalpur, India; his MTech (Postharvest Engineering) degree from the

Indian Institute of Technology, Kharagpur, India; and his PhD from the Indian Institute of Technology in Dairy and Food Engineering. In his PhD research, he worked on development of double wall basket centrifuge for pressing and chilling for batch production of Indian cheese (paneer).

ABOUT THE SENIOR EDITOR-IN-CHIEF

Megh R. Goyal, PhD, PE
Retired Professor in Agricultural and Biomedical Engineering, University of Puerto Rico, Mayaguez Campus Senior Acquisitions Editor, Biomedical Engineering and Agricultural Science, Apple Academic Press, Inc.

Megh R. Goyal, PhD, PE, is a Retired Professor in Agricultural and Biomedical Engineering from the General Engineering Department in the College of Engineering at the University of Puerto Rico–Mayaguez Campus; and Senior Acquisitions Editor and Senior Technical Editor-in-Chief in Agriculture and Biomedical Engineering for Apple Academic Press, Inc.

He has worked as a Soil Conservation Inspector and as a Research Assistant at Haryana Agricultural University and Ohio State University. He was the first agricultural engineer to receive the professional license in Agricultural Engineering in 1986 from the College of Engineers and Surveyors of Puerto Rico. On September 16, 2005, he was proclaimed as "Father of Irrigation Engineering in Puerto Rico for the Twentieth Century" by the ASABE, Puerto Rico Section, for his pioneering work on micro irrigation, evapotranspiration, agroclimatology, and soil and water engineering. During his professional career of 45 years, he has received many prestigious awards, including Scientist of the Year, Blue Ribbon Extension Award, Research Paper Award, Nolan Mitchell Young Extension Worker Award, Agricultural Engineer of the Year, Citations by Mayors of Juana Diaz and Ponce, Membership Grand Prize for ASAE Campaign, Felix Castro Rodriguez Academic Excellence, Rashtrya Ratan Award and Bharat Excellence Award and Gold Medal, Domingo Marrero Navarro Prize, Adopted son of Moca, Irrigation Protagonist of UPRM, Man of Drip Irrigation by Mayor of Municipalities of Mayaguez/Caguas/Ponce, and Senate/Secretary of Agriculture of ELA, Puerto Rico. On March 1, 2017:

Tamil Nadu Agricultural University conferred on him "Distinguished Scientist Award" at National Congress on Sustainable Microirrigtion by Water Technology Centre.

A prolific author and editor, he has written more than 200 journal articles and textbooks and has edited over 48 books including: *Elements of Agroclimatology* (Spanish) by UNISARC, Colombia; and two bibliographies on drip irrigation.

He received his BSc degree in engineering from Punjab Agricultural University, Ludhiana, India; his MSc and PhD degrees from Ohio State University, Columbus; and his Master of Divinity degree from Puerto Rico Evangelical Seminary, Hato Rey, Puerto Rico, USA.

Apple Academic Press, Inc. (AAP) has published his books, namely, *Management of Drip/Trickle or Micro Irrigation*, and *Evapotranspiration: Principles and Applications for Water Management*, his ten-volume set on *Research Advances in Sustainable Micro Irrigation*. During 2016–2020, AAP will be publishing book volumes on emerging technologies/issues/challenges under two book series, *Innovations and Challenges in Micro Irrigation*, and *Innovations in Agricultural and Biological Engineering*. Readers may contact him at: goyalmegh@gmail.com.

BOOK ENDORSEMENTS

Dairy and food engineering combines the fundamentals of basic branches of engineering for specific applications to the dairy and food industry. The present book is an amalgamation of review chapters on various emerging topics of milk processing and dairy products manufacturing. The comprehensive text, written by various technocrats, is very useful for all persons engaged in various facets of dairying. The source book of this type is needed urgently, particularly for dairy professionals. It would fill the information gap between textbooks and findings of present investigations.

—S. P. Agarwala, PhD
Ex-Head, Principal Scientist and Emeritus Scientist,
Dairy Engineering Division,
National Dairy Research Institute, Karnal – 132001, India

Agricultural processing and food engineering is the integral portion of agricultural engineering curricula. The undergraduate program is comprised of basic courses while the postgraduate program consists of advanced courses of dairy and food engineering. This volume presents several chapters that would be helpful in understanding the recent developments in this emerging field. This is a well-written book that gives a lucid exposition and addresses the needs of the professionals of both the academic and the industrial obligations. I have no doubt in saying that the undergraduate and post-graduate courses of the various universities engaged in fields like agricultural processing, milk and milk product processing, food processing, etc. would find this book as a useful contribution toward dissemination of knowledge. I extend my warm greetings and best wishes for all success.

—V. K. Pandey, PhD
Dean, S.V. College of Agric. Eng. and Tech. and Research Station,
Raipur – 492012, India

The book edited by Dr. A. K. Agrawal and Dr. M. R. Goyal has covered various aspects of dairy and food engineering in a most useful manner for professionals, entrepreneurs, scientists and research scholars who are working in this field. This book will be useful as a reference book for future dairy and food engineering research. In the production of dairy and food products, newer and better techniques are continuously evolved. This book presents information on these aspects. I extend my warm greetings and best wishes to the publisher for future success.

—Sudhir Uprit, PhD
Dean, College of Dairy Science and Food Technology,
Raipur – 492012, India

This book is a valuable asset for all dairy plants engineers, managers, professionals as well as for the decision makers working in dairy and food processing industries. This book gives a glimpse of advance knowledge in various topics of dairy and food processing. It would serve as a reference book for all investigators working in these fields.

—B. P. Shah, PhD
Former Principal and Dean,
SMC College of Dairy Science, Anand, India

BOOKS ON AGRICULTURAL AND BIOLOGICAL ENGINEERING FROM APPLE ACADEMIC PRESS, INC.

Management of Drip/Trickle or Micro Irrigation
Megh R. Goyal, PhD, PE, Senior Editor-in-Chief

Evapotranspiration: Principles and Applications for Water Management
Megh R. Goyal, PhD, PE, and Eric W. Harmsen, Editors

Book Series: Research Advances in Sustainable Micro Irrigation
Senior Editor-in-Chief: Megh R. Goyal, PhD, PE
  Volume 1: Sustainable Micro Irrigation: Principles and Practices
  Volume 2: Sustainable Practices in Surface and Subsurface Micro Irrigation
  Volume 3: Sustainable Micro Irrigation Management for Trees and Vines
  Volume 4: Management, Performance, and Applications of Micro
            Irrigation Systems
  Volume 5: Applications of Furrow and Micro Irrigation in Arid and
            Semi-Arid Regions
  Volume 6: Best Management Practices for Drip Irrigated Crops
  Volume 7: Closed Circuit Micro Irrigation Design: Theory and Applications
  Volume 8: Wastewater Management for Irrigation: Principles and Practices
  Volume 9: Water and Fertigation Management in Micro Irrigation
Volume 10: Innovation in Micro Irrigation Technology

Book Series: Innovations and Challenges in Micro Irrigation
Senior Editor-in-Chief: Megh R. Goyal, PhD, PE
Volume 1: Principles and Management of Clogging in Micro Irrigation
Volume 2: Sustainable Micro Irrigation Design Systems for Agricultural
          Crops: Methods and Practices
Volume 3: Performance Evaluation of Micro Irrigation Management:
          Principles and Practices
Volume 4: Potential Use of Solar Energy and Emerging Technologies in
          Micro Irrigation

Volume 5: Micro Irrigation Management: Technological Advances and
          Their Applications
Volume 6: Micro Irrigation Engineering for Horticultural Crops: Policy
          Options, Scheduling, and Design
Volume 7: Micro Irrigation Scheduling and Practices
Volume 8: Engineering Interventions in Sustainable Trickle Irrigation:
          Water Requirements, Uniformity, Fertigation, and Crop
          Performance

Book Series: Innovations in Agricultural and Biological Engineering
Senior Editor-in-Chief: Megh R. Goyal, PhD, PE
• Dairy Engineering: Advanced Technologies and their Applications
• Developing Technologies in Food Science: Status, Applications, and
  Challenges
• Emerging Technologies in Agricultural Engineering
• Engineering Interventions in Agricultural Processing
• Engineering Interventions in Foods and Plants
• Engineering Practices for Agricultural Production and Water
  Conservation: An Interdisciplinary Approach
• Flood Assessment: Modeling and Parameterization
• Food Engineering: Modeling, Emerging Issues and Applications.
• Food Process Engineering: Emerging Trends in Research and Their
  Applications
• Food Technology: Applied Research and Production Techniques
• Modeling Methods and Practices in Soil and Water Engineering
• Novel Dairy Processing Technologies: Techniques, Management, and
  Energy Conservation
• Processing Technologies for Milk and Milk Products: Methods,
  Applications, and Energy Usage
• Soil and Water Engineering: Principles and Applications of Modeling
• Soil Salinity Management in Agriculture: Technological Advances
  and Applications
• State-of-the-Art Technologies in Food Science: Human Health,
  Emerging Issues and Specialty Topics
• Sustainable Biological Systems for Agriculture: Emerging Issues in
  Nanotechnology, Biofertilizers, Wastewater, and Farm Machines
• Technological Interventions in Dairy Science: Innovative Approaches
  in Processing, Preservation, and Analysis of Milk Products
• Technological Interventions in Management of Irrigated Agriculture
• Technological Interventions in the Processing of Fruits and Vegetables

EDITORIAL

Apple Academic Press, Inc., (AAP) is publishing book volumes in the specialty areas as part of the *Innovations in Agricultural and Biological Engineering* book series over a span of 8 to 10 years. These specialty areas have been defined by *American Society of Agricultural and Biological Engineers* (http://asabe.org).

The mission of this series is to provide knowledge and techniques for Agricultural and Biological Engineers (ABEs). The series aims to offer high-quality reference and academic content in Agricultural and Biological Engineering (ABE) that is accessible to academicians, researchers, scientists, university faculty, and university-level students and professionals around the world. The following material has been edited/modified and reproduced below *"Goyal, Megh R., 2006. Agricultural and biomedical engineering: Scope and opportunities. Paper Edu_47 at the Fourth LACCEI International Latin American and Caribbean Conference for Engineering and Technology (LACCEI' 2006): Breaking Frontiers and Barriers in Engineering: Education and Research by LACCEI University of Puerto Rico – Mayaguez Campus, Mayaguez, Puerto Rico, June 21–23."*

WHAT IS AGRICULTURAL AND BIOLOGICAL ENGINEERING (ABE)?

"Agricultural Engineering (AE) involves application of engineering to production, processing, preservation and handling of food, fiber, and shelter. It also includes transfer of technology for the development and welfare of rural communities," according to http://isae.in." *ABE is the discipline of engineering that applies engineering principles and the fundamental concepts of biology to agricultural and biological systems and tools, for the safe, efficient and environmentally sensitive production, processing, and management of agricultural, biological, food, and natural resources systems,"* according to http://asabe.org. *"AE is the branch of engineering involved with the design of farm machinery, with soil management,*

land development, and mechanization and automation of livestock farming, and with the efficient planting, harvesting, storage, and processing of farm commodities," definition by: http://dictionary.reference.com/browse/agricultural+engineering.

"AE incorporates many science disciplines and technology practices to the efficient production and processing of food, feed, fiber and fuels. It involves disciplines like mechanical engineering (agricultural machinery and automated machine systems), soil science (crop nutrient and fertilization, etc.), environmental sciences (drainage and irrigation), plant biology (seeding and plant growth management), animal science (farm animals and housing) etc.," by: http://www.ABE.ncsu.edu/academic/agricultural-engineering.php.

"According to https://en.wikipedia.org/wiki/Biological_engineering: *"BE (Biological engineering) is a science-based discipline that applies concepts and methods of biology to solve real-world problems related to the life sciences or the application thereof. In this context, while traditional engineering applies physical and mathematical sciences to analyze, design and manufacture inanimate tools, structures and processes, biological engineering uses biology to study and advance applications of living systems."*

SPECIALTY AREAS OF ABE

Agricultural and Biological Engineers (ABEs) ensure that the world has the necessities of life including safe and plentiful food, clean air and water, renewable fuel and energy, safe working conditions, and a healthy environment by employing knowledge and expertise of sciences, both pure and applied, and engineering principles. Biological engineering applies engineering practices to problems and opportunities presented by living things and the natural environment in agriculture. BA engineers understand the interrelationships between technology and living systems, have available a wide variety of employment options. *"ABE embraces a variety of following specialty areas,"* http://asabe.org. As new technology and information emerge, specialty areas are created, and many overlap with one or more other areas.

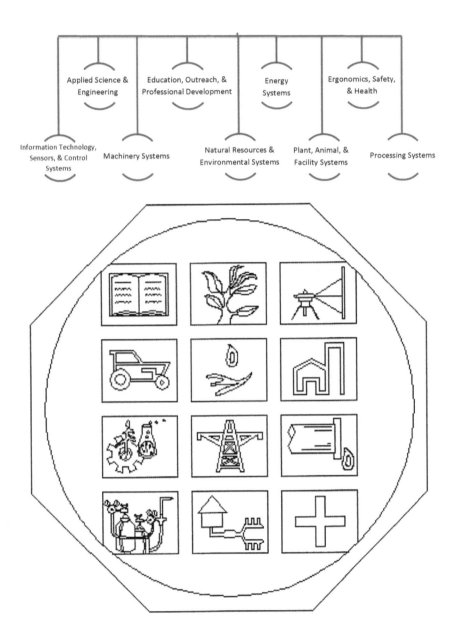

1. **Aquacultural Engineering**: ABEs help design farm systems for raising fish and shellfish, as well as ornamental and bait fish. They specialize in water quality, biotechnology, machinery, natural resources, feeding and ventilation systems, and sanitation. They

seek ways to reduce pollution from aquacultural discharges, to reduce excess water use, and to improve farm systems. They also work with aquatic animal harvesting, sorting, and processing.

2. **Biological Engineering** applies engineering practices to problems and opportunities presented by living things and the natural environment.

3. **Energy:** ABEs identify and develop viable energy sources – biomass, methane, and vegetable oil, to name a few – and to make these and other systems cleaner and more efficient. These specialists also develop energy conservation strategies to reduce costs and protect the environment, and they design traditional and alternative energy systems to meet the needs of agricultural operations.

4. **Farm Machinery and Power Engineering**: ABEs in this specialty focus on designing advanced equipment, making it more efficient and less demanding of our natural resources. They develop equipment for food processing, highly precise crop spraying, agricultural commodity and waste transport, and turf and landscape maintenance, as well as equipment for such specialized tasks as removing seaweed from beaches. This is in addition to the tractors, tillage equipment, irrigation equipment, and harvest equipment that have done so much to reduce the drudgery of farming.

5. **Food and Process Engineering:** Food and process engineers combine design expertise with manufacturing methods to develop economical and responsible processing solutions for industry. Also food and process engineers look for ways to reduce waste by devising alternatives for treatment, disposal and utilization.

6. **Forest Engineering**: ABEs apply engineering to solve natural resource and environment problems in forest production systems and related manufacturing industries. Engineering skills and expertise are needed to address problems related to equipment design and manufacturing, forest access systems design and construction; machine-soil interaction and erosion control; forest operations analysis and improvement; decision modeling; and wood product design and manufacturing.

7. **Information and Electrical Technologies Engineering** is one of the most versatile areas of the ABE specialty areas, because it is

applied to virtually all the others, from machinery design to soil testing to food quality and safety control. Geographic information systems, global positioning systems, machine instrumentation and controls, electromagnetics, bioinformatics, biorobotics, machine vision, sensors, spectroscopy: These are some of the exciting information and electrical technologies being used today and being developed for the future.

8. **Natural Resources:** ABEs with environmental expertise work to better understand the complex mechanics of these resources, so that they can be used efficiently and without degradation. ABEs determine crop water requirements and design irrigation systems. They are experts in agricultural hydrology principles, such as controlling drainage, and they implement ways to control soil erosion and study the environmental effects of sediment on stream quality. Natural resources engineers design, build, operate and maintain water control structures for reservoirs, floodways and channels. They also work on water treatment systems, wetlands protection, and other water issues.

9. **Nursery and Greenhouse Engineering**: In many ways, nursery and greenhouse operations are microcosms of large-scale production agriculture, with many similar needs – irrigation, mechanization, disease and pest control, and nutrient application. However, other engineering needs also present themselves in nursery and greenhouse operations: equipment for transplantation; control systems for temperature, humidity, and ventilation; and plant biology issues, such as hydroponics, tissue culture, and seedling propagation methods. And sometimes the challenges are extraterrestrial: ABEs at NASA are designing greenhouse systems to support a manned expedition to Mars!

10. **Safety and Health:** ABEs analyze health and injury data, the use and possible misuse of machines, and equipment compliance with standards and regulation. They constantly look for ways in which the safety of equipment, materials and agricultural practices can be improved and for ways in which safety and health issues can be communicated to the public.

11. **Structures and Environment:** ABEs with expertise in structures and environment design animal housing, storage structures, and

greenhouses, with ventilation systems, temperature and humidity controls, and structural strength appropriate for their climate and purpose. They also devise better practices and systems for storing, recovering, reusing, and transporting waste products.

CAREERS IN AGRICULTURAL AND BIOLOGICAL ENGINEERING

One will find that university ABE programs have many names, such as biological systems engineering, bioresource engineering, environmental engineering, forest engineering, or food and process engineering. Whatever the title, the typical curriculum begins with courses in writing, social sciences, and economics, along with mathematics (calculus and statistics), chemistry, physics, and biology. Student gains a fundamental knowledge of the life sciences and how biological systems interact with their environment. One also takes engineering courses, such as thermodynamics, mechanics, instrumentation and controls, electronics and electrical circuits, and engineering design. Then student adds courses related to particular interests, perhaps including mechanization, soil and water resource management, food and process engineering, industrial microbiology, biological engineering or pest management. As seniors, engineering students team up to design, build, and test new processes or products.

For more information on this series, readers may contact:

Ashish Kumar, Publisher and President Sandy Sickels, Vice President Apple Academic Press, Inc. Fax: 866-222-9549 E-mail: ashish@appleacademicpress.com http://www.appleacademicpress.com/pub- lishwithus.php	Megh R. Goyal, PhD, PE Book Series Senior Editor-in-Chief *Innovations in Agricultural* *and Biological Engineering* E-mail: goyalmegh@gmail.com

PART I

INNOVATIVE TECHNIQUES IN DAIRY ENGINEERING

CHAPTER 1

INNOVATIVE TECHNIQUES IN MILK PROCESSING: A REVIEW

BHUSHAN D. MESHRAM, P. G. WASNIK, K. K. SANDEY,
SHAKEEL ASGAR, and A. K. AGRAWAL

CONTENTS

1.1 INTRODUCTION

The thermal food processing is a classic technique for ensuring the microbiological safety of foods [77, 193]. This technique leads to unwanted changes in the sensory attributes foods (by overheating) or to low nutritional value of the food products [206]. The increased interest of consumers in high quality foods with higher nutritive value and fresh-like sensory attributes led to the development of a number of non-thermal food processing technologies as alterative to conventionally heat treatments [193, 167]. Among these novel technologies, microwave heating, high pressure processing (HPP), ohmic heating, atmospheric pressure plasma (APP), ultrasonic, high hydrostatic pressure (HHP) and pulsed electric field (PEF) are the most investigated ones [77].

HHP is an innovative technology for food preservation that protects the foods' sensory attributes and produces minimal quality loss [19]. In addition, HHP has the potential to improve energy efficiency and sustainability of food production [158]. PEF is a non-thermal technology that provides minimally processed, safe, nutritious and like-fresh foods to consumers [212, 158]. PEF has been commercially applied for preservation of liquid foods, as pre-step for solid food processes such as drying and for extraction [63]. These two technologies rely on the lethal effect of HHPs and strong electric fields, respectively and are entrusted to result in better quality retention and longer shelf-life [212].

In the recent years, Microwave heating, the HVAD, the cold plasma (CP), ultraviolet light (UV), ultrasound (US), pulsed light (PL) and ionizing radiation are also proposed as alternative non-thermal processing for foods. HVAD consists in application of electricity to pasteurize fluids by rapidly discharging electricity through an electrode gap, generating intense waves and electrolysis, thereby inactivating the microorganisms [149]. The use of arc discharge for foods is unsuitable largely because electrolysis and the formation of highly reactive chemicals occur during the discharge. CP is a relatively unexplored decontamination technology which does not require extreme process conditions compared to HVAD treatment [174]. Gamma irradiation has long been developed and researched and it has high potential in producing safe and nutritious food. Unfortunately, its development and commercialization has been hampered in the past by unfavorable public perceptions. Although several researches have demonstrated the effectiveness of plasmas for killing microorganisms, yet further studies into the nutritional and chemical changes in plasma treated food are required to accurately assess the effect of plasma treatment on product quality and shelf-life and to confirm that no harmful by-products are generated [67]. Although HHP, PEF, HVAD and CP offer great opportunities for food preservation, yet they are often technically difficult to apply into production practice, expensive and require specialized equipment and trained personnel [77]. Moreover, consumer acceptance and safety issues should be considered. The majority of European food producers are small companies with few resources and limited expertise to develop and implement novel emerging technologies.

The aim of this chapter is to present some general aspects about HHP, PEF, HVAD, and CP and to explore the opportunities and drawbacks for the food and milk industry.

1.2 HIGH HYDROSTATIC PRESSURE PROCESSING (HHPP)

In these days, in the emerging field of functional foods, minimal processed foods have increased in popularity along with organic foods. One of the promising technology which could serve as an alternative method for food preservation is the application of HHPP. For the first time in history, this technique was proposed by Royer in 1895 to kill bacteria and Hite in 1899 explored HHPP effects on milk, meat, fruits and vegetables processing.

1.2.1 THE PROCESS

Typically, HHPP of food is performed at 300–600 MPa at room temperature for 2–30 min. During pressurization, an increase in temperature (3–9°C per 100 MPa, depending on the pressure-transmitting fluid) occurs due to adiabatic heating; and a corresponding decrease occurs during depressurization [9]. These changes can be minimized by temperature control of the high-pressure equipment by circulating water. Conversely, this temperature rise can be used to achieve desired effects in the treated foods. During HHPP, the pressure is instantaneously and uniformly transmitted in all directions, regardless of the shape or volume (based on Pascal's principle) of the food in question. The large bio-molecules such as proteins, nucleic acids and polysaccharides that depend on non-covalent bonding to maintain structure and function are most affected. On the other hand, smaller organic molecules such as those responsible for colors, flavors, and nutrients are hardly affected. Milk treated at 400 MPa resulted in no significant loss of vitamins B1 and B6 [185]. The low temperature at which high-pressure treatments are usually performed ensures little or no heat induced changes in these components. However, recent developments in the use of pressure assisted thermal processing (PATP) and pressure assisted thermal sterilization (PATS) utilize both heat and pressure and hence cause some heat-induced changes.

1.2.2   BASIC PRINCIPLES

Hydrostatic pressure is generated by increasing the free energy by physical compression during pressure treatment in the closed system by mechanical volume reduction. Usually HHPP accompanied by a moderate increase in temperature, called the adiabatic heating, depends on the composition of the food product being processed [59]. There are three fundamental operational principles that govern the behavior of foods under underlying HHPP: Le-Chatelier's principle [74], isostatic principle [90], and principle of microscopic ordering [115, 116].

1.2.3   ADVANTAGES OF HHPP

• Significant reduction of heating, this will minimize thermal degradation of food components.
• Inactivation of microorganisms, spores and enzymes.
• High retention of flavor, color and nutritional value.
• Pressure is transmitted uniformly and instant so that food product retains its shape.

- Potential for the design of new products due to the creation of new textures, tastes and functional properties.
- Clean technology, flexible system for number of products and operation.
- Process time is less dependence of product shape and size.
- Reduced requirement of chemical additives, and Increased bioavailability.
- Positive consumer appeal.

1.2.4   DISADVANTAGES OF HHPP

- Food must contain water, as the whole phenomenon is based on compression.
- Some enzymes are very pressure resistant.
- May not inactivate spores.
- Structurally fragile foods needs special attention, and
- High installation cost.
- Foods should have approximately 40% free water for anti-microbial effect.
- Batch processing.
- Limited packaging options.
- Regulatory issues to be resolved.

1.2.5   EFFECTS OF HHPP ON DAIRY PRODUCTS AND PROCESSES

HHPP inactivates most of spoilage and pathogenic bacteria present in milk. Resistance of microorganisms to pressure varies considerably depending on the applied pressure range, temperature and treatment duration, and type of microorganism [188, 199]. Heat resistant groups of microorganisms were usually pressure resistant [59]. The number of yeasts, molds, psychrotrophs and coliforms were decreased more rapidly with pressure than that of acidic and heat-resistant bacteria and proteolytic microorganisms [117]. The lower resistance of Gram-negative bacteria compared with Gram-positive bacteria is due to their lack of teichoic acid, which strengthens the cell wall of Gram-positive bacteria. Bacteria in the log phase of growth are more sensitive than those in the stationary phase.

1.2.6 IMPACT OF HHPP ON PHYSICO-CHEMICAL PROPERTIES OF MILK

White color of milk is due to scattering of light particles by fat globules and casein Micelles. Hunter Luminance value (L-value) of milk [88] was reported to be reduced by HHPP treatment, due to disintegration of casein micelles, thus leading to decrease in the turbidity of milk. Treatment of milk at 200 MPa showed slight effect on L-value; while at 250–450 MPa significant decreased in the L-value was observed. When skim milk treated at 600 MPa for 30 min, L value decreased from 78 to 42 and skim milk becomes almost translucent or semi-transparent [54].

1.2.7 IMPACT OF HHPP ON CREAM, BUTTER AND ICE CREAM

HHPP treatment induces fat crystallization, shortens the time required to achieve a desirable solid fat content and thereby reduces the aging time of ice-cream mix and also enhances the physical ripening of cream for butter making [25]. The cream was subjected to HHPP of 100 to 150 MPa at 23°C for pasteurization and then studied for freeze fracture and transmission electron microscopy [24]. It was observed that pasteurization of cream induced fat crystallization within the small emulsion droplets mainly at the globule periphery [24]. Fat crystallization increased with the length of pressure treatment and was maximal after processing at 300–500 MPa. Moreover, the crystallization proceeded during further storage, after the pressure was released. Two potential applications of HHPP were fast aging of ice cream mix and physical ripening of dairy cream for butter making. Whipping properties were improved when cream was treated at pressure of 600 MPa for up to 2 min [61] probably due to better crystallization of milk fat. When treatment conditions exceeded the optimum conditions, an excessive denaturation of whey protein occurred which resulted in longer whipping time and destabilization of whipped cream. Study with modified whey protein concentrate added at a concentration of less than 10% in ice-cream mix exhibited enhanced overrun and foam stability, confirming the effect of HHPP on foaming properties of whey proteins in a complex system [127].

1.2.8 IMPACT OF HHPP ON YOGURT

Yoghurt suffers from common defects of syneresis and low viscosity. Preservation and rheological properties of yogurt can be improved by pressure treatment. Skim milk treated with combined treatments of HHPP (400–500 MPa) and thermal treatment (85°C for 30 min) showed increased yield stress, resistance to normal penetration, elastic modulus and reduced syneresis [88].

Pressure treatment at 200–300 MPa at 10–20°C for 10 min can be used to control 'post- acidification' of yogurt without decreasing the number of viable lactic acid bacteria (LAB) or modifying the yogurt texture. Treatment at higher pressures destroyed LAB. When exceeding 400 MPa for 15 min, *Lactobacillus delbrueckii* subsp. *bulgaricus* was inactivated, whereas *Streptococcus thermophilus* was more resistant but it lost its acidifying capacity. An extended shelf life probiotic yogurt has been developed using pressure of 350–650 MPa at 10–15°C. The process inactivated yeasts and molds but not specially selected pressure-resistant probiotics, extending the shelf life of yogurt up to 90 days.

1.2.9 IMPACT OF HHPP ON CHEESE

High-pressure treatment of milk at ≤300 MPa decreased the rennet coagulation time (RCT) and increased the curd-firming rate, curd firmness and curd yield during cheese manufacturing. At such pressures, the casein micelle was largely intact or increased slightly in size, and the extent of denaturation of β-lactoglobulin was modest. At higher pressures (>300 MPa), the RCT remained unaffected or increased when compared to that of untreated milk. The effect of high pressure appeared to be the result of two mechanisms with opposite effects: (i) disintegration of the casein micelle and (ii) denaturation of β-lactoglobulin. The quality of Cheddar cheese manufactured from high-pressure treated (31-min cycle at 586 MPa) milk was not significantly different in sensory quality to that made from high-temperature, short-time (HTST) pasteurized milk. However, the pressurized milk cheese had higher moisture content, which led to pasty and weak texture defects, due to the increased water-holding capacity of milk proteins [58, 159].

Milk treated at 300–400 MPa significantly increased wet curd yield up to 20% and reduced both the loss of protein in whey and the volume of whey, due to the denaturation of β-Lg and thus its incorporation in the curd. This leads to high cheese yield to the extent of 7%. Quick maturation and stronger flavor development has also been reported when treated at 400–600 MPa/5–15 min cycle [5, 94]. This can help in accelerating cheese-ripening process and provides better opportunity for improving cheese prepared from low fat milk. The cheese curd obtained from milk treated by HP gives dense network of fine strands thereby having a great potential for the design of new products due to the creation of modified textures, tastes and functional properties [50].

1.2.10 COMMERCIAL APPLICATIONS OF HHPP

A potential application of HHPP is the tenderization of meat. Commercially produced products also include pressure-processed salted raw squid and fish sausages [89]. Other possible applications are improved microbiological safety and elimination of cooked flavors from sterilized meats and pate [102]. Processing at 103 MPa and 40–60°C for 2.5 min improved the eating quality of meat and reduced cooking losses. The extent of tenderization depends on three factors: pressure, temperature and holding time. Starch molecules are similarly opened and partially degraded, to produce increased sweetness and susceptibility to amylase activity. Other research indicated that the appearance, odor, texture and taste of soybeans and rice did not change after processing, whereas root vegetables (potato and sweet potato) became softer, more pliable, sweeter and more transparent [74].

Fruit products are reported to retain the flavor, texture and color of the fresh fruit. Other applications include tempering chocolate, where the high pressures transform cocoa butter into the stable crystal form, preservation of honey and other viscous liquids, sea foods, dairy products such as unpasteurized milk and mold ripened cheese [74]. Compared to thermal and chemical alternatives, HHPP is an effective non-thermal technology. The effects of HHPP (600 MPa, 3 min) and storage (40 days, 4°C) were positive on the stability of avocado paste (*Persea americana* cv. Hass) carotenoids. Likewise, the effects of HHPP and storage on hydrophilic and lipophilic oxygen radical absorbance capacities (ORAC) of the product were studied [47]. Pressurization induced a significant increase (approx. 56%) in concentrations of total extractable carotenoids. Highest increases for individual carotenoids were observed for neoxanthin-β (513%), followed by α-cryptoxanthin (312%), α-carotene (284%), β-cryptoxanthin (220%), β-carotene (107%), and lutein (40%). Carotenoid levels declined during storage, but at the end of the sensory shelf-life of product were higher than those initially present in unprocessed avocado paste. Interestingly, ORAC-values followed a different trend than carotenoids; they decreased immediately after HPP and increased during storage, therefore indicating that carotenoids appear to be minor contributors to the total antioxidant capacity of the fruit [125].

1.2.11 PACKAGING REQUIREMENTS OF HHPP

Packaging technology for HHPP involves different considerations, based on whether a product is processed in-container or packaged after processing. For batch in-container process, flexible or partially rigid packaging is best suited. On the other hand, fluid products require continuous or semi-continuous systems, which are aseptically packaged after pressure treatment. The effectiveness of HHPP is greatly influenced by the physical and mechanical properties of the packaging material. The packaging material must be able to withstand the operating pressures, have good sealing properties and the ability to prevent quality deterioration during the application of pressure. At least one interface of the package should be flexible enough to transmit the pressure. Thus, rigid metal, glass or plastic containers cannot be used [170].

The most common packaging materials used for HP processed food are polypropylene (PP), polyester tubes, polyethylene (PE) pouches, and nylon cast PP pouches. Plastic packaging materials are best suited for HHPP packaging applications, because of reversible response to compression, flexibility and resiliency. The head-space must be also minimized as much as possible [169] in order to control the deformation of packaging materials while sealing the package in order to ensure efficient utilization of the package as well as space within the pressure vessel. Packaging materials for HHPP must be flexible to withstand a 15% increase in volume followed by a return to original size, without losing physical integrity, sealing or barrier properties. Sufficient headspace also minimizes the time taken to reach the target pressure. Film barrier properties and structural characteristics of polymer based packaging material were unaffected when subjected to pressures of 400 MPa for 30 min at 25°C [153].

1.2.12 LEGAL AND SAFETY CONCERNS OF HHPP FOODS

HHPP treated food items falls under the category of novel food as per the definition of Novel Food by European Union (EU) countries (Regulation (EC) No 258/97). The 'Novel Foods Regulation' defines novel food as a food that does not have a significant history of consumption within the EU before the 15th of May, 1997 [90] and such foods are subject to a pre-market safety assessment. Further, the legislations defines Novel foods as follows: foods and food ingredients to which has been applied a production process not currently used, where that process gives rise to significant changes in the composition or structure of the foods or food ingredients which affect their nutritional value, metabolism or level of undesirable substances. If a food falls under the definition of novel food, the person responsible for placing it on the market has to apply for an authorization. In EU countries for introducing novel foods to the market, food companies must have an approval that such products are in compliance with the food law. Food safety issues, the achievable extension of shelf-life and the legislative situation need to be inspected.

High pressure treated foodstuffs have been marketed in Japan since 1990, in Europe and United States since 1996. Information relating to the adverse effects of HHPP on toxins, allergens, and nutrients are rare. There are no published reports available on health and safety issues of HHPP foods. In the developing nations, where there is no such regulations established, for them supportive data on validation is required, and to be sorted out before the marketing of high pressure treated foods. Further, markers/indicators of the effectiveness of HHPP treatment needs to be worked out before the enforcement of legal requirements for such processed dairy foods.

1.3 PULSED ELECTRIC FIELD (PEF)

Non-thermal processes have gained importance in recent years due to the increasing demand for foods with a high nutritional value and fresh-like characteristics, representing an alternative to conventional thermal treatments. PEF are an emerging technology that has been extensively studied for non-thermal food processing. PEF processing has been studied by a number of researchers across a wide range of liquid foods.

Apple and orange juices are among the foods most often treated in PEF studies. The sensory attributes of juices are reported to be well preserved, and the shelf life is extended. Yogurt drinks, apple sauce, and salad dressing have also been shown to retain a fresh-like quality with extended shelf life after processing. Other PEF-processed foods include milk, tomato juice, carrot juice, pea soup [59], liquid whole egg [58], and enzymes in liquid foods such as milk [90]. The conventional HTST processing technique can affect the organoleptic and nutritional properties of milk to varying degrees [140, 190].

PEF is a non-thermal method of food preservation that uses short pulses of electricity for microbial inactivation and causes minimal detrimental effect on food quality attributes. PEF technology aims to offer consumers high-quality foods. For food quality attributes, PEF technology is considered superior than the traditional thermal processing methods, because it is energy efficient and avoids or greatly reduces detrimental changes in the sensory and physical properties of foods. PEF technology aims to offer high-quality foods [78, 102, 104].

1.3.1   TYPICAL COMPONENTS OF PEF EQUIPMENT

- **Power supply**: this may be an ordinary direct current power supply or a capacitor charging power supply (latter can provide higher repetition rates).
- **Energy storage element**: this can be either electric (capacitive) or magnetic (inductive).
- **Switch:** Either for closing or opening. Devices suitable for use as the discharge switch include a mercury ignitron spark gap, a gas spark gap, thyratron, a series of Switch Circuit diode, a magnetic switch or a mechanical rotary switch.
- **Pulse shaping and triggering circuit** in some cases is being used.
- **Treatment chamber:** in wide variety of designs.
- **Pump:** to supply a feed of product to the chamber.
- **Cooling system:** to control the temperature of the feed and/or output material.

1.3.2   PRINCIPLE OF PEF

The basic principle of the PEF technology is the application of short pulses of high electric fields with duration of microseconds to milliseconds and intensity in the order of 10–80 kV/cm. The processing time is calculated by multiplying the number of pulses times with effective pulse duration. The applied high voltage results in an electric field that causes microbial inactivation. When an electrical field is applied, electrical current flows into the liquid food and is transferred to each point in the liquid because of the charged molecules present [223]. The electric field may be applied in the form of exponentially decaying, square wave, bipolar, or oscillatory pulses and at ambient, sub-ambient, or slightly above-ambient temperature. After the treatment, the food is packaged aseptically and stored under refrigeration.

1.3.3   MECHANISMS OF MICROBIAL INACTIVATION

Two mechanisms have been proposed for the mode of PEF action on microbial membrane: electroporation (cell exposed to high voltage electric field pulses temporarily destabilizes the lipid bilayer and proteins of cell membranes); and electrical breakdown (normal resisting potential difference across the bacterial membrane is 10 mV which leads to the build-up of a membrane potential difference due to charge separation across the membrane). In both cases, the phenomena starts by electroporation, by which the cell wall is perforated and cytoplasm contents leak out resulting in cell death ([135, 200].

1.3.4   APPLICATION OF PEF IN MILK PROCESSING

It has been observed that raw skim milk under PEF treatment (intensity of 35 kV/cm; pulse with width 3 μs and time of 9 μs) did not show any significant difference in color, pH, proteins, moisture and particle size [141].

Application of PEF treatment (35 kV/cm; 2.3 μs of pulse width at 65°C for <10 s) immediately after HTST pasteurization extended the shelf life of milk to 78 days at 4°C [184]. PEF treatment of bovine immunoglobulin enriched soymilk (at 41 kV/cm for 54 μs) did not cause any significant change in bovine IgG activity but resulted in a 5.3 log reduction of initial microbial flora [126].

1.3.4.1 Microbial Inactivation and Shelf Life Enhancement of Milk

PEF treatment results in less flavor degradation of milk than equivalent heat treatments, and does not cause any chemical or physical changes in milk fat, protein integrity, and casein structure. Hence, PEF may be used to preserve heat-sensitive dairy products such as whey protein concentrates. An extension of shelf life of raw milk to 2 weeks at refrigeration temperature using PEF (two steps of seven pulses, and one step of six pulses at 40 kV/ cm) was noted. There was no significant difference in sensory quality of PEF treated and heat pasteurized milk. High intensity PEF (HIPEF) treatment of whole milk at 35.5 kV for 1000 μs with 7 μs bipolar pulses at 111 Hz reduced the mesophilic aerobic count from 3.2 log to 1 log cycle ensuring microbial stability for 5 days at 4°C without significant change in acidity, pH and FFA; proteolysis and lipolysis was not observed [156]. The combination of PEF and heat treatment reduced the *salmonella enteritidis* count in skim milk by 2.3 log cycle compared to 1.2 log cycle obtained by PEF treatment alone [156]. The shelf of skim milk under HIPEF treatment (40 kV/cm, 60 μs; 36 kV/cm, 84 μs) was enhanced to 14 days when stored at 4°C [68]. Whole milk processed with HIPEF (35.5 kV/cm, 1000 μs) had a shelf life of 5 days at 4°C and change in acidity was not observed during storage [156]. Subjecting chocolate milk to HIPEF (30 kV/cm, 45 μs) prior to heating at 105°C and 112°C for 31 s resulted in product having shelf life of 119 days at ambient (37°C) storage temperature [135].

1.3.5 INACTIVATION OF ENZYMES BY PEF

The alkaline phosphatase enzyme activity was reduced by 65% in simulated milk ultra-filtrate (SMUF) subjected to electric field strength 22 kV/ cm and seventy pulses using static chamber [31]. The 90% reduction in plasmin activity in SMUF at 30 kV/cm after 50 pulses was observed [199]. Destruction of plasmin by non-thermal PEF is of significance since its activity can cause bitterness and gelation in UHT milk. PEF treatment at 21.5 kV/cm achieved 60% reduction in the milk lipase activity too. Inactivation of protease and lipases produced by psychrotropic microorganism such as *Bacillus* and *Pseudomonas* by use of PEF is of significance in dairy industry since these enzymes are heat-stable and some remain active even after high-temperature treatment. The inactivation of enzymes was increased with field strength, treatment time, input energy and pulse frequency and was dependent on the composition of milk. The inactivation of enzymes in skim milk was higher than that of whole milk. PEF treatment at 21.5 kV/cm and high-energy input reduced the lipase and peroxidase activity of raw milk by 65% and 25%, respectively with negligible effect on alkaline phosphatase activity [82].

1.3.6   APPLICATION OF PEF IN JUICE PROCESSING

Juice extraction from Chardonnay white grape using PEF with two pressure conditions was studied by Grimi [83]. A PEF treatment of 400 V/cm was applied. The PEF pre-treatment increased the juice yield by 67 to 75% compared to the control sample without any adverse effects. The juice volume, total phenolics, betacyanins, betaxanthins concentrations and antioxidant activity were increased ($p = 0.01$) for PEF treated beetroot juice compared to non-PEF treated. Whole beetroots were treated with 1.5 kV/cm electric field, 0.66 μF capacitance and 20 pulses [207].

1.3.7   EFFECT OF PEF ON OSMOTIC DEHYDRATION

Carrot slices (2 cm in diameter, 1 cm in thickness) were pre-treated with PEF at different levels with exponential pulses, 5 pulses and a total specific energy input range of 0.04–2.25 kJ/kg [170]. The electric field strength range was 0.22 to 16 kV/cm and pulse duration between 378 and 405 μs. Both PEF-treated and untreated samples were osmotically dehydrated (immersion in 50° B sucrose solution at 40°C for 5 h). PEF pre-treated samples showed the decrease in moisture content and the increase in solid content during osmotic dehydration.

1.3.8   APPLICATION OF PEF IN PROCESSING OF EGGS

The liquid whole egg with 0.15% citric acid processed at 35 kV/cm for 20 μs and at a maximum temperature of 45°C was preferred over a commercial brand in an acceptance test. Also, scrambled eggs under PEF were not distinguished from a control in a triangle test and did not detect significant changes in the viscosity, °Brix, and color parameters between untreated liquid whole egg controls and samples treated with PEF (25 kV/cm, 250 μs) plus a following heat treatment at 55°C for 3.5 min. These treatment conditions were chosen to obtain a product with a long shelf life in refrigeration temperatures (more than 60 days), since PEF alone resulted insufficient [164].

1.3.9   APPLICATION OF PEF IN EXTRACTION OF BIOACTIVE COMPOUNDS

Thirteen percent increase in total polyphenolic (TP) content in fresh pressed grape juice was reached in comparison to the referent sample simultaneously with 24% increase of TP content in grape residue, under treatment conditions of 0.5 kV/cm, 50 pulses, 0.1 kJ/kg 2.4 kV/cm, and 50 pulses 2.3 kJ/kg [10]. There was a higher juice yield (75%) of PEF-treated vine grapes in comparison to referent sample (70%). With 3 kV/cm and 50 pulses, total anthocyanin content was almost 3 times higher than that in the untreated grapes [197].

1.4   ULTRASOUND TECHNOLOGY

Use of ultrasound in food processing includes extraction, drying, crystallization, filtration, de-foaming, homogenization, meat tenderization and preservation. Ultrasound is one of the non-thermal methods that are used for foods in the last decades. It can be applied to solid, liquid and gas systems for different purposes. Its instrumentation can be fully automated to allow precise measurements [57]. The principle aim of ultrasound technology is to reduce the processing time, save energy and improve the shelf-life and quality of food products [37]. The advantages of ultrasound over the heat treatment include; minimization of flavor loss, greater homogeneity and significant energy savings [60, 202].

Ultrasound refers to sound waves, mechanical vibrations, which propagate through solids, liquids, or gases with a frequency greater than the upper limit of human hearing. The range of human hearing is from about 16 Hz to 18 kHz. When these waves propagate into liquid media, alternating compression and expansion cycles are produced. During the expansion cycle, high intensity ultrasonic waves make small bubbles to grow in liquid. When they attain a volume at which they can no longer absorb enough energy, they implode violently. This phenomenon is known as cavitation. During implosion, very high temperatures (approximately 5000 K) and pressures (estimated at 50 MPa) are reached inside these bubbles.

1.4.1 ULTRASOUND GENERATION

Ultrasonic wave producing system consists of generator, transducer and the application system. Generator produces electrical or mechanical energy; and transducer converts this energy into the sound energy at ultrasonic frequencies.

There are three main types of transducers: fluid-driven, magnetostrictive and piezoelectric [151]. The fluid-driven transducer produces vibration at ultrasonic frequencies by forcing liquid to thin metal blade which can be used for mixing and homogenization systems. The magnetostrictive transducer is made from ferromagnetic materials which change in dimension upon the application of a magnetic field and these changes produce mechanical vibrations. The efficiency of system is about for conversion into an acoustic energy [123]. The piezoelectric transducers produce acoustic energy by changes in size produced by electrical signals in piezo-ceramic materials such as lead zirconate titanate, barium titanate and lead metaniobate. The piezoelectric transducers are most commonly used devices and are more efficient (80–95% transfer to acoustic energy) [123, 151].

1.4.2 CLASSIFICATION OF ULTRASOUND APPLICATIONS

Methods of ultrasound applications can be divided: (i) direct application to the product, (ii) coupling with the device, and (iii) submergence in an ultrasonic bath [3]. Also, ultrasonic applications in the food industry are divided into two distinct categories according to the energy generated by sound field: Low and high energy ultrasounds, which are classified by the sound power (W), sound energy density (Ws/m^3) and sound intensity (W/m^2).

Low energy (low power, low-intensity) ultrasound applications are performed at frequencies higher than 100 kHz and below 1 W/cm^2 intensities. Small power level is used for low intensity ultrasound so that it is non-destructive and no change occurs in the physical or chemical properties of food. Low intensity ultrasound in the food industry is generally used for analytical applications to get information on the physicochemical properties of foods such as composition, structure and physical state [101].

High energy (high power and high-intensity) ultrasonic applications are performed generally at frequencies between 18 and 100 kHz and at intensities higher than 1 W/cm^2 (typically in the range 10–1000 W/cm^2) [138]. At this power, destruction can be observed due to the physical, mechanical or chemical effects of ultrasonic waves (e.g., physical disruption, acceleration of certain chemical reactions). High-intensity ultrasound has been used for many years to generate emulsions, disrupt cells and disperse aggregated materials. More recently, it is used for many purposes such as: modification and control of crystallization processes, degassing of liquid foods, enzyme inactivation, enhanced drying and filtration and the induction of oxidation reactions [138].

1.4.3 METHODS OF ULTRASOUND

Ultrasound can be used for food preservation in combination with other treatments by improving its inactivation efficacy. There have been many studies combining ultrasound with either pressure, temperature, or pressure and temperature, some of them are briefly discussed below:

1. Ultrasonication (US) is the application of ultrasound at low temperature. Therefore, it can be used for the heat sensible products. However, it requires longer treatment time to inactivate stable enzymes and/or microorganisms, which may cause high energy requirement. During ultrasound application, there may be a rise in

temperature depending on the ultrasonic power and time of application; and needs control to optimize the process [224].

2. Thermosonication (TS) is a combined method of ultrasound and heat. The product is subjected to ultrasound and moderate heat simultaneously. This method produces a greater effect on inactivation of microorganisms than heat alone. When TS is used for pasteurization or sterilization purpose, lower process temperatures and processing times are required to achieve the same lethality values as with conventional processes [137, 213].

3. Manosonication (MS) is a combined method in which ultrasound and pressure are applied together. MS provides to inactivate enzymes and/or microorganisms by combining ultrasound with moderate pressures at low temperatures. Its inactivation efficiency is higher than ultrasound alone at the same temperature.

4. Manothermosonication (MTS) is a combined method of heat, ultrasound and pressure. MTS treatments inactivate several enzymes at lower temperatures and/or in a shorter time than thermal treatments at the same temperatures [36]. Applied temperature and pressure maximizes the cavitation or bubble implosion in the media, which increases the level of inactivation. Microorganisms that have high thermos tolerance can be inactivated by MTS. Also some thermos – resistant enzymes (such as: lipoxygenase, peroxidase and poly-phenoloxidase and heat labile lipases and proteases from pseudomonas) can be inactivated by MTS [133].

1.4.4 APPLICATIONS IN DAIRY INDUSTRY

1.4.4.1 Applications in Cheese

When sonication was applied to milk to study the proteolytic activity of the enzymes related to curdling, the main observable effect was that ultrasound speeds up the hardening of the curd and the final product showed a better firmness because of the activity on the chymosin, pepsin and other related enzymes [214]. When ultrasound (20 kHz) was used to enhance the extraction not only the yield and enzyme activity were increased considerably, but also extraction times were much shorter than without sonication, due to the destruction of cellular structure because of the action of ultrasound, increasing the activity of the substances contained in the cells and the migration of proteins and minerals from the cells to the solution. The activity of the chymosin was increased with sonication and the nitrogen content of the extract was decreased at the same time. During cheese making, the methods used to test curd firmness are destructive methods (penetrometers, suspended bodies, torsion viscometers, and rotational viscometers) that are not easy to automate. A pulse-echo technique has been used to determine variations in ultrasonic attenuation and velocity during the coagulation process [12].

The changes during enzymatic coagulation were planned by ultrasound using 1-MHz transducer and the pulse-echo technique. Ultrasonic velocity did not show any variation during coagulation, but ultrasonic dilution was decreased when coagulation progressed, due to changes in viscosity, which increases the viscous attenuation of ultrasound [7]. Resonant techniques (sonic frequencies) have been used to classify defects in cheese based on the differences found in the spectrum of cheeses with and without defect. The pulse-echo technique was used to estimate the size and number of voids in kamaboko by counting the number of ultrasonic echo pulses on the oscillograms [195]. Using this technique, cracked cheeses can be identified. Furthermore, it is also possible to determine the distance of the crack from the surface by assuming a range of velocities that include the maximum and minimum values found for the particular type of cheese. The calculated range for the cheese was 1.84–1.98 cm (at velocity range 1620–1740 m/s), which coincided with the distance measured with a digital gauge (1.9 cm).

1.4.4.2 Applications in Lactose-Free Milk

Lactose-free milk without lactose can be produced by fermentation of lactose-hydrolyzed milk or by the simultaneous addition of β-galactosidase and lactic acid bacteria. These bacteria produce β-galactosidase, which hydrolyses the lactose in fermented milk. Ultrasound has the capability of raising the reaction activity of cells or to stimulate a new action into the cells, for example, in sterol synthesis with baker's yeast or in lactose hydrolyzed fermented milk. Using ultrasound in the processing of lactose-free milk, the lactose hydrolysis was around 55% compared to 36% with traditional methods to produce lactose-free milk (fermentation) [215].

1.4.4.3   Applications in Ice Cream

A narrow ice crystal size distribution is necessary for production of high quality ice-cream with smooth texture and desired sensory characteristics [176]. High power ultrasound treatment of ice cream inside the scraped surface freezer induces crystal fragmentation by cavitation bubbles and also prevents accumulation on the cold surface due to the high heat transfer rate [136]. Increasing the ultrasound pulse time significantly decreased the freezing process time of ice cream, and improved sensory flavor, texture and mouth feel [150].

1.4.4.4   Applications in Homogenization and Emulsification

Sonication of fresh cow milk at 20 kHz resulted in a reduction in the size of fat globules. Homogenization at a power level of 40 for 10 min was similar to conventional homogenization [213]. Ultrasonic emulsification is mainly driven by cavitation, wherein the bubbles collapse at the interface of two immiscible continuous and dispersed phases [136]. High amplitude homogenization also improved the water-holding capacity and viscosity and also reduced syneresis of yogurt produced from ultrasonicated milk [145].

1.4.4.5   Applications in Milk Adulteration

Milk adulteration has usually consisted of adding water. To detect this fraud, the variation of ultrasonic velocity in two types of milk, cow and buffalo, adulterated with different percentages of water was measured [18]. Ultrasonic velocity of cow and buffalo milk was found to be different due to the differences in composition. In both cases, velocity was decreased in line with the water addition and was dependent on temperature, density and viscosity of the samples [18].

1.4.4.6   Applications in Fouling Detection

When dairy products are processed in continuous high temperature processing plants, the internal walls of the plant can become fouled by burnt on or chemically deposited material. The fouling layer will affect the flow rate and also heat flow to or from the product. An ultrasonic sensor has been developed to detect and measure the thickness of these films in a dairy plant [217]. The sensor was operated by transmitting a pulse of ultrasound across the pipe being tested. The received signal was analyzed in the time domain to determine film presence and thickness. Thickness measurement was possible over a range of 0.5–6.0 mm. Product temperature compensation over a temperature range of 20–140°C was implemented. Changes in product flow rate from 0 to 25 l/min and pressure from 0 to 3 bars had no effect on the ultrasonic measurements.

1.4.5 EFFECT OF ULTRASONICATION ON MICROBIAL INACTIVATION

Combined effect of power ultrasound and heat (thermosonication) has proved to be more efficient method of microbial inactivation than either of the two methods alone [172]. Microbial inactivation of ultrasound treatment accounts for generation of acoustic cavitations, resulting in increased permeability of membranes, selectivity loss, cell membrane thinning [180], confined heating [194] and singlet electron transfer in cooling phase [124]. Ultrasonic power of 100 W was found to be optimal for maximum microbial inactivation [222] and US has been found to be effective method for microbial inactivation in *Escherichia coli*, *Listeria monocytogenes*, and other pathogens [73]. Efficiency of ultrasonic treatment as antimicrobial tool depends on the physical (size, hydrophobicity) and biological (gram-status, growth phase) characteristics of the micro-organisms. It has been demonstrated that micro-organisms with "soft" and thicker capsule are extremely resistant to ultrasonic treatment [76].

1.4.6 EFFECT OF ULTRASONICATION ON ENZYME INACTIVATION

Heat treatment to eliminate enzymes is the commonly used method but it also destroys nutrients and may cause loss of food quality. For this reason, non-thermal technologies are being tested as an option for reducing the enzymatic activities in foods [66]. Ultrasound is an effective method in the inactivation of enzymes when it is used alone or with temperature and pressure. There are many enzymes inactivated with ultrasound such as: glucose oxidase [84], peroxidase [49, 177], pectin methyl esterase [171], protease and lipase [208, 209], watercress peroxidase [45] and poly-phenoloxidase [168]. TS is used as a means for enzyme inactivation such as lipoxygenase, peroxidase, lipase, and protease, and tomato or orange pectin methylesterase [171].

The ultrasound stability of individual proteins varies between the enzymes [41, 132, 160, 210, 211] and also depends on ultrasound treatment conditions [172], the composition of treatment medium, treatment pH, and whether they are bound (e.g., membrane-bound proteins) or free (e.g., cytoplasmic proteins). Enzyme inactivation generally increases with increasing ultrasound power, ultrasound frequency, exposure time, amplitude level, cavitation intensity, processing temperature and processing pressure, but decreases as the volume being treated increases [172, 211].

1.4.7   ULTRASONICATION IN MEAT TECHNOLOGY

A large number of applications of ultrasonic treatment are reported in meat technology like, reduction of meat toughness due to large proportion of connective tissue [100], examining the composition of fish, poultry, raw, and fermented meat products by supporting genetic enhancement programs in case of livestock [75] and in the tenderization of meat products.

1.4.8   ULTRASONICATION IN FRUIT AND VEGETABLE PROCESSING

US is used to maintain both pre- and post-harvest quality attributes in fresh fruits and vegetables [75] and is considered a substitute for washing of fruit and vegetable in the food industry [14]. In an attempt to meet the consumers' needs of not only maintaining but also improving the nutritional value of fruit juices [15, 16], US has proved to be one such technique [1] and is described to retain fresh quality, nutritional value, and microbiological safety in guava juice [38], orange juice [205], and tomato juice [221].

Ultrasound treatment can also be used to recover the nutrient loss occurred during blanching, resulting in achieving the collaborative benefit of both the techniques [98]. US cleaners (20–400 kHz) have been efficiently used to produce fruits and vegetables free of contamination [129]. At 40 kHz, it has been applied on strawberry fruits, in which decay and infection were considerably reduced along with quality maintenance [29].

1.4.9   ULTRASONICATION IN EMULSIFICATION

US is relatively cheaper technique for emulsion formation with significant effect on emulsion droplet size and structure. In ultrasonic emulsification application of high energy, researchers observed viscosity decrease and lesser particle size distribution in sub-micron oil-droplets emulsions. However, change in sonication parameters caused remarkable change in stability and oil droplet size of the emulsion formed [108]. Ultrasonically produced W/O emulsions are used by emulsion liquid membrane for the separation and recapture of cationic dyes, and the stability is governed by operating variables such as emulsification time, carrier, ultrasonic power, surfactant and internal phase concentrations, volume ratios of internal phase to organic phase and of external phase to W/O emulsions, stirring speed, contact time, and diluents [56].

1.4.10   ULTRASONICATION IN OIL TECHNOLOGY

US stimulates the mixing and desired reaction for conversion of soybean oil to biodiesel and can achieve optimum yield using 9:1 oil to methanol ratio [181]. Ultrasonic irradiation is also used to increase the rate of trans-esterification [53].

1.4.11   ULTRASONICATION IN HONEY

Ultrasound applications in honey include use of velocity of ultrasonic wave propagation as a means to differentiate between different types of honey determination of adulteration in honey and evaluation of the type of protein, aggregation state, and size [75].

1.4.12 ADVANTAGES AND DISADVANTAGES OF ULTRASONICATION

- Ultrasound waves are non-toxic, safe, and environmentally friendly [112].
- US in combination with other non-thermal methods is considered an effective means of microbial inactivation [211].
- US involves lower running cost, ease of operation, and efficient power output.
- US does not need sophisticated machinery and wide range of technologies [75].
- Use of ultrasound provides more yield and rate of extraction as compared to other conventional methods of extraction [9].
- US involves minimum loss in flavor, superior consistency (viscosity, homogenization) and significant savings in energy expenditure [33].
- Ultrasound has gained huge applications in the food industry such as processing, extraction, emulsification, preservation, homogenization, etc. [36].

1.4.13 DISADVANTAGES OF ULTRASONICATION

- Ultrasound due to shear stress developed by swirls from the shock waves (mechanical effects) cause inactivation of the released products [122].
- Ultrasound application needs more input of energy which makes industrialists to think over while using this technique on commercial scale [222].
- Ultrasound induces physicochemical effects which may be responsible for quality impairment of food products by development of off – flavors, alterations in physical properties, and degradation of components.
- US leads to the formation of radicals as a result of critical temperature and pressure conditions that are responsible for changes in food compounds. The radicals (OH and H) produced in the medium deposit at the surface of cavitation bubble that stimulates the radical chain reactions which involve formation of degradation products and thus lead to considerable quality defects in product [46].

- Frequency of ultrasound waves can impose resistance to mass transfer [62].
- Ultrasonic power is considered to be responsible for change in materials based on characteristics of medium. So, this power needs to be minimized in the food industry in order to achieve maximum results [65].

1.5 OHMIC HEATING

Heating technologies for processing and preservation of foods have observed marvelous advancements with the development of technologies such as ohmic heating, dielectric heating (which includes microwave heating and radio frequency heating) and inductive heating. All the advanced methods of processing are highly energetic and efficient as the heat is generated directly inside the food. These are called novel thermal processing technologies.

Ohmic heating is also-called Joule heating, electrical resistance heating, direct electrical resistance heating, electro heating or electro conductive heating. It is a process where heat is internally generated due to electrical resistance, when electric current is passed through it [4]. Ohmic heating is distinguished from other electrical heating methods as the electrodes are in contact with the foods unlike in microwave and inductive heating where electrodes are absent, the applied frequency is lower as compared to radio or microwave frequency range and also the waveform is unrestricted, although typically sinusoidal. A successful application of electricity in food processing was developed in the 19th century to pasteurize milk called "electropure process" [81]. But this application was dejected apparently due to high processing costs [72]. Also, other applications were abandoned because of the short supply of inert materials needed for the electrodes [144]. However recent research has been carried out by various scientists worldwide on fruits, vegetables and meat products, flours and starches, etc. [32, 161, 216]. The ohmic heating system helps in the production of highly shelf-stable products with proper maintenance of the color and nutritional value of food. Ohmic heating is one of these alternative-processing techniques to emerge in the last 20 years [17].

1.5.1 FUNDAMENTAL PRINCIPLE OF OHMIC HEATING

Ohmic heating method is one of the several electromagnetic based methods such as capacitive dielectric, radiative, dielectric, inductive and radiative magnetic heating. Ohmic heating is somewhat similar to microwave heating but with very different frequencies [48]. Ohmic heating (direct resistance heating) is a process in which food liquids and solids are heated simultaneously bypassing an electric current through them.

The applicability of ohmic heating is dependent on product electrical conductivity. Most food preparations contain a moderate percentage of free water with dissolved ionic salts and therefore conduct sufficiently well for the ohmic effect to be applied. The ohmic heater column typically consists of seven-electrode housing machined from a solid block of polytetrduoroethylene and encased in stainless steel, each containing a single cantilever electrode. The electrode housings are connected using stainless-steel spacer tubes lined with a generally recognized as safe electrically insulating plastic. The column is mounted in a vertical or near-vertical position, with upward flow of product. A vent valve positioned at the top of the heater ensures that the column is always full. The column is configured so that each heating section has the same electrical impedance. Hence, the interconnecting tubes generally increase in length toward the outlet because the electrical conductivity of food products usually increases with increase in temperature. For aqueous solutions of ionized salts, there is a linear relationship between temperature and electrical conductivity, due to increased ionic mobility with increase in temperature and it applies to most food products. Exceptions could be products in which viscosity increases markedly at a higher temperatures, such as those containing un-gelatinized starches [162].

1.5.2 ADVANTAGES OF OHMIC HEATING

- Continuous production without heat-transfer surfaces;
- Rapid and uniform treatment of liquid and solid phases with minimal heat damage; and
- Nutrient losses (e.g., unlike microwave heating, which has a finite penetration depth into solid materials);

- Ideal process for shear-sensitive products because of low flow velocity;
- Optimization of capital investment and product safety as a result of high solids loading;
- Reduced fouling when compared to conventional heating;
- Better and simpler process control with reduced maintenance costs;
- Environmentally friendly system;
- Maintenance of the color and nutritional value of food;
- Less cleaning requirements;
- Heating of particulate foods and liquid–particle mixtures;
- Low risk of product damage due to burning;
- High energy conversion efficiency.

1.5.3   DISADVANTAGES OF OHMIC HEATING

The installation and operation cost of ohmic food processing systems was found to be more costly as compared to that of conventional retorting, freezing and heating in a conventional tubular heat exchanger and ohmic heating [6].

The food containing fat globules is not effectively heated during ohmic heating process, as it is non-conductive due to lack of water and salt [166]. If these globules are present in a highly electrical conductive region where currents can bypass them, they may heat slower due to lack of electrical conductivity. Any pathogenic bacteria that may be present in these globules may receive less heat treatment than the rest of the sub-stance [182].

Also there is the possibility of 'runaway' heating [69]. As the tempera-ture of a system rises, the electrical conductivity also increases due to the faster movement of electrons.

1.5.4   APPLICATIONS OF OHMIC HEATING

1.5.4.1   Applications of Ohmic Heating Extraction

The application of ohmic heating in conjunction with extraction processes increased the extraction efficiency of sucrose from sugar beets [110]. Diffusion of beet dye from beetroot into a carrier fluid was increased in ohmic heating and the amount of dye extracted was proportional to the strength of electrical field [128]. Ohmic heating improved the extraction of soymilk from soybeans [113].

1.5.4.2 Applications of Ohmic Heating Enzyme Inactivation

In addition to improvement food quality (e.g., texture and flavor) and for the recovery of by-products, enzymes may also have negative effects on food quality such as production of off –odors, tastes and altering textural properties. Therefore, control of enzymatic activity is required in many food processing steps to promote/inhibit enzymatic activity.

The effects of electric field on important food processing enzymes have been studied and reported in literature [33]. The tested enzymes were poly-phenoloxidase (PPO), lipoxygenase, pectinase, alkaline phosphatase and β-galactosidase and the inactivation assays were performed under conventional and ohmic heating conditions. All the enzymes followed 1st-order inactivation kinetics for both conventional and ohmic heating treatments. The presence of an electric field did not cause an enhanced inactivation to alkaline phosphatase, pectinase, and β-galactosidase. However, lipoxygenase and PPO kinetics were significantly affected by the electric field, reducing the time needed for inactivation. Fresh grape juice was ohmically heated at different voltage gradients (20, 30, and 40 V/cm) from 20°C to temperatures of 60, 70, 80 or 90°C and the change in the activity of PPO enzyme was measured [97]. The critical deactivation temperatures were at 60°C or lower for 40 V/cm, and 70°C for 20 and 30V/cm. Various kinetic models for the deactivation of PPO by ohmic heating at 30 V/cm were fitted to the experimental data. The simplest kinetic model involving one step first-order deactivation was better than more complex models. The activation energy of the PPO deactivation at the temperature range of 70–90°C was 83.5 kJ/mol.

Ohmic heating using continuous alternating current electric field was applied to orange juice containing *Bacillus subtilis* spores to examine its inactivation. An effective inactivation of spores was achieved using a pressurized electric sterilization system, using a combination of high temperature and high electric field, in a shorter time. Also the loss in the ascorbic acid and development of peculiar smell was minimized in ohmic heating [201].

1.5.4.3   Applications of Ohmic Heating in Blanching

The peroxidize inactivation and color changes during ohmic blanching of pea puree were studied by application of four different voltage gradients in the range of 20–50 V/cm; the puree samples were heated from 30 to 100°C to achieve adequate blanching. The conventional blanching was performed at 100°C water bath. The ohmic blanching was applied by using 30 V/cm and above. Voltage gradient inactivated peroxidase enzyme in lesser time than the water blanching. The ohmic blanching at 50 V/cm gave the shortest critical inactivation time of 54 s with the best color quality. First order reaction kinetics adequately described the changes in color values during ohmic blanching. Hue angle is the most appropriate combination ($R^2 = 0.954$), which describes closely the reaction kinetics of total color changes of pea puree for ohmic blanching at 20 V/cm [96].

1.5.4.4   Applications of Ohmic Heating in Pasteurization and Sterilization

Ohmic heating is very often used in pasteurization/sterilization of food products resulting in excellent quality it also enhanced the air drying rate [128, 192]. In a recent study, the use of ohmic heating on sterilization of guava juice has reported [62]. The resistance heating technique was used for milk pasteurization in the early 20th century [165]. Ohmic heating can be used for ultra-high temperature (UHT) sterilization of foods. A reusable pouch with electrodes for long term space missions was developed [106]. The pouch permits reheating and sterilization of its internal contents. The 3-D model design ensured sterility and permitted identification of cold spots over the entire pouch. This 3-D model was observed to be useful tool to optimize electrode configurations and to assure adequate sterilization process.

1.5.4.5   Applications of Ohmic Heating in Fruit and Vegetable Products

The nutritional quality of most of the fruits and vegetable products is altered during conventional thermal processing. This necessitates the search for alternative processing technologies to achieve better quality of end products. Several strawberry based products were tested by ohmic heating. The results proved that high heating rates could be achieved for most of the products. Also, the increase of the applied electric field could increase the heating rate. The suitability of ohmic heating for strawberry products having different solids concentrations was also evaluated [33]. Electrical conductivity was observed to be decreased with the increase in solid contents in a mixture of particles, but the decrease was more significant for the bigger particles. The results also suggested that for higher solid content (> 20% w/w) and sugar contents over 40.0° Brix, electrical conductivity was too low to use in the conventional ohmic heaters and a new design is required.

1.5.4.6   Applications of Ohmic Heating in Milk Fouling

The influence of material (stainless steel. tin, and graphite electrodes), flow rate, electric current density (at constant frequency 50 Hz) and temperature (in a limited temperature range 65–75°C) on the fouling of skimmed milk during direct ohmic heating was studied and it was observed that the stainless steel electrodes are worst while the graphite electrodes, where no fouling was observed, are the best, thus confirming the significant role of corrosion and electrical phenomena [189].

 While studying the hydrodynamic behavior of milk in a flat ohmic cell, it was found that fouling of fluid occurs due to greater quantity deposited in the zone where the temperature is lowest (entrance zone) and velocity is non-uniform. During continuous ohmic heating, there is a chance of slightest hydrodynamic disturbance which results in a thermal and electric disturbance and there by creates zones, which are subjected to fouling [8].

1.5.4.7   Applications of Ohmic Heating in Waste Water Treatment

Waste water treatment is one of the problems in surimi production due to high volume and high biological oxygen demand (BOD) of water. Protein coagulation by heating and subsequent separation is the method to reduce the BOD of waste water having high protein concentration. Ohmic heating is an efficient heating method to heat the fluid. Thus ohmic heating might be a viable alternative for waste water treatment in surimi production plants [182].

A continuous ohmic heating system to coagulate protein from surimi waste water to reduce the biological oxygen demand of the waste water were developed and a simple model, based on the energy conservation equation, was used to predict the temperature profiles of the waste water. Samples were diluted and NaCl solution (10% by wt.) was added to make them suitable for testing in the developed device. All samples were heated under different conditions (electric field strength of 20, 25 and 30 V/cm; flow rates of 100, 200 and 300 cc/min). After heating, the samples were centrifuged and the remaining protein in supernatants was measured and compared with the results from the previous batch experiments. Heating under higher electric field strength and lower flow rates resulted in higher temperatures of samples. The predicted temperature values agreed well with the experimental results. The amount of the remaining protein was also in agreement with that of the previous work. The lab scale ohmic heating system possessed good performance to coagulate protein (60%) from surimi waste water [109].

1.6   FOOD IRRADIATION

Food irradiation is a process in which food products are exposed to ionizing radiation in form of gamma radiation, X-rays and electron beams in a controlled amount to destroy pathogenic microorganisms in order to increase its safety and shelf life [219]. It can be used to replace chemical preservatives as well as thermal treatment. It is considered as cold pasteurization of food.

The use of gamma irradiation in dairy products is one of the most important peaceful application [218]. There was no hazard caused by irradiation up to 10 kilo gray which did not cause cancer, genetic mutation or tumors [139]. Therefore, hospitals use irradiated food for patients with severely impaired immune system [118]. In 1981, the United Nation Food and Agricultural Organization (UN-FAO-WHO) endorsed irradiation doses up to 10 k Gray as a major technology for the prevention of food borne illness and for the reduction in food losses due to spoilage by microorganisms and vermin. Ionizing radiation is now approved for use in more than 41 countries for over 35 specified foods, and the list is growing [30, 131, 198]. Approximately 26 countries currently employ radiation on a commercial scale for food application [191]. Many consumers are not adequately educated about the safety of irradiated foods. Investment for a commercial irradiator facility is high. As a result, it is very challenging for a food preserved with an unconventional technology to enter and compete in the market place [70].

1.6.1 FOOD IRRADIATION METHODS

Three principal types of radiation source can be used in food irradiation according to the *Codex Alimentations General Standard* [69].

Gamma radiation source is from radionuclides such as 60 Cu (copper) or 137 C_s (cesium), Machine sources of electron beams with energies up to 10 MeV, Machine sources of bremsstrahlung (X rays) with electron energies up to 5 MeV. Because of their greater penetrating capability, gamma rays and X-rays may be used for processing of relatively thick or dense products. Ionizing radiation for food processing is limited to high energy photons (gamma rays) of radio nuclides 60 Cu or C_s, X-rays from machine sources with energies up to 5 MeV and accelerated electrons with energies up to 10 MeV generated by electron accelerating machines [95]. These kinds of Ionizing radiation are preferred due to: the suitable food preservative effects do not generate radioactivity in foods or packaging materials and available at costs as commercial use of the irradiation process [64].

1.6.2 ADVANTAGES OF IRRADIATION

To ensure hygienic quality of food, the use of ethylene oxide fumigation for decontaminating the ingredients has been increasing restricted in recent years. The European community is used a directive, which prohibited the use of ethylene oxide on food starting effective January 1991 [55]. Most microorganisms and all insects cause damage to fresh commodities such as fish, meat, fruit, vegetable, etc., and their products are sensitive to low dose irradiation. Thus, irradiating the food with dose between 1 and 5 kGy resulted in insect's disinfestation and a several fold reduction of spoilage microorganisms, thereby extending the shelf life of the food [204].

1.6.3   LIMITATIONS OF IRRADIATION

Food treatment adds cost to the product. Like other physical food processes, irradiation has high capital costs and requires a critical capacity and product volume for economic operation. A number of market tests of irradiated food have been carried out in the past five years with interesting results [203], by consumer critical education.

1.6.4   PROCESS OF IRRADIATION

During the irradiation process, food is exposed to the energy source in such a way that a precise and specific dose is absorbed. To do that, it is necessary to know the energy output of the source per unit of time, to have a defined spatial relationship between the source and the target and to expose the target material for a specific time. The radiation dose ordinarily used in food processing ranges from 50 Gy to 10 kGy, and depends on the kind of food being processed and the desired effects [220].

The actual dose of radiation employed in any food processing application represents a balance between the amount needed to produce a desired result and the amount the product can tolerate without suffering unwanted change. High radiation dose can cause organoleptic changes (off-flavors or changes in texture), especially in foods of animal origin, such as dairy products. In fresh fruits and vegetables, radiation may cause softening and increase the permeability of tissue [219].

1.6.5   IRRADIATION OF DAIRY PRODUCTS

Shelf life extension and/or sterilization of dairy products for making it shelf stable using radiation treatment are not a widely accepted practice. The reason for its limited use is that ionizing energy, through the formation of radiolytic products especially in high lipid-based foods, generates unacceptable off-odors and flavors via oxidation, or complementary with use of other preservation techniques including refrigeration and/or preservatives such as sorbic acid. Cheddar cheese developed off-flavors when irradiated at 0.5 kGy. However, none was detected when the dose was reduced to 0.2 kGy [21]. A dose > 1.5 kGy when applied to Turkish Kashar cheese, not only resulted in off-flavor development but also contributed to color deterioration [103]. By decreasing the dose to 1.2 kGy, the sensory problems were eliminated and the mold-free shelf life was extended 12 to 15 days when stored at room temperature. In contrast, non-irradiated cheese became moldy within 3 to 5 days. When combined with refrigeration storage, radiation increased the shelf-life period of the cheese five-folds. With Gouda cheese, however, no taste difference was reported between irradiated (3.3 kGy) and non-irradiated samples [175]. In order to stabilize the cheese by preventing additional growth of *Penicillium roqueforti*, a minimal dose of 2.0 kGy was recommended. Results from a subsequent study, however, reported that full fat Camembert cheese suffered no off-flavor development up to a dose of 3 kGy [39] and treatment at 2.5 kGy was sufficient to eliminate initial populations of 103 to 104 colony forming units (cfu)/g of the pathogen *Listeria monocytogenes* [15]. In contrast, flavor changes were quite noticeable when radiation treatment was applied to cottage cheese, the minimal threshold dose being 0.75 kGy. At this dosage, the cheese was described as having a slight bitter, cooked, or foreign taste. However, in order to reduce spoilage by psychrotrophic bacteria by at least three logs, the applied dose would have to be nearly doubled [103]. This resulted in cheese with a definite burnt off-flavor. Using electron beam irradiation and doses of 0.21 and 0.52 kGy, the shelf life of vacuum packaged cheddar cheese at 10°C containing 101 cfu/cm², *Aspergillus ochraceus* spores was extended by approximately 42 and 52 days, respectively [20].

Sterilization of yogurt bars, ice cream, and non-fat dry milk by gamma irradiation using a dose of 40 kGy at −78°C resulted in an overall decrease in acceptance [103]. Although the use of MAP or the inclusion of antioxidants appeared to reduce the level of off-flavors, yet the effects were product specific. Irradiation of fluid milk also resulted in unacceptable flavor scores. Off-flavors and browning originating from chemical reactions involving lactose were identified. Irradiation preservation of yogurt was similarly investigated. Left at room temperature, plain yogurt reached a population of 109 cfu/g by 6 days and was judged unacceptable. However, when treated with gamma irradiation using a dose of 1 kGy, this population level was not reached until 18 days of incubation. Irradiation combined with refrigeration further extended the shelf life of yogurt to 30 days. In comparison, the shelf life of the refrigerated controls was only 15 days [120].

1.7   DEGREES OF FOOD PROCESSING

1.7.1   MINIMALLY PROCESSED FOODS

Not all foods undergo the same degree of processing. In this section, processed foods are classified in three categories: minimally processed food, processed food ingredients, and highly processed food [147].

Fruits, vegetables, legumes, nuts, meat and milk are often sold in minimally processed forms. Foods sold as such are not substantially changed from their raw, unprocessed form and retain most of their nutritional properties. Minimal forms of processing include washing, peeling and slicing, juicing and removing inedible parts [157].

1.7.2   PROCESSED FOOD INGREDIENTS

This group includes flours, oils, fats, sugars, sweeteners, starches and other ingredients. High fructose corn syrup, margarine and vegetable oil are common examples. Processed food ingredients are rarely eaten alone; and these are typically used in cooking or in the manufacture of highly processed foods [146].

1.7.3   HIGHLY PROCESSED FOODS

Highly processed foods are made from combinations of unprocessed food, minimally processed food and processed food ingredients. They are often portable, can be eaten anywhere (while driving), working at the office and watching TV and require little or no preparation. Highly processed foods include: snacks and desserts, such as cereal bars, biscuits, chips, cakes and pastries, ice cream and soft drinks; as well as breads pasta, breakfast cereals and infant formula [187].

1.8   MICROWAVE HEATING

Microwave heating refers to the use of electromagnetic waves of certain frequencies to generate heat in material. For food applications, most commonly used microwave frequencies are 2450 MHz and 915 MHz. When a microwavable container with food is placed in a microwave oven, a temperature gradient develops between the center and the edges. Meat, fish, fruit, butter and other food-stuffs can be tempered from cold store temperature to around $-3°C$ for ease of further processing (i.e., grinding of meat in production of burgers, blending and portioning of butter packs). Food products, such as bread, precooked foods have been processed using microwaves for pasteurization or sterilization or simply to improve their digestibility.

1.8.1   BASIC COMPONENTS OF MICROWAVE HEATING

These include: power supply and control; magnetron; waveguide; stirrer; turntable; cooking cavity; and door with choke.

1.8.2   PRINCIPLE OF MICROWAVE HEATING

Microwave heating in foods occurs due to coupling of electrical energy from an electromagnetic field in a microwave cavity with the food and its subsequent dissipation within food product. This results in a sharp increase in temperature within the product. Microwave energy is delivered at a molecular level through the molecular interaction with the electromagnetic field, in particular, through molecular friction resulting from dipole rotation of polar solvents and from the conductive migration of dissolved ions. Water in the food is the primary dipolar component responsible for the *dielectric heating*. In an alternating current electric field, the polarity of the field is varied at the rate of microwave frequency and molecules attempt to align themselves with the changing field. Heat is generated rapidly as a result of internal molecular friction.

The second major mechanism of heating with microwaves is through the *polarization of ions* as a result of the back and forth movement of the ionic molecules trying to align themselves with the oscillating electric field [52].

1.8.3   ADVANTAGES OF MICROWAVE TECHNOLOGY

• Microwave penetrates inside the food materials and therefore, cooking takes place throughout the whole volume of food internally, uniformly and rapidly, which significantly reduces the processing time and energy.
• Since the heat transfer is fast, nutrients and vitamins contents, as well as flavor, sensory characteristics, and color of food are well preserved.
• Minimum fouling depositions, because of the elimination of the hot heat transfer surfaces, since the piping used is microwave transparent and remains relatively cooler than the product.
• High heating efficiency (80% or higher efficiency can be achieved).
• Suitable for heat-sensitive, high-viscous and multiphase fluids.
• Low cost in system maintenance.

1.8.4   DISADVANTAGES OF MICROWAVE TECHNOLOGY

The rather slow spread of food industrial microwave applications has a number of reasons: there is the conservatism of the food industry [52] and its relatively low research budget. Linked to this, there are difficulties in moderating the problems of microwave heating applications. One of the main problems is that, in order to get good results, they need a high input of engineering intelligence. Different from conventional heating systems, where satisfactory results can be achieved easily by perception, good microwave application results often need a lot of knowledge or experience to understand and moderate effects like uneven heating or the thermal runway. Another disadvantage of microwave heating as opposed to conventional heating is the need for expensive electrical energy, high initial capital investment, more complex technology devices and microwave radiation leakage problem.

1.8.5 APPLICATIONS OF MICROWAVE TECHNOLOGY IN THE FOOD INDUSTRY

1.8.5.1 Applications of Microwave Technology in Thawing-Tempering

By using microwaves mostly with 915 MHz due to their larger penetration depth, the tempering time can be reduced to minutes or hours and the required space is diminished to one sixth of the conventional system [82]. Another advantage is the possibility to use the microwaves at low air temperatures, thus reducing or even stopping microbial growth.

Without a doubt, thawing and tempering are the most industrially widespread applications of microwave heating. There are about 400 systems in use in the United States for vegetables and fruits; and at least four in the United Kingdom for the tempering of butter. Most of the studies carried out [6, 13] have analyzed the behavior and final characteristics of diverse types of meat during microwave tempering. Tempered meat shows good final characteristics with less process time. This technology was attempted for stretching Mozzarella curd was not successful [28].

Recently, other possible applications of this technology to products such as rice balls [105] mashed potatoes [91] or cereal pellets or pieces [183] have been studied, with a few encouraging results in terms of the good physical and sensory properties. On the other hand, some studies [34, 107] were mainly focused on reaching a better understanding of the relationship between the equipment (applied powers and cycles of work) and the product (dielectric properties, loads, and geometry).

1.8.5.2 Applications of Microwave Technology in Heating of Precooked Products

The heating of precooked products is the principal practical application of microwave ovens, both in domestic use and in the catering industry, since a rapid, safe, and hygienic heating of the product is obtained [26]. The objective of pre-cooking operations is to reduce preparation time for the consumer. In the case of cereals, these operations consist basically of treating starch to reduce its gelatinization time during the final preparation of the food product. Pre-cooked rice and wheat flour with good sensory and nutritional characteristics can be prepared with microwave [35]. The precooking process can be accelerated with the help of microwaves as has been established for precooking of poultry, meat patty and bacon [22, 52].

1.8.5.3   Applications of Microwave Technology in Cooking

Microwave ovens are now being used in about 92% of homes in the United States. Microwave ovens are very popular home appliances for the food processing applications. Cooking is one of the major applications of microwave. Microwave heating is so rapid; it takes the product to the desired temperature in such a short time that product cooking does not take place; the product is hot, but has the appearance and flavor of the raw product.

There are several products used in the continuous study of this technology, such as fish [178], beans [154], egg yolk [152], and shrimp [85]. The nutritious characteristics of the food are quite well retained, but it does not reach the typical flavor of the cooked dish; thus it is necessary to combine microwave treatment with conventional technologies [41]. It is reported that there was no significant change on the loss of B-complex vitamins during microwave boiling of cow and buffalo milk in comparison to conventional heating [115]. Microwave cooking reduces cooking times of common beans and chickpeas [134]. In addition microwave treatment reduced cooking losses, increased the soluble to insoluble and soluble to total dietary fiber ratio, but did not modify in-vitro starch digestibility. A higher protein concentration in soya milk was obtained by microwave heating of soya slurry than by the conventional methods of heating such as the use of boiling water [2]. Microwave oven heating of soya slurry, which was effective for protein extraction, also made the prepared tofu more digestible.

1.8.5.4   Applications of Microwave Technology in Baking

Baking process includes three phases: expansion of dough and moisture loss initiates in the first phase; the second phase, in which expansion and the rate of moisture loss becomes maximal. The changes that continue to take place in the third phase of baking include rise in product height and decrease in rate of moisture loss because the structure of the air cells within the dough medium collapses as a result of increased vapor pressure. Baking using microwave energy has been limited due to poor product quality compared to products baked by using conventional energy sources, which can be a reflection of the differences in the mechanism of heat and mass transfer [179]. In products such as breads, cakes and cookies, microwave baking can affect texture, moisture content and color of the final product, which represents a great challenge for research. Some researchers suggested adjustments in formulation and alterations in the baking process, while others studied the interactions between microwave energy and the ingredients of the formulation [163]. A process was patented to obtain a sponge cake free from bake shrinkage and good-looking voluminous appearance, through a batter prepared by adding a thermo-coagulation protein to a sponge cake premix containing main ingredient as a cereal powder consisting of starch and a pre-gelatinized starch cooked under heat with a microwave oven [196].

1.8.5.5   Applications of Microwave Technology in Blanching

Blanching is generally used for color retention and enzyme inactivation, which is carried out by immersing food materials in hot water, steam or boiling solutions containing acids or salts. Blanching has additional benefits, such as the cleansing of the product, the decreasing of the initial microbial load, exhausting gas from the plant tissue, and the preheating before processing. A moderate heating process such as blanching may also release carotenoids and make them more extractable and bio-available.

Blanching with hot water after the microwave treatment compensates for any lack of heating uniformity that may have taken place, and also prevents desiccation or shriveling of delicate vegetables. And while microwave blanching alone provides a fresh vegetable flavor, the combination with initial water or steam blanching provides an economic advantage. This is because low-cost hot water or steam power is used to first partially raise the temperature, while microwave power, which costs more, does the more difficult task of internally blanching the food product. Microwave blanching of marjoram and rosemary was carried out by soaking the herbs in a minimum quantity of water and ex-posed to microwaves [186].

1.8.5.6   Applications of Microwave Technology in Food Sterilization and Pasteurization

Pasteurization and sterilization are done with the purpose of destroying or inactivating microorganisms to enhance the food safety and storage life [52, 51]. Solid products are usually sterilized after being packed, so no metallic materials should be used in packaging when microwaves are used in this process. This factor limits the use of this technique in food sterilization. Possible non-thermal effects on destruction of microorganisms under microwave heating has been reported: The polar and/or charged moieties of proteins (i.e., COO– and NH_4+) can be affected by the electrical component of the microwaves [52]; and the disruption of non-covalent bonds by microwaves is a more likely cause of speedy microbial death [119]. Academic and industrial approaches to microwave pasteurization or sterilization cover the application for precooked food like yogurt or pouch packed meals as well as the continuous pasteurization of fluids like milk [51, 52].

1.8.5.7   Applications of Microwave Technology in Food Dehydration

In drying of food materials, the goal is to remove moisture from food materials without affecting their physical and chemical composition. It is also important to preserve the food products and enhance their storage stability which can be achieved by drying. Dehydration of food can be done by various drying methods such as solar (open air) drying, smoking, convection drying, drum drying, spray drying, fluidized-bed drying, freeze drying, explosive puffing and osmotic drying [42].

The application of microwaves to food drying has also received widespread attention recently [52, 80]. The heat generated by microwaves induces an internal pressure gradient that involves vaporization and expelling of the water toward the surface. This greatly accelerates the process, when compared to hot air or infrared dehydration [190], in which the drying rate is dependent on the diffusion of water inside the product toward the surface.

1.8.6   APPLICATIONS OF MICROWAVE TECHNOLOGY IN DAIRY INDUSTRY

Milk is traditionally pasteurized in a heat exchanger before distribution. The application of microwave heating to pasteurize milk has been well studied and has been a commercial practice for quite a long time. The success of microwave heating of milk is based on established conditions that provide the desired degree of safety with minimum product quality degradation. Since the first reported study on the use of a microwave system for pasteurization of milk [87], several studies on microwave heating of milk have been carried out. The majority of these microwave-based studies have been used to investigate the possibility of shelf-life enhancement of pasteurized milk, application of microwave energy to inactivate milk pathogens, assess the influence on the milk nutrients or the non-uniform temperature distribution during the microwave treatment [121].

Microwave application allows pasteurization of glass, plastic, and paper products, which offers a useful tool for package treatment. The food products that best respond to MW pasteurization treatment are pastry, prepared dishes, and soft cheeses [27]. The technique has also been tested on milk [79] and fruit juices [71] in devices suitable for continuous treatment and domestic microwave. It is reported that the microwave pasteurization has no effect on amino acid composition [2].

1.9   COLD PLASMA

CP is a novel non-thermal food processing technology that uses energetic, reactive gases to inactivate contaminating microbes on meats, poultry, fruits, and vegetables. This flexible sanitizing method uses electricity and a carrier gas, such as air, oxygen, nitrogen, or helium; antimicrobial chemical agents are not required. The primary modes of action are due to UV light and reactive chemical products of the CP ionization process [155]. Reductions of greater than 5 logs can be obtained for pathogens such as *Salmonella, Escherichia coli O157:H7, Listeria monocytogenes, and Staphylococcus aureus.* Effective treatment times can range from 120 s to as little as 3 s, depending on the food treated and the processing conditions. Key limitations for CP are the relatively early state of technology development, the variety and complexity of the necessary equipment, and the largely unexplored impacts of CP treatment on the sensory and nutritional qualities of treated foods.

Plasma is often called the *"Fourth State of Matter,"* the other three being solid, liquid and gas. Plasma is a distinct state of matter containing a significant number of electrically charged particles, a number sufficient to affect its electrical properties and behavior. In an ordinary gas each atom contains an equal number of positive and negative charges; the positive charges in the nucleus are surrounded by an equal number of negatively charged electrons, and each atom is electrically "neutral." A gas becomes plasma when the addition of heat or other energy causes a significant number of atoms to release some or all of their electrons. The remaining parts of those atoms are left with a positive charge, and the detached negative electrons are free to move about. Those atoms and the resulting electrically charged gas are said to be "ionized." When enough atoms are ionized to significantly affect the electrical characteristics of the gas, it is called plasma.

Physical plasma is defined as a gas in which part of the particles are present in ionized form. This is achieved by heating a gas which leads to the dissociation of the molecular bonds and subsequently ionization of the free atoms. Thus, plasma consists of positively and negatively charged ions and negatively charged electrons as well as radicals, neutral and excited atoms and molecules. Plasma not only occurs as a natural phenomenon as seen in the universe in the form of fire, in the polar aurora borealis and in the nuclear fusion reactions of the sun but also can be created artificially which has gained importance in the fields of plasma screens or light sources.

There are two types of plasma: thermal and non-thermal or cold atmospheric plasma. Thermal plasma has electrons and heavy particles (neutral and ions) at the same temperature. Cold Atmospheric Plasma (CAP) is said to be non-thermal because it has electron at a hotter temperature than the heavy particles that are at room temperature. CAP is a specific type of plasma that is less than 104°F at the point of application.

1.9.1 METHODS OF CP GENERATION

CP can be produced by a variety of means, some of which have been the subject of research since the earliest years of inquiry into electrical phenomena [11]. There are three basic forms of CP discharge systems: The *glow discharge* has electrodes at either end of a separating space, which may be partially evacuated or filled with a specific gas. The *radio frequency discharge* uses pulsed electricity to generate CP within the center of the electrical coil. The *barrier discharge* uses an intervening material with high electrical resistance (the dielectric material) to distribute the flow of current and generate the plasma.

A simple form of the barrier discharge systems is shown here as an example. These systems may use one or two layers of dielectric material, arranged in various configurations [92]. These may also be arranged in an annular or tubular form, with one electrode entirely within the other. In those designs, the CP is generated in the space between the electrodes. These designs allow for gas movement across the zone of plasma generation and delivery of the CP to the target [93].

1.9.2 ACTION OF PLASMA ON MICROORGANISMS

The reactive species in plasma have been widely associated to the direct oxidative effects on the outer surface of microbial cells. As an example, commonly used oxygen and nitrogen gas plasma are excellent sources of reactive oxygen-based and nitrogen-based species, such as O, O_2, O_3, $OH\bullet$, $NO\bullet$, NO_2, etc. Atomic oxygen is potentially a very effective sterilizing agent, with a chemical rate constant for oxidation at room temperature of about 106 times that of molecular oxygen [44]. These act on the unsaturated fatty acids of the lipid bilayer of the cell membrane, thereby impeding the transport of biomolecules across it. The double bonds of unsaturated lipids are particularly vulnerable to ozone attack [86]. Membrane lipids are assumed to be more significantly affected by the reactive oxygen species (ROS) due to their location along the surface of bacterial cell, which allows them to be bombarded by these strong oxidizing agents [148]. The proteins of the cells and the spores are equally vulnerable to the action of these species, causing denaturation and cell leakage. Oxidation of amino acids and nucleic acids may also cause changes that result in microbial death or injury [44].

1.9.3 APPLICATIONS OF COLD PLASMA

CP can be used to coat the surface of foods (dairy or non-dairy) with a film of vitamins or sensitive bioactive compounds. Equally known as the fourth state of matter, CP can be used to disinfect, but it does not penetrate deeply. It effectively disinfects the irregular surfaces of equipment and packaging.

1.9.3.1 Applications of CP in Surface Decontamination

CP can be used for decontamination of products where micro-organisms are externally located. Unlike light (UV decontamination), plasma flows around objects, ensuring all parts of the product are treated.

1.9.3.2 Applications of CP in Mild Surface Decontamination

For products like cut vegetables and fresh meat, there is no mild surface decontamination technology available currently; CP could be used for this purpose.

1.9.4 CP USE IN FOOD PACKAGING TECHNOLOGY

- It was originally developed to increase the surface energy of polymers, enhancing adhesion, printability and sealability.
- It has recently emerged as a powerful tool for surface decontamination food packaging materials. Gas plasma reactions establish efficient inactivation of micro-organisms (bacterial cells, spores, yeasts and molds) adhering to polymer surfaces within short treatment times. *Packaging materials such as plastic bottles, lids and films can be rapidly sterilized using CP,* without adversely affecting their bulk properties or leaving any residues.
- New trends aim to develop in package decontamination, offering non-thermal treatment of foods post packaging. (active packaging technique) it is effective and quick method to destroy microbes.

1.9.5   IN-PACKAGE PLASMA TECHNOLOGY

Recently DBDs have been employed for generation of plasma inside sealed packages containing bacterial samples [43, 143], fresh produce [114], fish [40] and meat [143]. The in-package plasma decontamination of foods and biomaterials relies on use of the polymeric package itself as a dielectric and has been studied using several packaging materials such as LDPE, HDPE, polystyrene (PS), etc. [111]. All these works have demonstrated significant reduction in microbial population on food products. Moreover, this approach is easy to scale-up to continuous industrial processing and could prevent post-packaging contamination [142]. For a complete assessment of the technology, it is essential to quantify all possible changes to the packaging, induced by the CP. For example, the migration limits of additives, monomers, oligomers and low molecular weight volatile compounds from the packaging material into the food (following in-package plasma) should be evaluated for food safety reasons, as well as water vapor and oxygen permeability.

1.9.6   ADVANTAGES OF USING COLD PLASMA

• Flexible sanitizing method uses electricity, for example, plasma torch.
• Antimicrobial chemical agents are not required to inactivation process.
• Cost effective method compared to other chemical and thermal decontamination methods.
• Less time consuming method (need few seconds to inactivate microbes within the food surface and mild surface).
• Green technology (since harmful effects to environment has not identified yet).

1.9.7   LIMITATIONS OF USING COLD PLASMA

CP are the relatively early state of technology development, the variety and complexity of the necessary equipment, and the largely unexplored impacts of CP treatment on the sensory and nutritional qualities of treated foods. Also, the antimicrobial modes of action for various CP systems vary depending on the type of CP generated. Optimization and scale up to commercial treatment levels require a more complete understanding of these chemical processes.

1.10   HIGH VOLTAGE ARC DISCHARGE (HVAD) TECHNOLOGY

The arc discharges have been used in many areas such as biochemistry, biology, medicine, microbial inactivation of food and also for biocompounds extraction from different products [23]. The arc discharge leads to a multitude of physical and chemical effects. The high pressure shock waves can induce bubble cavitation, which can create strong secondary shocks with very short duration. These shocks can interact with structures of the cells. The phenomena result in mechanically rupture of the cell membranes that accelerate the extraction of intracellular compounds [130]. The voltage arc discharge prompts the formation of highly reactive free radicals from chemical species in foods, such as oxygen. The free radicals are toxic compounds that serve to inactivate certain intracellular components required for cellular metabolism. The bacterial inactivation was not due to heating, but mainly to irreversible loss of membrane function as a semipermeable barrier between the bacterial cell and the environment. Moreover, the formation of toxic compounds (oxygen radicals and other oxidizing compounds) was noticed. The major drawbacks of this method are the contamination of the treated foods by chemical products of electrolysis and disintegration of food particles by shock waves. The method based on continuous HVADs may be unsuitable for use in the food industry [99].

1.11   CONCLUSIONS

The concern behind the thermal processing of food is loss of volatile compounds, nutrients, and flavor. To overcome these problems, innovative methods are being developed in food industries to increase the production rate and profit. The non-thermal processing is used for all foods for its better quality, acceptance, and for its shelf life also reduces the operational cost. Innovative methods have better potential than other conventional methods and still is an evolving challenging field. The cost of equipment used in the non-thermal processing is high when compared to equipment used in thermal processing. After minimizing the investment costs and energy saving potential of non-thermal processing methods, it can also be employed in small scale industries.

1.12 SUMMARY

Preservation is the most important process related to all food products. Preservation of food products can be achieved by various ways like addition of salt, sugars, preservatives, antioxidants, naturally occurring antimicrobial substances and also by the processes like drying, freezing, refrigerated storage and Hurdle technology. Novel technologies like microwave heating, PEF technology, HPP, PL technology, ohmic heating, ultrasonics, pulsed X-rays are also applied for the preservation of food products. The main problem with the thermal processing method is loss of color, flavor, vitamins and other nutrients in food products. A detailed review is presented for different non-thermal processing methods and its merits and demerits are analyzed and illustrated for applications in various industries. This chapter investigates different non-thermal processing methods and its suitability to different food processing industries which deal with foods like meat, milk, fish, egg and ready-to-eat foods.

KEYWORDS

- atmospheric pressure
- baking
- blanching
- cold plasma

- cooking
- emulsification
- enzyme inactivation
- food dehydration
- fruit and vegetable processing
- high pressure processing
- high voltage arc discharge
- honey
- irradiation
- isostatic
- meat technology
- microscopic ordering
- microwave heating
- milk fouling
- monosonication
- non thermal
- ohmic heating
- oil
- pasteurization
- plasma
- preservation
- pulsed electric field
- pulsed light
- sterilization
- tempering
- thawing
- thermosonication
- ultra sonication
- ultrasonic
- ultrasound
- ultraviolet light

REFERENCES

1. Abid, M., Jabbar, S., Wu, T., Hashim, M. M., Hu, B., Lei, S., & Zhang, X. (2013). Effect of ultrasound on different quality parameters of apple juice. *Ultrasonics Sonochemistry, 20*, 1182–1187.

2. Albert, Cs., Mandoki, Zs., Csapo-Kiss, Zs., & Csapo, J. (2009). The effect of microwave pasteurization on the composition of milk. *Acta Univ. Sapientiae Alimentaria, 2*(2), 153–165.

3. Allen, K., Eidman, V., & Kinsey, J. (1996). An economic engineering study of ohmic food processing. *Food Tech., 50*, 269–273.

4. Alwis, A. A. P., Halden, K., & Fryer, P. J., (1989). Shape and conductivity effects in the ohmic heating of foods. *Chemical Engineering Research and Design, 67*, 159–168.

5. Arias, M., Lopez-Fandino, R., & Olano, A. (2000). Influence of pH on effect of high pressure on Milk. *Milchwissenschaft., 55*(4), 191–194.

6. Aronowicz, J. (1975). In-line microwave tempering upgrades quality of sliced meats. *Food Process., 36*(12), 54–55.

7. Ay, C., & Gunasekaran, S. (1994). Ultrasonic attenuation measurements for estimating milk coagulation time, *Transactions of de ASAE, 37*(3), 857–862.

8. Ayadi, M. A., Leuliet, J. C., Chopard, F., Berthou, M., & Lebouche, M. (2005). Experimental study of hydrodynamics in a flat ohmic cell impact on fouling by dairy products. *Journal of Food Engineering, 70*, 489–498.

9. Balachandran, S., Kentish, S. E., Mawson, R., & Ashokkumar, M. (2006). Ultrasonic enhancement of the supercritical extraction from ginger. *Ultrasonics Sonochemistry, 13*, 471–479.

10. Balasa, A., Toepfl, S., & Knorr, D. 2006. Pulsed electric field treatment of grapes. *Food Factory of the Future, 3*, Gothenburg, Sweden.

11. Becker, K. H., Kogelschatz, U., Schoenbach, K. H., & Barker, R. J. (2005). Generation of cold plasmas. In: *Non-Equilibrium Air Plasma at Atmospheric Pressure*, edited by Becker, K. H., Kogelschatz, U., Schoenbach, K. H., & Barker, R. J., Inst. Phys. Pub., Bristol, UK, pp. 19–24.

12. Benguigui, L., Emery, J., Durand, D., & Busnel, J. P. (1994). Ultrasonic study of milk clotting, *Lait, 74*(3), 197–206.

13. Bezanson, A. (1975). Thawing and tempering frozen meat. In: *Proceedings of the Meat Industry Research Conference*, Raytheon Co., Waltham, Massachusetts, USA and American Meat Science Association: Illinois, USA, pp. 51–62.

14. Bhat, R., Ameran, S. B., Voon, H. C., Karim, A. A., & Tze, L. M. (2011). Quality attributes of starfruit (*Averrhoa carambola* L.) juice treated with ultraviolet radiation. *Food Chemistry, 127*, 641–644.

15. Bhat, R., & Karim, A. A. (2009). Effects of radiation processing on phytochemicals and antioxidants in plant produce. *Trends Food Sci. Technol., 5*, 201–212.

16. Bhat, R., Kamaruddin, N. S. B. C., Min-Tze, L., & Karim, A. A. (2011). Sonication improves kasturi lime (*Citrus microcarpa*) juice quality. *Ultrasonics Sonochemistry, 18*, 1295–1298.

17. Bhat, R., Sridhar, K. R., & Karim, A. A. (2010). Microbial quality evaluation and effective decontamination of nutraceutically valued lotus seeds by electron beams and gamma irradiation. *Radiation Physics and Chemistry, 79*, 976–981.

18. Bhatti, S. S., Bhatti, R., & Singh, S. (1986). Ultrasonic testing of milk. *Acoustica,* 62, 96–99.
19. Bilbao-Sáinz, C., Younce, F. L., Rasco, B., & Clark, S. (2009). Protease stability in bovine milk under combined thermal-high hydrostatic pressure treatment, *Food Science and Emerging Technologies, 10*(3), 314–320.
20. Blank, G., Shamsuzzaman, K., & Sohal, S. (1992). Use of electron beam irradiation for mold decontamination on Cheddar cheese. *J. Dairy Sci., 75,* 13.
21. Bongirwar, D. R., & Kumta, U. S. (1967). Preservation of cheese with combined use of gamma-rays and sorbic acid. *Int. J. Appl. Rad. Isotopes, 18,* 133.
22. Bookwalter, G. N., Shukla, T. P., & Kwolek, W. F. (1982). Microwave processing to destroy *Salmonellaein* corn-soy-milk blends and effect on product quality. *J. Food Sci., 47*(5), 1683–1686.
23. Boussetta, N., Vorobiev, E., Reess, T., De Ferron, A., Pecastaing, L., Ruscassié, R., & Lanoisellé, J. L. (2012). Scale-up of high voltage electrical discharges for polyphenols extraction from grape pomace: Effect of the dynamic shock waves. *Innovative Food Science and Emerging Technologies, 16,* 129–136.
24. Buchheim, W., & Nour, A. E. (1992). Introduction of milk fat crystallization in emulsified state by high hydrostatic pressure. *Fett Wissenschaft Technologie, 94,* 369–373.
25. Buchheim, W., & Frede, E. (1996). Use of high pressure treatment to influence the crystallization of emulsified fats. *DMZ Lebensm Ind Milchwirtsch, 117,* 228–237.
26. Burfoot, D., & Foster, A. M. (1991). Microwave reheating of ready meals. *MAFF Microwave Science Series, 2,* 1–43.
27. Burfoot, D., & James, S. J. (1992). Developments in microwave pasteurization systems for eadymeals. *Process Technol.,* 6–9.
28. Cadeddu, S. (1981). Using microwave techniques in the production of Mozzarella cheese. In: *Proceedings from the Second Biennial Marshall International Cheese Conference,* Madison, Wisconsin, pp. 176–179.
29. Cao, S., Hu, Z., Pang, B., Wang, H., Xie, H., & Wu, F. (2010). Effect of ultrasound treatment on fruit decay and quality maintenance in strawberry after harvest. *Food Control, 21,* 529–532.
30. CAST (1989). Ionizing energy in food processing and pest control: II. Applications. Report 115, Council for Agricultural Science and Technology, Ames, Iowa.
31. Castro, A. J. (1994). Pulsed electric field modification of activity and denaturation of alkaline phosphate. *PhD Dissertation,* Washington State University, Washington.
32. Castro, I., Teixeira, J. A., & Vicente, A. A. (2002). The influence of food additives on the electrical conductivity of a strawberry pulp. In: *Proceedings of the 32nd Annual Food Science and Technology Research Conference,* University College Cork, Cork, Ireland.
33. Castro, I., Teixeira, J. A., Salengke, S., Sastry, S. K., & Vicente, A. A. (2003). The influence of field strength, sugar and solid content on electrical conductivity of strawberry products. *Journal of Food Process Engineering, 26,* 17–29.
34. Chamchong, M., & Datta, A. K. (1999). Thawing of foods in a microwave oven, I: Effect of power levels and power cycling. *J. Microw. Power Electromagn. Energy, 34*(1), 9–21.
35. Chavan, R. S., & Chavan, S. R. (2010). Microwave baking in food industry: A review. *International Journal of Dairy Science, 5*(3), 113–127.

36. Chemat, F., Huma, Z., & Khan, M. K. (2011). Applications of ultrasound in food technology: Processing, preservation and extraction. *Ultrasonics Sonochemistry, 18*, 813–835.

37. Chen, H. T., Bhat, R., & Karim, A. A. (2010). Effects of sodium dodecyl sulfate and sonication treatment on physicochemical properties of starch. *Food Chemistry, 120*, 703–709.

38. Cheng, L. H., Soh, C. Y., Liew, S. C., & Teh, F. F. (2007). Effects of sonication and carbonation on guava juice quality. *Food Chemistry, 104*, 1396–1401.

39. Chincholle, R. C. (1991). Action of the ionization treatment on the soft cheeses made from unpasteurized milk. *CR. Acad. Agric. Fr., 77*, 26.

40. Chiper, A. S., Chen, W., Mejlholm, O., Dalgaard, P., & Stamate, E. (2011). Atmospheric pressure plasma produced inside a closed package by a dielectric barrier discharge in Ar/CO(2) for bacterial inactivation of biological samples. *Plasma Sources Science and Technology, 20*, 10.

41. Chung, J. C., Shu, H. H., & Der, S. C. (2000). The physical properties of steamed bread cooked by using microwave-steam combined heating. Taiwanese. *J. Agric. Chem. Food Sci., 38*(2), 141–150.

42. Cohen, J. S., & Yang, T. C. S. (1995). Progress in food dehydration. *Trends in Food Science and Technology, 6*, 20–25.

43. Connolly, J., Valdramidis, V. P., Byrne, E., Karatzas, K. A. G., & Cullen, P. J. (2013). Characterization and antimicrobial efficacy against *E. coli* of a helium/air plasma at atmospheric pressure created in a plastic package. *Journal of Physics D: Applied Physics, 46*, 035401–035412.

44. Critzer, F., Kelly-Wintenberg, K., South, S., & Golden, D. (2007). Atmospheric plasma inactivation of foodborne pathogens on fresh produce surfaces. *J Food Protec, 70*, 2290.

45. Cruz, R. M. S., Vieira, M. C., & Silva, C. L. M. (2006). Effect of heat and thermosonication treatments on peroxidase inactivation kinetics in watercress (*Nasturtium officinale*). *Journal of Food Engineering, 7*, 8–15.

46. Czechowska-Biskup, R., Rokita, B., Lotfy, S., Ulanski, P., & Rosiak, J. M. (2005). Degradation of chitosan and starch by 360 kHz ultrasound. *Carbohydrate Polymers, 60*, 175–184.

47. Daniel, A. J., & C. Hernández-Brenes (2012). Stability of avocado paste carotenoids as affected by high hydrostatic pressure and storage. *Innovative Food Science and Emerging Technologies, 16*, 121–128.

48. De Alwis, A. A. P., & Fryer, P. J. (1990). The use of direct resistance heating in the food industry. *Journal of Food Engineering, 11*, 3–27.

49. De Gennaro, L., Cavella, S., Romano, R., & Masi, P. (1999). The use of ultrasound in food technology, I: Inactivation of peroxidase by thermosonication. *Journal of Food Engineering, 39*, 401–407.

50. De La Fuente, M. A., B. Carazo, & M. Juarez (1997). Determination of major minerals in dairy products digested in closed vessels using microwave heating. *J. Dairy Sci, 80*, 806–811.

51. Dealler, S., Rotowa, N., & Lacey, R. (1990). Microwave reheating of convenience meals. *Br. Food J., 92*(3), 19–21.

52. Decareau, R. V. (1985). *Microwaves in the Food Processing Industry.* Academic Press: New York.

53. Deshmane, V. G., Gogate, P. R., & Pandit, A. B. (2009). Ultrasound-assisted synthesis of biodiesel from palm fatty acid distillate. *Industrial and Engineering Chemistry Research, 48*, 7923–7927.

54. Desobry-Banon, S., Richard, F., & Hardy, J. (1994). Study of acid and rennet coagulation of high pressurized milk. *J. Dairy Sci., 77*(11), 3267–3274.

55. Dicman, S., (1991). Compromise eludes EC. *Nature, 349*, 273.

56. Djenouhat, M., Hamdaoui, O., Chiha, M., & Samar, M. H. (2008). Ultrasonication assisted preparation of water-in-oil emulsions and application to the removal of cationic dyes from water by emulsion liquid membrane. *Separation and Purification Technology, 62*, 636–641.

57. Dolatowski, J. Z., Stadnik, J., & Stasiak, D. (2007). Application of ultrasound in food technology. *Acta Scientiarum Polonorum Technologia Alimentaria, 6*, 89–99.

58. Drake, M. A., Harrison, S. L., Asplund, M., Barbicosa, C. G., & Swanson, B. G. (1997). High pressure treatment of milk and effects on microbiological and sensory quality of Cheddar cheese. *J Food Sci, 62*, 843–845.

59. Dring, J. G. (1976). Some aspects of the effects of hydrostatic pressure on microorganisms. In: *Inhibition and Inactivation of Microorganisms*, edited by Skinner, F. A., & Hugo, W. B. London: Academic Press, pp. 257–277.

60. Earnshaw, R. G., Appleyard, J., & Hurst, R. M. (1995). Understanding physical inactivation processes: Combined preservation opportunities using heat, ultrasound and pressure. *International Journal of Food Microbiology, 28*, 197–219.

61. Eberhard, P., Strahm, W., & Eyer, H. (1999). High pressure treatment of whipped cream. *Agrarforschung, 6*, 352–354.

62. Elzubier, A. S., Thomas, C. S. Y., Sergie, S. Y., Chin, N. L., & Ibrahim, O. M. (2009). The effect of buoyancy force in computational fluid dynamics simulation of a two-dimensional continuous ohmic heating process. *Am J Appl Sci, 6*(11), 1902–1908.

63. Ersus, S., & Barrett, D. M. (2010). Determination of membrane integrity in onion tissues treated by pulsed electric fields: Use of microscopic images and ion leakage measurements. *Innovative Food Science and Emerging Technologies, 11*(4), 598–603.

64. Farkas, J., (2004). Charged particle and photon interactions with matter. In: *Food Irradiation*, Mozumder, A., & Hatano, Y. (eds.). Marcel Dekker, New York, pp. 785–812.

65. Fazilah, A., Azemi, M. N. M., Karim, A. A., & Norakma, M. N. (2009). Physico-chemical properties of hydrothermally treated hemicellulose from oil palm frond. *Journal of Agricultural and Food Chemistry, 57*(4), 1527–1530.

66. Feng, H., Barbosa-Canovas, G. V., & Weiss, J. (2011). *Ultrasound technologies for food and bioprocessing*, Springer, New York.

67. Fernández, A., & Thompson, A. (2012). The inactivation of Salmonella by cold atmospheric plasma treatment. *Food Research International, 45*(2), 678–684.

68. Fernandez-Molina, J. J., Fernandez-Gutierrez, S. A., Altunakar, B., Bermudez-Aguirre, Swanson, B. G., & Barbosa-Canovas, G. V. (2005). The combined effect of pulsed electric fields and conventional heating on the microbial quality and shelf life of skim milk. *Journal of Food Processing and Preservation, 29*, 390–406.

69. Food and Agriculture Organization (FAO), World Health Organization (1984). *Codex General Standard for Irradiated Foods and Recommended International Code of Practice for the Operation of Radiation Facilities used for the Treatment of Food.* Codex Aliment Arius, volume 15, FAO/WHO, Rome.

70. Fox, J. A. (2002). Influence on purchase of irradiated foods. *J. Food Technol.*, *56*(11), 34.
71. Fox, K. (1994). Innovations in citrus processing. *Fruit Process*, *4*(11).
72. Fryer, P. J., & Li, Z. (1993). Electrical resistance heating of foods. *Journal of Food Science and Technology*, *4*, 364–369.
73. Furuta, M., Yamaguchi, M., Tsukamoto, T., Yim, B., Stavarache, C. E., Hasiba, K., & Maeda, Y. (2004). Inactivation of *Escherichia coli* by ultrasonic irradiation. *Ultrasonics Sonochemistry, 11*, 57–60.
74. Galazka, V. B., & Ledward, D. A. (1995). Developments in high pressure food. In: *Food Technology International Europe*, edited by Turner, A. Sterling Publications International, London, pp. 123–125.
75. Gallego-Juárez, J., Rodriguez, G., Acosta, V., & Riera, E. (2010). Power ultrasonic transducers with extensive radiators for industrial processing. *Ultrasonics Sonochemistry, 17*, 953–964.
76. Gao, S., Lewis, G. D., Ashokkumar, M., & Hemar, Y. (2014). Inactivation of microorganisms by low frequency high power ultrasound, I: Effect of growth phase and capsule properties of the bacteria. *Ultrasonics Sonochemistry, 21*, 446–453.
77. Garcia-Gonzalez, L., Geeraerd, A. H., Spilimbergo, S., Elst, K., Van Ginneken, L, Debevere, J., VanImpe, J. F., & Devlieghere, F. (2007). High pressure carbon dioxide inactivation of microorganisms in foods: The past, the present and the future. *International Journal of Food Microbiology, 117*(1), 1–28.
78. Gaudreau, M., Hawkey, T., Petry, J., & Kempkes, M. (2008). Pulsed electric field processing for food and waste streams. *Food Australia, 60*, 323–325.
79. Geczi, G., Nagy, P. I., & Sembery, P. Primary Processing of the animal food products with microwave heat treatment. http://www.agir.ro/buletine/1311.pdf
80. George, M. (1997). Industrial microwave food processing. *Food Rev., 24*(7), 11–13.
81. Getchel, B. E. (1935). Electric pasteurization of milk. *Agriculture Engineering, 16*(10), 408–410.
82. Grahl, T., & Märkl, H. (1996). Killing of microorganisms by pulsed electric fields. *Applied Microb. Biotec., 45,* 148–157.
83. Grimi, N., Lebovka, N. I., Vorobiev, E., & Vaxelaire, J. (2009). Effect of a pulsed electric field treatment on expression behavior and juice quality of chardonnay grape. *Food Biophysics, 4*(3), 191–198.
84. Guiseppi-Elie, A., Choi, S. H., & Geckeler, K. E. (2009). Ultrasonic processing of enzymes: Effect on enzymatic activity of glucose oxidase. *Journal of Molecular Catalysis, 58*, 118–123.
85. Gundavarapu, S., Hung, Y. C., & Reynolds, A. E. (1998). Consumer acceptance and quality of microwave-cooked shrimp. *J. Food Qual., 21*(1), 71–84.
86. Guzel-Seydim, Z. B., Greene, A. K., & Seydim, A. C. (2004). Use of ozone in the food industry. *Lebensmittel-Wissenschaftund-Technologie, 37*, 453–460.
87. Hamid, M. A. K., Boulanger, R. J., Tong, S. C., Gallop, R. A., & Pereira, R. R. (1969). Microwave pasteurization of raw milk. *J. Microwave Power, 4*, 272–275.
88. Harte, F. M., Luedecke, L., Swanson, B. G., & Barbosa-Canovas, G. V. (2003). Low-fat set yogurt made from milk subjected to combinations of high hydrostatic pressure and thermal processing. *J Dairy Sci, 86*, 1074–1082.

89. Hayashi, R. (1995). Advances in high pressure in Japan. In: *Food: Recent Developments*, edited by Gaonkar, A. G., Elsevier, London, pp. 85.

90. Heinz, V., & Buckow, R. (2010). Food preservation by high pressure. *J. Consum. Protect. Food Saf., 5*(1), 73–81.

91. Hoke, K., Klima, L., Gree, R., & Houska, M. (2000). Controlled thawing of foods. Czech *J. Food Sci., 18*(5), 194–200.

92. http://dx.doi.org/10.1016/j.apenergy.2013.08.085

93. http://www.researchgate.net/publication/258344120_Applications_of_cold_plasma_technology_in_food_packaging.

94. Huppertz, T., Alan, L., & Fox, P. F. (2002). Effect of high pressure on constituents and properties of milk. *Int. Dairy Journal, 12*, 561–572.

95. Hvizdzak, A. L., Beamer, S., Jaczynski, J., & Matak, K. E. (2010). Use of electron beam radiation for the reduction of Salmonella enteric Serovars Typhimurium and Tennessee in peanut butter. *J. Food Protec., 73*(2), 353–357.

96. Icier, F., Yildiz, H., & Baysal, T. (2006). Peroxidase inactivation and color changes during ohmic blanching of pea puree. *Journal of Food Engineering, 74*, 424–429.

97. Icier, F., Yildiz, H., & Baysal, T. (2008). Polyphenoloxidase deactivation kinetics during ohmic heating of grape juice. *Journal of Food Engineering, 85*, 410–417.

98. Jabbar, S., Abid, M., Hu, B., Wu, T., Hashim, M. A., Lei, S., & Zeng, X. (2014). Quality of carrot juice as influenced by blanching and sonication treatments. *LWT – Food Science and Technology, 55*, 16–21.

99. Jayaram, S., Castle, G. S. P., & Margaritis, A. (1991). Effects of high electric field pulses on Lactobacillus brevis at elevated temperatures, *IEEE Industrial Applications in Society Annual Meeting, 5*, 674–681.

100. Jayasooriya, S. D., Torley, P. J., D'Arcy, B. R., & Bhandari, B. R. (2007). Effect of high power ultrasound and aging on the physical properties of bovine Semitendinosus and Longissimus muscles. *Meat Science, 75*, 628–639.

101. Jayasooriya, S. D., Bhandari, B. R., Torley, P., & Darcy, B. R. (2004). Effect of high power ultrasound waves on properties of meat: A review. *International Journal of Food Properties, 2*, 301–319.

102. Johnston, D. E. (1995). High pressure effects on milk and meat. In: *High Pressure of Foods*. D. A. Ledward, D. E. Johnson, R. G. Earnshaw, & A. P. M. Hasting (eds.). Nottingham University Press, pp. 99–122.

103. Jones, T. H., & Jelen, P. (1988). Low dose gamma irradiation of camembert, cottage cheese and cottage cheese whey. *Milchwissenschaft, 43*, 233.

104. Jose, A., Sepulveda, D. R., Gongora-Nieto, M. M., Swanson, B., & Barbosa-Canovas, G. V. (2010). Milk thermization by pulsed electric fields and electrically induced heat. *J Food Engg, 100*, 56–60.

105. Juliano, B. O. (1985). Production and utilization of rice. In: *Rice Chemistry and Technology*. (2nd Ed.), edited by Juliano, B. O. St. Paul: American Association of Cereal Chemists, pp. 1–16.

106. Jun, S., & Sastry, S. K. (2007). Reusable pouch development for long term space missions: A 3D ohmic model for verification of sterilization efficacy. *Journal of Food Engineering, 80*, 1199–1205.

107. Jun, S., Chang, H. L., & Ouk, H. (1988). Effects of height for microwave defrosting on frozen food. *J. Kor. Soc. Food Sci. Nutr., 27*(1), 109–114.

108. Kaltsa, O., Michon, C., Yanniotis, S., & Mandala, I. (2013). Ultrasonic energy input influence on the production of sub-micron o/w emulsions containing whey protein and common stabilizers. *Ultrasonics Sonochemistry, 20,* 881–891.

109. Kanjanapongkul, K., Tia, S., Wongsa-Ngasri, P. and yoovidhya, T., (2009). Coagulation of protein in surimi wastewater using a continuous ohmic heater. *Journal of Food Engineering, 91,* 341–346.

110. Katrokha, L., Matvienko, A., Vorona, L., Kolchak, M., & Zaets, V., (1984). Intensification of sugar extraction from sweet sugar beet cossettes in an electric field. *Sakharnaya Promyshlennost, 7,* 28–31.

111. Keener, K. M., Jensen, J., Valdramidis, V., Byrne, E., Connolly, J., Mosnier, J., & Cullen, P. (2012). Decontamination of Bacillus subtilis spores in a sealed package using a non-thermal plasma system. In: *NATO Advanced Research Workshop: Plasma for Bio-Decontamination, Medicine and Food Security,* edited by Hensel, K., & Machala, Z., Jasna, Slovakia, pp. 445–455.

112. Kentish, S., & Ashokkumar, M. (2011). The physical and chemical effects of ultrasound. In: *Ultrasound Technologies for Food and Bioprocessing,* edited by Feng, H., Barbosa-Canovas, G. V., & Weiss, J., London: Springer, pp. 1–12.

113. Kim, J., & Pyun, Y., (1995). Extraction of soymilk using ohmic heating. *9th Congress of Food Sci. Tech,* Budapest, Hungary.

114. Klockow, P. A., & Keener, K. M. (2009). Safety and quality assessment of packaged spinach treated with a novel ozone-generation system. *LWT – Food Science and Technology, 42,* 1047–1053.

115. Knorr, D. (1995). High pressure effects on plant derived foods. In: *High Pressure of Foods,* edited by Ledward, D. A., Johnson, D. E., Earnshaw, R. G., & Hasting, A. P. M., Nottingham University Press, pp. 123–136.

116. Knorr, D., Zenker, M., Heinz, V., & Lee, D. U. (2004). Applications and potential of ultrasonics in food processing. *Trends in Food Science and Technology, 15,* 261–266.

117. Kolakowski, P., Reps, A., Dajnowiec, F., Szczepek, & J., Porowski, S. (1997). Effect of high pressures on the microflora of raw cow's milk. In: *Process Optimization and Minimal Processing of Foods, Proceedings of the Third Main Meeting, Volume 4,* edited by Jorge C. Oliveira & Oliveira, F. A. R. Leuven Catholic University Press, Belgium. pp. 46–50.

118. Konteles, S., Sinanoglou, V. J., Batrinou, A., & Sflomos, K. (2009). Effects of gamma irradiation on Listeria monocytogenes population, color, texture and sensory properties of Feta cheese during cold storage. *Food Microbiol., 26*(2), 157–165.

119. Koutchma, T., Le Bail, A., & Ramaswamy, H. S. (2001). Comparative experimental evaluation of microbial destruction in continuous-flow microwave and conventional heating systems. *Can Biosyst Eng., 43,* 3.1–3.8.

120. Kunstadt, P. (2001). Economic and technical considerations in food irradiation. In: *Food Irradiation: Principles and Applications by Molins,* R. A. (Ed). Wiley – Interscience: New York, pp. 415–442.

121. Kutchma, T. (1998). Synergistic effect of microwave heating and hydrogen peroxide on in activation of microorganisms. *J. Microw. Power Electromagn. Energy, 33*(2), 77–87.

122. Lateef, A., Oloke, J. K., & Prapulla, S. G. (2007). The effect of ultrasonication on the release of fructosyl-transferase from *Aureobasidium pullulans* CFR 77. *Enzyme and Microbial Technology, 40,* 1067–1070.

123. Leadley, C. E., & Williams, A. (2006). Pulsed electric field processing, power ultrasound and other emerging technologies. In: *Food Processing Handbook*, edited by Brennan, J. G., Wiley-VCH, Weinheim, pp. 214–218.
124. Lee, H., & Feng, H. (2011). Effect of power ultrasound on food quality. In: *Ultrasound Technologies for Food and Bioprocessing*, edited by Feng, H., Barbosa-Canovas, G. V., & Weiss, J., London: Springer. pp. 559–582.
125. Lexandre, E. M. C., Brandao, T. R. S., & Silva, C. L. M. (2013). Impact of non-thermal technologies and sanitizer solutions on microbial load reduction and quality factor retention of frozen red bell peppers. *Innovative Food Science and Emerging Technologies, 17*, 199–205.
126. Li, S. Q., Zhang, Q. H., Lee, Y. Z., & Pham, T. V. (2003). Effects of pulsed electric field and thermal processing on the stability of bovine IgG in enriched soymilk. *J Food Sci, 68*, 1201–1207.
127. Lim, S. Y., Swanson, B. G., Ross, C. F., & Clark, S. (2007). High hydrostatic pressure modification of whey protein concentrate for improved body and texture of low fat ice-cream. *J Dairy Sci, 91*, 1308–1316.
128. Lima, M., & Sastry, S. K., (1999). The effects of ohmic heating frequency on hot air drying rate and juice yield. *Journal of Food Engineering, 41*, 115–119.
129. Lin, I., & Erel, D. (1992). *Dynamic ultrasonic cleaning and disinfecting device and method*. US Patent No. 5113881A. Washington, DC: U. S. Patent and Trademark Office.
130. Liu, D., Vorobiev, E., Savoire, R., & Lanoisellé, J. L. (2011). Intensification of polyphenols extraction from grape seeds by high voltage electrical discharges and extract concentration by dead-end ultrafiltration. *Separation and Purification Technology, 81*(2), 134–140.
131. Loaharanu, P., (2005). Irradiation as a cold pasteurization process of food. *Journal of Veterinary and Parasitology, 64*(2), 171–182.
132. Lopez, P., Sala, F. J., de la Fuente, J. L., Condon, S., Raso, J., & Burgos, J. (1994). Inactivation of peroxidase, lipoxygenase and polyphenol oxidase by monothermoso-nication. *Journal of Agricultural and Food Chemistry, 42*, 252–256.
133. Manas, P., Munoz, B., Sanz, D., & Condon, S. (2006). Inactivation of lysozyme by ultrasonic waves under pressure at different temperatures. *Enzyme and Microbial Technology, 39*, 1177–1182.
134. Marconi, E., Ruggeri, S., Paoletti, F., Leonardi, D., & Carnovale, E. (1998). Physicochemical and Structural modifications in chickpea and common bean seeds after traditional and microwave cooking processes. In: *3rd European Conference on Grain Legumes. Opportunities for High Quality, Healthy and Added Value Crops to European Demands.* Valladolid, Spain, 14–19 November. pp. 358–359.
135. Martin-Belloso, O., & Martinez, P. (2005). Food safety aspects of pulsed electric field. In: *Emerging Technologies for Food Processing*, edited by Da-Wen Sun, Elsevier Academic Press, London, pp. 184–217.
136. Mason, T. J. (1998). Power ultrasound in food processing – The way forward. In: *Ultrasound in Food Processing*, edited by Povey, M. J., & Mason, T. J., London, UK: Thomson Science. pp. 105–126.
137. Mason, T. J., Paniwnyk, L., & Lorimer, J. P. (1996). Uses of ultrasound in food technology. *Ultrasonics Sono Chemistry, 3*, 253–260.

138. McClements, D. J. (1995). Advances in the application of ultrasound in food analysis and processing. *Trends in Food Science and Technology*, *6*, 293–299.
139. Mehran, N. T., Tawfeak, Y., & Hewedy, M. (2005). Incidence of pathogens in kareash cheese. *Egyptian Journal of Dairy Science*, *26*(1), 295–300.
140. Mertens, B., & Knorr, D. (1992). Developments of non-thermal processes of food preservation. *Food Technol*, *5*, 124–133.
141. Michalac, S. L., Alvarez, V. B., & Zhang, Q. H. (2003). Inactivation of selected microorganisms and properties of pulsed electric field processed milk. *J Food Process Preser*, *27*, 137–151.
142. Misra, N. N., Tiwari, B. K., Raghavarao, K. S. M. S., & Cullen, P. J. (2011). Nonthermal plasma inactivation of food-borne pathogens. *Food Engineering Reviews, 3*, 159–170.
143. Misra, N. N., Ziuzina, D., Cullen, P. J., & Keener, K. M. (2012). Characterization of a novel cold atmospheric air plasma system for treatment of packaged liquid food products. Paper 121337629 Presented at Meeting by American Society of Agricultural and Biological Engineers, Dallas, Texas, July 29–August 1.
144. Mizrahi, S., Kopelman, I., & Perlaman, J. (1975). Blanching by electroconductive heating. *Journal of Food Technology, 10*, 281–288.
145. Mongenot, N., Charrier, S., & Chalier, P. (2000). Effect of ultrasound emulsification on cheese aroma encapsulation by carbohydrates. *J Agric Food Chem.*, *48*, 861–867.
146. Monteiro, C. A. (2009). Nutrition and health: The issue is not food, nor nutrients, so much as processing. *Public Health Nutrition*, *12*(5), 729–731.
147. Monteiro, C. A. (2010). A new classification of foods based on the extent and purpose of their processing. Public Health Nutrition, *26*(11), 2039–2049.
148. Montie, T. C., Kelly, K., & Roth, J. R. (2002). An overview of research using the one atmosphere uniform glow discharge plasma for sterilization of surfaces and materials. *Plasma Sci IEEE Transactions, 28*, 41–50.
149. Morris, C., Brody, A. L., & Wicker, L. (2007). Non-thermal food processing/ preservation technologies: a review with packaging implications. *Packaging Technology and Science*, *20*(4), 275–286.
150. Mortazavi, A., & Tabatabaie, F. (2008). Study of ice cream freezing process after treatment with ultrasound. *World Applied Sciences Journal*, *4*(2), 188–190.
151. Mulet, A., Carcel, J., Benedito, C., Rossello, C., & Simal, S. (2003). Ultrasonic mass transfer enhancement in food processing. Chapter 18, In: *Transport Phenomena of Food Processing*, edited by Welti-Chanes, J., Vélez-Ruiz, F., & Barbosa-Cánovas, G. V., Boca Raton.
152. Murcia, M. A., Martınez Tome, M., del Cerro, I., Sotillo, F., & Ramırez, A. (1999). Proximate composition and vitamin E levels in egg yolk: losses by cooking in a microwave oven. *J. Sci. Food Agric.*, *79*(12), 1550–1556.
153. Nachamansion, J. (1995). Packaging solutions for high quality foods processed by high hydrostatic pressure. *Proceedings of Europack.*, *7*, 390–401.
154. Negi, A., Boora, P., & Khetarpaul, N. (2001). Effect of microwave cooking on the starch and protein digestibility of some newly released moth bean (*Phase olusaconitifolius Jacq.*) cultivars. *J. Food Compos. Anal.*, *14*(5), 541–546.
155. Niemira, B. A. (2012). Cold plasma decontamination of foods. *Annual Rev Food Sci Technol*, *3*, 125–142.

156. Odriozola-serrano, I., Bendicho, S., & Martin, O. (2006). Comparative study on shelf life of whole milk processed by high intensity PEF or heat treatment. *J Dairy Sci*, *89*, 905–911.

157. Ohlsson, T., (2002). *Minimal Processing Technologies in the Food Industry*. Boca Raton, FL: CRC Press.

158. Oms-Oliu, G., Odriozola-Serrano, I., Soliva- Fortuny, R., & Martín-Belloso, O. (2009). Effects of high intensity pulsed electric field processing conditions on lycopene, vitamin C and antioxidant capacity of watermelon juice. *Food Chemistry*, *115*(4), 1312–1319.

159. Orlandini, I., & Annibaldi, S. (1983). New techniques in evaluation of the structure of parmesan cheese: ultrasonic and X-rays. *Sci. Latiero-Caseria*, *34*, 20–30.

160. Ozbek, B., & Ulgen, K. O. (2000). The stability of enzymes after sonication. *Process Biochemistry*, *35*, 1037–1043.

161. Palaniappan, S., & Sastry, S. K., (1991). Electrical conductivity of selected juices: Influences of temperature, solids content, applied voltage and particle size. *Journal of Food Process Engineering*, *14*(4), 247–260.

162. Parrott, D. L. (1992). Use of OH for aseptic processing of food particulates. *Food Technology*, *45*(12), 68–72.

163. Picouet, R. A., A. Fernandez, X. Serra, J. J. Sunol, & J. Arnau (2007). Microwave heating of cooked pork patties as a function of fat content. *J. Food Sci.*, *72*(2), E57–E63.

164. Qin, B. L., Pothakamury, U. R., Barbosa-Cánovas, G. V., & Swanson, B. G. (1996). Nonthermal pasteurization of liquid foods using high-intensity pulsed electric fields. *Critical Reviews in Food Science and Nutrition*, *36*(6), 603–627.

165. Quarini, G. L. (1995). Thermal hydraulic aspects of the ohmic heating process. *J Food Eng.*, *24*, 561–574.

166. Rahman, M. S. (1999). *Handbook of Food Preservation*. CRC Press.

167. Rajkovic, A., Smigic, N., & Devlieghere, F. (2010). Contemporary strategies in combating microbial contamination in food chain. *International Journal of Food Microbiology*, *141*(1), S29–S42.

168. Raso, J., & Barbosa-Canovas, G. V. (2003). Nonthermal preservation of foods using combined processing techniques. *Critical Reviews in Food Science and Nutrition*, *43*, 265–285.

169. Rastogi, N. K., Raghavarao, K. S. M. S., Balasubramaniam, V. M., Niranjan, K., & Knorr, D. (2007). Opportunities and challenges in high pressure processing of foods. *Crit. Rev. Food Sci. Nutr.*, *47*, 69–112.

170. Rastogi, N. K., Eshtiaghi, M. N., & Knorr, D. (1999). Accelerated mass transfer during osmotic dehydration of high intensity electrical field pulse pre-treated carrots *Journal of Food Science*, *64*, 1020–1023.

171. Raso, J., & Barbosa-Canovas, G. V. (2003). Non-thermal preservation of foods using combined processing techniques. *Critical Reviews in Food Science and Nutrition*, *43*, 265–285.

172. Raso, J., Palop, A., & Condon, S. (1998). Inactivation of Bacillus subtilis spores by combining ultrasound waves under pressure and mild heat treatment. *Journal of Applied Microbiology*, *85*, 849–854.

173. Raviyan, P., Zhang, Z., & Feng, H. (2005). Ultrasonication for tomato pectin methylesterase inactivation: Effect of cavitation intensity and temperature on inactivation. *Journal of Food Engineering*, *70*, 189–196.

174. Rod, S. K., Hansen, F., Leipold, F., & Knochel, S. (2012). Cold atmospheric pressure plasma treatment of ready-to-eat meat: Inactivation of Listeria innocua and changes in product quality. *Food Microbiology, 30*, 233–238.

175. Rosenthal, I., Martinot, M., Linder, P., & Juven, B. J. (1983). A study of ionizing radiation of dairy products. *Milchwissenschaft, 38*, 467.

176. Russell, A. B., Cheney, P. E., & Wantling, S. D. (1999). Influence of freezing conditions on ice crystallization in ice cream. *Journal of Food Engineering, 39*(2), 179–191.

177. Şahin, S., & Soysal, Ç. (2011). Effect of ultrasound and temperature on tomato peroxidase. *Ultrasonics Sonochemistry, 18*, 689–695.

178. Sahin, S., & Sumnu, G. (2002). Effects of microwave cooking on fish quality. *Int. J. Food Prop., 4*(3), 501–512.

179. Sakonidoua, E. P., Karapantsiosa, T. D., & Raphaelides, S. N. (2003). Mass transfer limitations during starch gelatinization *Carbohydrate Polymers, 53*(1), 53–61.

180. Sams, A. R., & Feria, R. (1991). Microbial effects of ultrasonication of broiler drumstick skin. *Journal of Food Science, 56*, 247–248.

181. Santos, F. F. P., Rodrigues, S., & Fernandes, F. A. N. (2009). Optimization of the production of biodiesel from soybean oil by ultrasound assisted methanolysis. *Fuel Processing Technology, 90*, 312–316.

182. Sastry, S. K., & Palaniappan, S. (1992). Mathematical modeling and experimental studies on ohmic heating of liquid-particle mixtures in a static heater. *Journal of Food Process Engineering, 15*, 241–226.

183. Schwab, E. C., & Brown, G. E.(1993). *Microwave Tempering of Cooked Cereal Pellets or Pieces. Minneapolis, MN: General Mills*, United States Patent 05182127.

184. Sepulveda-Ahumada, D. R. (2003). Preservation of fluid foods by pulse electric fields in combination with mild thermal treatments. PhD Thesis at Pullman WA: Washington State University.

185. Sharma, R., & Lal, D. (1998). Influence of various heat processing treatments on some B vitamins in buffalo and cow's milk. *J Food Sci. Technol., 35*(6), 524–526.

186. Singh, M., Raghavan, B., & Abraham, K. O. (1996). Processing of marjoram (*Marjonahortensis Moench.*) and rosemary (*Rosmari nusofficinalis* L.): Effect of blanching methods on quality. *Nahrung, 40*, 264–266.

187. Slimani, N., Deharveng, G., & Southgate, D. T. (2009). Contribution of highly industrially processed foods to the nutrient intakes and patterns of middle-aged populations in the European prospective investigation into cancer and nutrition study. *European Journal of Clinical Nutrition, 63*(4), S206–S225.

188. Smelt, J. M. (1998). Recent advances in the microbiology of high pressure processing. *Trends Food Sci. Technol., 9*, 152–158.

189. Stanel, J., & Zitny, R., (2010). Milk fouling at direct ohmic heating. *Journal of Food Engineering, 99*, 437–444.

190. Steele, R. J. (1987). Microwave in the food industry. *CSIRO Food Res. Q., 47*(4), 73.

191. Stevenson, M. H. (1994). Nutritional and other implications of irradiating meat. *Proceedings of The Nutrition Society, 53*(2), 317–325.

192. Stirling, R. (1987). Ohmic heating: a new process for the food industry. *J Power Eng, 6*, 365–371.

193. Stoica, M., Bahrim, G., & Cârâc, G. (2011). Factors that Influence the Electric Field Effects on Fungal Cells. In: *Science against microbial pathogens: communicating current research and technological advances*, Formatex Research Center, Badajoz, pp. 291–302.

194. Suslick, K. S. (1998). Sonochemistry. In: *Kirk-Othmer Encyclopedia of Chemical Technology* (4th ed., Vol. 26, 517–541). New York, NY: Wiley.

195. Takai, R., Suzuki, T., Mihori, T., Chin, S., Hocchi, Y., & Kozima, T. (1994). Nondestructive evaluation of voids in kamaboko by an ultrasonic pulse-echo technique. *J. Japanese Society of Food Sci. Technol.*, *41*(12), 897–903.

196. Takashima, H. (2005). *Sponge Cake Premix and Method of Manufacturing Sponge Cake by Using Said Premix*. United States Patent 6,884,448.

197. Tedjo, W., Eshtiaghi, M. N., & Knorr, D. (2002). Use, non-thermal method for cell permeabilization of grapes and extraction of content materials (*Einsatz, nichtthermischer verfahren zur zellpermeabilisierung von weintrauben ung gewinnung von inhaltsstoffen*). *Fluessiges Obst*, *9*, 578–583.

198. Thayer, D. W. (2005). Food irradiation: benefits and concerns. *Journal of Food Quality*, *13*(1), 147–169.

199. Timson, W. J., & Short, A. J. (1965). Resistance of microorganisms to hydrostatic pressure. *Biotechnol. Bioengin.*, *7*, 139–159.

200. Tsong, T. (1990). Reviews on electroporation of cell membranes and some related phenomena. *Bioelectrochem Bioenergetics*, *24*, 271.

201. Uemura, K., & Isobe, S. (2003). Developing a new apparatus for inactivating Bacillus subtilis spore in orange juice with a high electric field AC under pressurized conditions. *Journal of Food Engineering*, *56*(4), 325–329.

202. Ulusoy, H. B. Colak, H., & Hampikyan, H. (2007). Use of ultrasonic waves in food technology. *Research Journal of Biological Science*, *2*, 491–497.

203. United States Food and Drug Administration (USFD) (1990). US Department of Health and Human Services, FDA, federal register 21 CFR part 179, May 2.

204. Urbain, J. H., (1983). Radurization and radicidation: fruits and vegetables. In: *Preservation of Foods by Iodine Radiation*. Vol. 3, edited by Josephson, E. S. & Peterson, M. Boca Raton, FL: CRC Press.

205. Valero, M., Recrosio, N., Saura, D., Munoz, N., Martí, N., & Lizama, V. (2007). Effects of ultrasonic treatments in orange juice processing. *Journal of Food Engineering*, *80*, 509–516.

206. Valizadeh, R., Kargarsana, H., Shojaei, M., & Mehbodnia, M. (2009). Effect of high intensity pulsed electric fields on microbial inactivation of cow milk, *Journal of Animal and Veterinary Advances*, *8*(12), 2638–2643.

207. Valli, K. (2011). *Extraction of Bioactive Compounds from Whole Red Cabbage and Beetroot using Pulsed Electric Fields and Evaluation of their Functionality*. MTech Thesis, The Graduate College at the University of Nebraska.

208. Vercet, A., Burgos, J., & Lopez-Buesa, P. (2002). Mana thermosonication of heat resistant lipase and protease from *Pseudomonas fluorescence*: Effect of pH and soni- cation parameters. *Journal of Dairy Research*, *69*, 243–254.

209. Vercet, A., Burgos, J., Crelier, S., & Lopez-Buesa, P. (2001). Inactivation of protease and lipase by ultrasound. *Innovation Food Science and Technologies*, *2*, 139–150.

210. Vercet, A., Lopez, P., & Burgos, J. (1997). Inactivation of heat resistant lipase and protease from Pseudomonas fluorescence by manothermosonication. *Dairy Science, 80,* 29–36

211. Vercet, A., Sanchez, C., Burgos, J., Montanes, L., & Lopez-Buesa, P. (2002). Effects of manothermosonication on tomato pectic enzymes and tomato paste rheological properties. *Journal of Food Engineering, 53,* 273–278.

212. Vervoort, L., Van der Plancken, I., Grauwet, T., Timmermans, R. A. H., Mastwijk, H. K., Matser, A. M., Hendrickx, M. E., & Loey, A. (2011). Innovative Comparing equivalent thermal, high pressure and pulsed electric field processes for mild pasteurization of orange juice, Part II: Impact on specific chemical and biochemical quality parameters. *Food Science and Emerging Technologies, 12*(4), 466–477.

213. Villamiel, M., & De Jong, P. (2000). Influence of high-intensity ultrasound and heat treatment in continuous flow on fat, proteins, and native enzymes of milk. *Journal of Agriculture and Food Chemistry, 48,* 472–478.

214. Villamiel, M., Hamersveld, V., & De Jong, P. (1999) Review: Effect of ultrasound processing on the quality of dairy products. *Milchwissenschaft, 54,* 69–73.

215. Wang, D., & Sakakibara, M. (1997). Lactose hydrolysis and β-galactosidase activity insonicated fermentation with lactobacillus strains. *Ultrasonics Sonochemistry, 4,* 255–61.

216. Wang, W. C., & Sastry, S. K., (1997). Starch gelatinization in ohmic heating. *Journal of Food Engineering, 34,* 225–242.

217. Withers, P. (1994). Ultrasonic sensor for the detection of fouling in UHT processing plants. *Food Control, 5*(2), 67–72.

218. World Health Organization (2005). http://www.who.int/media center/factsheets/.

219. World Health Organization (1981). Food Irradiation: Use of Irradiation to Ensure Hygienic Quality. Technical Report Series 659.

220. Desobry-Banon, S., Richard, F., & Hardy, J. (1994). Study of acid and rennet coagulation of high pressurized milk. *Journal of Dairy Science, 77*(11), 3267–3274.

221. Wu, J., Gamage, T. V., Vilkhu, K. S., Simons, L. K., & Mawson, R. (2008). Effect of thermosonication on quality improvement of tomato juice. *Innovative Food Science and Emerging Technologies, 9,* 186–195.

222. Yusaf, T., & Al-Juboori, R. A. (2014). Alternative methods of microorganism disruption for agricultural applications. *Applied Energy, 114,* 909–923.

223. Zhang, Q. H., Barbosa-Cánovas, G. V., & Swanson, B. G. (1995). Engineering aspects of pulsed electric field pasteurization. *J Food Eng, 25,* 261–281.

224. Zheng, L., & Sun, D. W. (2006). Innovative applications of power ultrasound during food freezing processes: a review. *Trends in Food Science and Technology, 17*(1), 16–23.

CHAPTER 2

DIGITAL IMAGE ANALYSIS: TOOL FOR FOOD QUALITY EVALUATION

PRASHANT G. WASNIK, REKHA RAVINDRA MENON, and BHUSHAN D. MESHRAM

CONTENTS

2.1   INTRODUCTION

Assurance and control of quality during and post processing is an integral part of the food processing industry. Conventionally, various subjective and objective methods are employed for the same. Subjective evaluation of quality stipulates the involvement of trained judges to visually identify, scale in linguistic terms or score in a numerical scale, the relevant quality parameters for the product. In addition to being a highly skill-oriented procedure, the subjective evaluation of quality is also time consuming, laborious and prone to human error and fatigue. Objective evaluation of quality is carried out using a range of measurement techniques, simple or sophisticated involving rigorous protocols of sampling and preparation steps of chemical analyzes and physical experiments. With the need for more rapid and economical objective measurement of quality, in recent times, computer vision technology or image analysis technique is garnering prominence as a relevant tool for the qualitative and quantitative assessment of quality parameters in food processing.

This chapter reviews image analysis as a tool for evaluation of food quality.

2.2   DIGITAL IMAGING

A digital image displayed on a computer screen is made up of pixels. These pixels are tiny picture elements arranged in matrix. The individual pixel can be seen by zooming up the image by magnifying tool. Width and height of the image are defined by the number of pixels in x and y direction. Each pixel has brightness or intensity value somewhere between black and white represented as a number. The intensity of the monochrome image is known as gray level. The limitations with regard to gray level are that it is positive and finite. The gray level interval varies from low intensity to high is called gray scale. A common practice is to represent this interval numerically as a value between 0 and L, where, the lowest value 0 represents pure black and the maximum value L is white. All intermediate values are shades of gray, varying continuously from black to white [23]. Digital image has a defined bit depth, such as 8-bit, 16-bit, or 32-bit, etc. When an 8-bit integer is used to store each pixel value, Grey levels range from 0 to 255 (i.e., 2^0-1 to 2^8-1).

2.2.1   READING THE IMAGE

In Figure 2.1, the name of the image is shown at the top of the image in title bar [87]. The image was scaled to 33.3% larger than the original and it

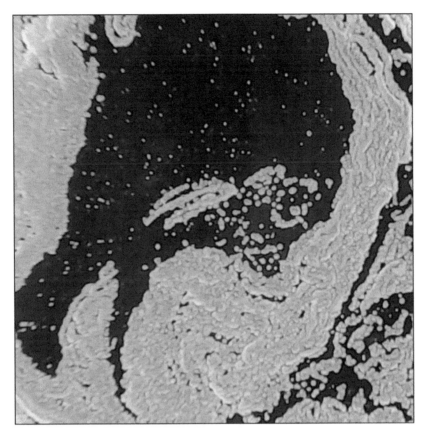

FIGURE 2.1   Digital image of cow ghee after crystallization at 28°C.

is saved in tagged image file (TIFF) format. The image properties are written below the title bar. The width and height of the image in mm and pixel is mentioned. The image is saved in 24 bit RGB (each color Red, Green and Blue) in 8-bit depth. The space occupied by the image is 8.3 MB.

2.2.2   IMAGE PROCESSING

We are familiar with the word food processing, where we are adding value to raw food material by applying various unit operations, in the same way in image processing we are adding value to image to extract meaningful information of our use from processed image by applying some mathematical operations on the raw image. In this process, we are not reducing the amount of data in the raw image but we are rearranging the data.

2.2.2.1 Purpose of Image Processing

The general objective of image processing for scientific and technical applications is to use radiation emitted by object. An imaging system collects the radiation to form an image. Then image-processing techniques are used to perform an area-extended measurement of the object features of interest. For scientific and technical applications area extended measurements constitute a significant advantage over point measurements, since also the spatial and not only the temporal structure of the signals can be acquired and analyzed [29].

The area of image analysis is in between image processing and computer vision. There is no clear-cut boundary in the continuum from image-processing at one end to computer vision at the other. However, one useful paradigm is to consider three types of computerized processes. Low-level processes involve primitive operations such as image pre-processing to reduce noise, contrast enhancement, and image sharpening. A low level process if characterized by the fact that both inputs and outputs are images.

Mid-level processing on images involves tasks such as segmentation (partitioning an image into regions or objects) descriptions of those objects reduce them to a form suitable for computer processing and classification (recognition) of individual objects. A mid-level process is characterized by the fact that both its inputs generally are images, but its outputs are attributes extracted from those images (e.g., edges, contours and the identity of individual objects). Finally higher level processing involves "making sense" of an ensemble of recognized objects, as in image analysis, and at the far end of continuum, performing the cognitive functions normally associated with visions [20].

Two types of errors are generally encountered while image processing: statistical and systematic errors. If one and the same measurement is repeated over and over again, the results will not be exactly same but rather scatter. The width of the scatter plot gives statistical error. The deviation of mean value of measured results from true value is called systematic error. In image analysis the statistical error may be large but the systematic error should be minimum [29].

2.3   DIGITAL IMAGE ACQUISITION

2.3.1   ILLUMINATION

Illumination is an important prerequisite of image acquisition for food quality evaluation. Acquisition of high quality image would naturally help to reduce the time and complexity of subsequent image processing steps, translating to a decreased cost of the image processing system and the quality of captured image is greatly affected by the lighting condition. By enhancing image contrast, a well-designed illumination system can improve the accuracy and lead to success of image analysis [23]. Also, the illumination strategy during image acquisition may depend on the intended application of the image processing technique. Most lighting arrangements can be grouped as front lighting, back lighting, and structured lighting [51]. The common lighting used in food research are: A (2856 K), C (6774 K), D65 (6500 K), and D (7500 K), etc. The illumination sources C, D65, and D are designed to mimic variations of daylight [36].

2.3.2   DIGITAL IMAGE ACQUISITION DEVICES

Image acquisition is capture of an image in digital form, and it is obviously the first step in any image processing system. During the last decades, considerable amount of research effort has been directed at developing techniques for image acquisition. A very intensive field of research in image acquisition is the development of sensors [14] and various configurations of sensors have been widely reported for the conversion of images into its digital form. In recent years, there have been attempts to develop non-destructive, non-invasive sensors for assessing composition and quality of food products. Different sensors such as charge coupled devices (CCD), ultrasound and magnetic resonance imaging (MRI), computed tomography (CT), and electrical tomography (ET) are been cited widely to obtain images of food products.

2.3.2.1 Scanner

In computing, an image scanner is a device that optically scans images (printed text, handwriting or an object) and converts it to a digital image which is then transferred to a computer. The scanner head (comprising of mirrors, lens, filter and Charge Coupled Device (CCD array) move over the document line by line by belt attached to steeper motor. Each line is broken down into "basic dots" which correspond to pixels. A captor analyzes the color of each pixel. The color of each pixel is broken down into three components (red, green, and blue). Each color component is measured and represented by a value. For 8-bit quantification, each component will have a value between 0 and 255 ($2^8-1 = 255$). The high-intensity light emitted is reflected by the document and converges towards a series of captors via a system of lenses and mirrors. The captors convert the light intensities received into electrical signals, which are in turn converted into digital data by an analog-digital converter. Scanners typically read red-green-blue color (RGB) data from the array. This data is then processed with some proprietary algorithm to correct for different exposure conditions, and sent to the computer via the device's input/output interface. Color depth varies depending on the scanning array characteristics, but is usually at least 24 bits. The other qualifying parameter for a scanner is its resolution, measured in pixels per inch (ppi), sometimes more accurately referred to as samples per inch (spi) [84].

The scanned result is a non-compressed RGB image, which can be transferred to a computer's memory. Once on the computer, the image can be processed with a raster graphics program (such as Photoshop or the GIMP) and saved on a storage device. During digital image processing each pixel can be represented in the computer memory or interface hardware (e.g., a graphics card) as binary values for the red, green and blue color components. When properly managed, these values are converted into intensities or voltages via gamma correction to correct the inherent non-linearity of some devices, such that the intended intensities are reproduced on the display [84]. Cross section of flat-bed scanner is shown in Figure 2.2 [11].

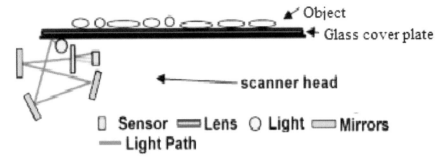

FIGURE 2.2   Cross section of a flatbed scanner used for scanning the food.

2.3.2.2   Digital Cameras

Digital cameras have a built-in computer, and all of them record images electronically. Firstly, light bouncing off an object passes into the camera, through a set of lenses, and onto a mirror. From there, the light bounces up and into a pentaprism. Once light enters the pentaprism, it bounces around in a complicated way until it passes through the eyepiece and enters our eye. Just like a conventional camera, it has a series of lenses that focus light to create an image of an object. But instead of focusing this light onto a piece of film, it focuses it onto a semiconductor device that records light electronically. A computer then breaks this electronic information down into digital data [10].

The fundamental technology of the digital camera is a light sensor and a program. The light sensor is most often a CCD and the program is firmware that is embedded right into the circuit board of the camera. The CCD is like a grid of millions of little squares, each one kind of like a solar cell. Each of those little squares on the CCD takes light energy and converts it to electrical energy. Each brightness and intensity condition of the light generates a very specific electrical charge. Those charges for each little square are then transported through an array of electronics to where it can be interpreted by the firmware. The firmware knows what each specific charge means and translates it to information that includes the color and other qualities of the light that the CCD picked up. This process is done for each of the squares in the grid of the CCD. The next step is for the firmware to record the information it saw into digital code. That code can be used to accurately reproduce the picture again and again. Now, that code can be passed to the view screen on the camera, or to a monitor or printer for reproduction [33]. Essentially, a digital image is just a long string of 1s and 0s that represent all the tiny colored dots – or pixels – that collectively make up the image. A digital camera with a minimum resolution of 1600 × 1200 pixels is recommended, which is equivalent to a 2.1 megapixels or higher camera for acquiring scientific images. The camera should also have macro and zoom feature. A memory card of at least 32 Mb and a digital film reader are also useful for storing the image files and transferring them to the computer [90].

A standard software can be employed/used to analyze a displayed image file created after digital scanning or photography both for color parameters and for the reflectance or luminosity. Modern color flatbed scanners, digital video cameras, and their image-processing software (graphics editors) provide wide possibilities for creating and editing images and their analysis [81, 82].

The colorimetric parameters of colored substances adsorbed on polyurethane foam can be measured using a desktop scanner and image-processing software [70]. In this method colored samples of polyurethane foam were scanned using a desktop scanner, the scanned color images were processed using Adobe Photoshop as a graphics editor, and the calibration plots of the luminosity of the selected channel (R, G, or B) as a function of the concentration of the test compound were found out using the Origin software. They prepared and scanned different colored chemical solutions. The separation of colors of images and the determination of the luminosity of R, G, and B channels were made using Adobe (R) Photoshop 8.0 software. The calibration plots were described by a first-order exponential decay function. It has been found that substances adsorbed on polyurethane foam can be determined with the use of a scanner and the corresponding image processing software with the same sensitivity as with the use of diffuse reflectance spectroscopy [76].

2.3.2.3 Digital Microscopy

Digital microscopy has been extensively used by many workers for characterization of food microstructure. Conventional optical microscope to obtain the images of bovine sodium caseinate (NaCAS) and soy protein isolate acid gels to access the possible changes in the microstructure can be used [27]. Polarization microscope equipped with a digital camera was used to acquire the hot stage images of starch/sodium chloride gels to study its microstructure [38]. The advantages of confocal scanning light microscopy (CSLM) over conventional microscopy that very thin focal plane can be observed by blocking the out-of-focus light through confocal optics [40]. They used a Bio-Rad 1024 CLSM system with Nikon Eclipse microscope for imaging the lipid microstructure. The images of low fat yogurt manufactured with microparticulated whey proteins acquired with confocal laser scanning microscopy used to investigate the difference in microstructure [83]. The images of concentrated milk suspensions and concentrated milk gels having similar casein composition acquired by transmission electron microscope equipped with CCD camera are also used to investigate the microstructure.

The basic tool for digital image processing consists of camera which acts as an electronic imagining system, having two main components lens and sensor. Lens collects radiation/refection to form image of the object features of interest and sensors converts irradiance at the image plane into an electric signal. Frame grabber converts electric signal into a digital image and stores it in the computer. Computer provides platform for digital processing of image and image processing software

1. **Input**
2. **Image acquisition**
3. **Pre processing**
4. **Feature extraction**
5. **Classification**
6. **Post processing**
7. **Decision**

FIGURE 2.3 The steps or components in the production of digital image.

provides algorithms to process and analyze contents of digital image [29]. A typical computer vision system can be divided into components (Figure 2.3). The components of machine vision system are described in following subsections.

2.3.3   IMAGE PRE-PROCESSING

Images captured by any image acquisition device are expected to experience various types of artifacts or noises. These artifacts may distort the quality of an image, which subsequently cannot provide correct information for successive image processing. In order to improve the quality of the image these artifacts needs to be removed by performing some operations on the image. The objective of the pre-processing operations is to improve and enhance the desired features for further specific applications. Generally two types of pre-processing operations are carried out for food quality evaluation: (a) pixel pre-processing; and (b) local pre-processing, according to the size of the pixel neighborhood that is used for the calculation of new pixel [14]. In pixel pre-processing method, input image is converted in such a way that each output pixel corresponds directly to the input pixel having the same coordinates. Pixel pre-processing may be viewed as a pixel-by-pixel copying operation, except that the values are modified according to specified transformation function. Color space transformation is the most common pixel pre-processing method for food quality evaluation [14].

Local pre-processing methods calculate the new value by averaging the brightness value of some neighborhood points, which have similar properties to the processed point [22]. Relatively simple and at the same time most complicated algorithms are available in ImageJ software to carry out this operation. Local pre-processing can be used to blur sharp edges, or to preserve edge in the image as per the demand of the specific problem. The median filter technique, a special group of filters called rank statistic filter, is observed to preserve the edges while filtering out the peak noise and is therefore the popular choice for filtering before applying an edge detection technique. For some special aims, more complex local pre-processing methods have been applied for food quality evaluation. The 'filter factor,', for example, modified unsharp filter transform, which is a Laplace transform of an image added to the same image, to enhance the detection cracks in the image of whole egg without overly enhancing other surface features and noise [22]. It is reported that this operation followed by a contrast stretch produced very satisfactory results as the sensitivity to translucent spots was decreased while sensitivity to cracks was increased. To smoothen a binary oyster image, shrink, expansion, and closing processes can be used. The shrink operation is applied to remove small objects (e.g., noise) in the image while the expansion process fills the holes and concavities in objects in the image [71].

2.3.4   IMAGE SEGMENTATION

Partitioning of an image into its constituent objects is known as Image segmentation which is a challenging task because of the richness of visual information in the image. The techniques of image segmentation developed for food quality evaluation can be divided into four different theoretical approaches, for example, thresholding-based, region-based, gradient-based, and classification-based segmentation. Current literature survey on image segmentation indicates that in most applications thresholding-based and region-based methods have been used for segmentation [14]. The other methods (gradient based and classification-based approaches) are used less frequently.

2.3.4.1   Thresholding-Based Segmentation

Thresholding-based segmentation is a particularly effective technique for images containing solid objects and having contrasting background, which distinguishes the object from the remaining part of an image with an optimal value. Among the thresholding-based segmentation methods for food quality evaluation, some perform segmentation directly by thresholding, and others combine the same with other techniques. Thresholding works well if the objects of interest have uniform interior gray level and rest upon a background of different, but uniform, gray level. For fast and computerized detection of bruises in magnetic resonance images of apples method is available in literature [93] in that computationally simple thresholding technique is used to distinguish between bright pixels representing the vascular system and those representing bruises. Whole pizza image is segmented from the white background using the RGB model and by setting the HSI values, segmentation of pizza sauce from pizza base can be achieved [76].

The defects of curved fruits can be segmented by generating the reference fruit image and it is normalized to get the normalized reference fruit image (NRFI) for inspection. The normalized original fruit image is subtracted from NFRI and a simple thresholding process is used to extract the defects [39]. It is recommended that the application of adaptive thresholding techniques instead of a fixed global threshold to segment an image for non–curvilinear fruits. Various automatic thresholding techniques are available in literature of which modified Otsu's algorithm can be used for successful automatic background segmentation of corn germplasm images [54].

2.3.4.2   Region-Based Segmentation

Region-based segmentation methods are divided into two basic classes: region growing-and-merging (GM) and region splitting-and-merging (SM) [91]. The former is a bottom-up method that groups pixels or subregions into larger regions according to a set of homogeneity criteria; and the latter is a top-down method that successively divides an image into smaller and smaller regions until certain criteria are satisfied. Region-based algorithms are computationally more expensive than the simpler techniques, for example, thresholding based segmentation, but region-based segmentation is able to utilize several image properties directly and simultaneously in determining the final boundary location. A region-based segmentation method, for example, the flooding algorithm, was developed to detect the apple surface feature [91].

Sun [76] developed a new region-based segmentation algorithm for processing the pizza images. This new algorithm adopted a scan line-based growing mode instead of the radial growing mode employed in traditional region growing algorithms. It uses the traditional region-based segmentation as a dominant method and combined the strengths of both thresholding and edge-based segmentation techniques.

2.3.4.3   Gradient-Based and Classification-Based Segmentation

Gradient-based approaches attempt to find the edges directly by their high gradient magnitudes and are similar to edge detection based on the gradient of an image. This image segmentation algorithm involving edge detection and boundary labeling and tracking is used to locate the position of whole fish [30]. However, the application of the gradient-based segmentation is limited because completed boundaries are difficult and sometimes impossible to trace in most food images. Classification-based methods attempt to assign each pixel to different objects based on classification techniques like statistical, fuzzy logic, and neural network methods. The Bayesian classification process can be used successfully to segment and detect apple defects [37].

2.3.5   PARTICLE ANALYSIS

Characterization of a population of granular objects is needed for many industrial processes, particularly in the food industry. According to its speed, accuracy and non-destructive operation abilities, image analysis is well-adapted tool for this purpose [52]. Quantitative analysis of morphological characteristics such as shape, size, count, area, perimeter, density, porosity, etc. can be estimated with particle analysis. Various software such as LUCIA, Image-pro, MATLAB, ImageJ, FIJI, etc., are available for particle analysis. Large number of workers worked on particle analysis protocol, selected work is briefly explained below for readers.

The microbial colonies in food matrices can be measured accurately by novel noise free method based on time-lapse shadow image analysis [52]. Flat bed scanner images of single layer rice kernel on a black background were used to measure length, width and area of rice kernels [11]. The distribution parameters, aspect ratio, density and number of broken kernels can be calculated from the measured results. The per cent area occupied by calcium lactate crystals on the surface of smoked cheddar cheese; size and shape distribution of fat particles in a model processed cheese product and the amount of surface area occupied by gas holes in cheeses using image analysis can be quantitatively analyzed [7, 61, 77].

The rapid particle size distribution analysis of Bacillus spore suspension can be measured with an image particle sizing analysis software and the dimensions such as length, breadth and perimeter of single spores can be determined [11]. For identifying shapes and determining particle size distribution, an Image-plug-in extracts the dimensions from digital image of disjoint particles [61].

The example of image processing steps involved in particle analysis of *ghee* (Clarified butter fat) particles is discussed below to understand the concept. The image acquired by web cam is calibrated in mm. Rectangular ROI of 24.52 × 24.52 mm (1476 × 1476 pixel) is selected and central portion of the image is cropped and converted to gray scale image by 8-bit ImageJ command. Cropped image in Figure 2.3 shows the typical region of interest. The brightness and contrast of the cropped image is adjusted by auto mode. The image is pre-processed by applying Gaussian Blur filter of Sigma 10. The blurred image is subtracted from original image and new image of 32 bit is obtained (shown as subtracted image). The resulted image again converted to 8 bit and the unsharp mask filter of radius 50 pixels and 0.9 mask weight was applied to remove the surface layer and sharpen the boundaries. Huang auto threshold protocol with upper and lower limits of gray scale intensities 228 and 255 is applied for image segmentation. The particle analysis is carried out with size range 0.01- infinity to avoid noise and very small particles with circularity 0.1 to 1 on the processed image. The transitions of images are shown in Figure 2.4 [87].

2.4   POROSITY

Porosity or void fraction is a measure of void spaces in a material and is a void fraction of volume of voids over total volume usually expressed as a per cent. This property is very important in case of baked products (such as bread, milk cake, *gulabjamun, rasogulla, pantua*, etc.), as it imparts

Original Image

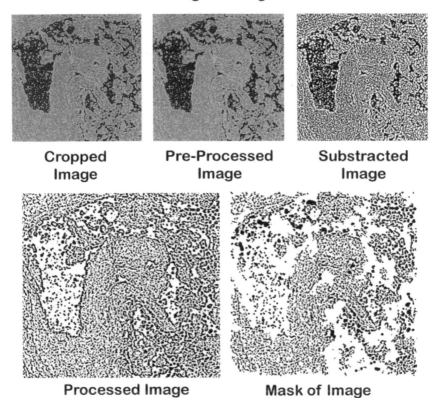

Cropped Image   **Pre-Processed Image**   **Substracted Image**

Processed Image   **Mask of Image**

FIGURE 2.4   Image processing steps.

sponginess to the product. Various methods are available for measurement of porosity of food products but image analysis technique provides the easiest way to measure the same. To elaborate the technique the acquired and segmented image of milk cake is shown in Figure 2.5. The image of section of milk cake was acquired by flat bed scanner. The portion of void spaces can be seen in segmented image. With the help of ImageJ the total area of void spaces and per cent area of voids can be measured with particle analysis protocol.

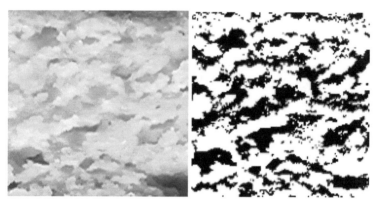

FIGURE 2.5   Acquired and segmented image of milk cake.

2.5   IMAGE TEXTURE ANALYSIS

Texture is an essential image attribute and has been applied greatly in the food industry for quality evaluation and inspection. A clear scientific definition, for image texture has not yet been available due to the limited understanding of texture properties as there is an infinite diversity of texture patterns of images [77]. The concept of texture in image processing/computer vision is totally different from the conventional concept of texture followed in the food industry. Currently in the food industry, food texture parameters, such as, hardness, cohesiveness, viscosity, elasticity, adhesiveness, brittleness, chewiness, and gumminess, is usually correlated to the manner in which human mouth perceives the mouth feel of the food while image texture descriptors, such as, fineness, coarseness, smoothness and graininess, are generally characterized by

IMAGE TEXTURE	Statistical texture	**GLCM**
		GL pixel run length matrix
		Neighboring GL dependence matrix
	Model based texture	**Fractal model**
		Auto regression model
	Transform based texture	**Convolution mask**
		Wavelets transform
		Fourier transform

FIGURE 2.6   The classification of image texture.

the spatial arrangement of the brightness values of the pixels in a region in the image [7].

There are four different types of image texture: statistical texture, structural texture, model-based texture, and transform- based texture [92]. Statistical approaches are employed to extract statistical textural features from the higher-order of pixel gray values of images while structural textural descriptors are acquired through certain structural primitives constructed from gray values of pixels. Transform-based texture is derived by using statistical measurements from transformed images. Model-based texture is obtained by calculating coefficients from a model based on the relationship of the gray values between a pixel and its neighboring pixels. Among these, statistical texture is the most widely reported during image analysis in the food industry due to its high accuracy and less computation time [92]. Transform-based texture and model-based texture are also used, although not as popular as statistical texture. However, structural texture is rarely used in the food industry. The selection of suitable method for image texture analysis has been restricted by the limited understanding of each of the methods suited for specific applications and the approach of employing different methods concurrently improved the final results [92]. The classification of image texture is depicted in Figure 2.6.

2.5.1 GREY LEVEL COOCCURANCE MATRIX

Grey Level Cooccurance Matrix (GLCM) is a statistical surface measuring algorithm, widely used for food texture evaluation. It can evaluate 14 different texture parameters in four different directions, by comparing the neighboring pixels. It tabulates how frequently different combinations of pixel brightness values occur in an image. At a time it considers the relation between two pixels, one considered as the reference and the other as neighbor pixel. Starting in the upper left corner and proceeding to the lower right of window each pixel within the window becomes the reference pixel in succession. Pixels along the right edge have no right hand neighbor, so they are not used for this count. Generally north, north east, east and south east directions are used to calculate GLCM texture attributes and then summing the counts. For evaluating the food texture one direction and one distance between the pixels is used. Angular second moment, contrast, correlation, inverse difference moment and entropy can

TABLE 2.1   Texture Parameters with Its Texture Features

Texture parameter	Texture feature
Angular second moment	Uniformity of an image
Contrast	Amount of local variations present in an image
Correlation	Pixel linear dependence
Entropy	Amount of the order in an image
Inverse difference moment	Homogeneity of an image
Sum of squares	Roughness of the image

FIGURE 2.7 The von Koch snow flake.

be calculated from GLCM matrix and used as descriptors of food texture. The commonly used six features are listed in Table 2.1.

2.5.2 FRACTAL ANALYSIS

Fractal geometry is a relatively new branch of mathematics proposed by a mathematician called Benoit Mandelbrot who introduced a generically different concept of dimension to account for the complicated geometrical figures with finite and infinite boundary. Fractal (derived from the Latin word *fractus* meaning broken) is applied to define broken or problematic shapes illustrated for example by the von Koch snow flake (Figure 2.7) [69].

The basic definition of a fractal indicates that it is a geometrical object consisting of mostly broken lines constructed by an iterative process using infinite repetitions of a set of operations such that geometrically it appears as infinite multiples of self-similar units [69]. A classical example of this geometry is demonstrated in all related literature in the form of the snow flake boundary; zooming into the boundary results in infinite number of triangles, each side of which could be further zoomed to appear as a set of self-similar triangles. A self-similar fractal is often defined using Hausdorff dimension (named after the mathematician bearing the same name):

$$sd = c \tag{1}$$

where, d is the Hausdorff dimension, c is the amount of new copies one gets after one iteration, and s is the scaling factor.

The dimension for non-self-similar fractals are designated in other ways, one of the methods commonly cited is the box-counting method. This method involves the "covering" of the image or superimposing the image with smaller objects, usually circles or squares, or hyper cubes or balls in whatever topological dimension the fractal set fits in. The minimum number of objects required to completely cover the fractal is calculated to obtain the fractal dimension. This is referred to as the box-counting dimension of a fractal [69].

The dimension of line, a square and a cube are one, two and three, respectively and its distance, area and volume, respectively are easily measurable. However, we cannot measure the dimension of complicated self-similar or dissimilar objects such as fat crystal network. The fractal dimension is a unit that could help to measure the dimension of complex objects [45]. The degree of complexity of such objects is measured by evaluating how fast the measurements increase or decrease as the scale becomes larger or smaller. To calculate the box- counting dimension, a grid is placed on a digital image. The reciprocal of the width of grid is taken as 's' and the count of number of blocks that touches the digitalized image is taken as N(s). The process is repeated by resizing the grid and the fractal dimension is measured as the slope of best fit line of log N(s) versus log (1/s) plot [45, 60].

The fractality of fat crystal networks is calculated by particle counting procedure based on counting the number of particles within boxes of increasing size corresponded to the fractal dimension of the network and a grid type box counting method can be used to determine the fractal dimension [45].

Various methodologies can be used for calculating the fractal dimensions. For calculating, the Fractal features can be extracted by traditional threshold method using ImageJ freeware or by applying "High Pass filter" in Photoshop software. For semisolid foods such as retrograded starch or yogurt images obtained with confocal microscopy will be useful for studying the fractality with suitable image processing algorithm. Fractal analysis in combination with PCA is a robust and objective tool to detect and quantify morphological changes in the microstructure of stirred yogurt manufactured with micro-particulated whey protein as ingredient.

The fractal dimension is very useful for measuring the quality of habitats. Area and perimeter fractal dimensions can be correlated with ecological index for habitat quality. The interior to edge ratio is also suitable for predicting interior habitat quality [26].

Bulk density particle diameter relation, Richardson's plot, gas adsorption and pore size distribution methods can be used for estimating the fractal dimensions of fine food particles of native and modified starch.

The perimeter and surface area of particles can be used effectively to calculate the natural fractal dimensions of caseinate structures [15]. The fractal dimensions can be used to relate morphological features to functional or even sensory properties, thus assisting food product development, process control and quality assurance [5].

2.5.3 ANGLE OF REPOSE

When bulk granular material is poured on to a horizontal surface, a conical shape of piled granular material will form. The internal angle between the surface of pile and the horizontal surface is known as the angle of repose. It is related to the density, surface area, shapes of particles, coefficient of friction. Materials with a low angle of repose form flatter piles than materials with a high angle of repose. In ImageJ a dedicated angle tool is provided for measurement of angles. Simply by drawing line over the angle of acquired image of pile the angle can be measured.

2.5.4 IMAGE COLOR ANALYSIS

Color measurement is not straightforward because it varies with the amount or intensity of mixture of primary colors which form innumerable color shades. Consequently, no single color measuring system encompasses these numerous combinations, leading to variations in color expressions. The Commission International de l'Eclairage (International Commission on Illumination, CIE) has been involved in defining a standard system to effectively express color in all possible combinations since 1930 and the proposed systems have been progressively refined and upgraded. In 1976, CIE recommended the CIELAB color scale, which measures color using three coordinates, the L*, a* and b* [18]. L* is the luminance or lightness component, which ranges from 0 to 100, and parameters a* (from green to red) and b* (from blue to yellow) are the two chromatic components, which range from −120 to 120 [91]. The L*a*b* color space is recommended as an absolute model to be used as a reference as it is perceptually uniform and device independent [55].

The CIE Color Systems utilizes a three coordinate system to locate any color in a color space; the commonly employed co-ordinate systems include CIE XYZ, CIE L*a*b* and CIE L*C*h. An alternate color scale commonly employed was developed by Hunter in 1958 and is also expressed as three coordinates, namely, L, a, and b with almost similar definitions to the CIE L*a*b* scale parameters but computed differently [43]. Most modern colorimeters automatically calculate either the CIELAB values or the Hunter values.

Color of food products is reported to be measured by various methods namely: spectrophotometers, colorimeters, color charts and computer vision methods. All these methods use one or the other expressions of color measuring systems viz. CIE XYZ, Lab, L*a*b* or Munsell color scale. Out of these color measuring systems, CIE L*a*b* is the most accepted and commonly used system because of its wide spectrum involving maximum number of color shades. However, CMYK mode is commonly used in printing systems [64]. The various equipment and techniques used for color measurement are briefly discussed in the following subsections.

2.5.4.1 Spectrophotometer

Spectrophotometry has been cited for boundless applications including corporate logo standardization, color control in paint industry, color testing of inks, color control for packaging material and labels, color control in plastics and textiles industry and monitoring of food color throughout the development, manufacturing and storage process. The principle of spectrophotometry is based on the capture of reflected light of well-defined wavelength incident on the sample. The commonly reported spectrophotometers employed for monitoring color include Spherical spectrophotometers, 0°/45° spectrophotometers and Multi-angle spectrophotometers [63].

The spectrophotometer was used to monitor the color of evaporated milk and related products and a convenient index for routine estimations of the darkening in color of evaporated milk was determined by noting changes in reflectance of light of 520 μm wave length [49]. Near infra-red (NIR) spectroscopy and Hunter-lab can be employed to predict color of European Emmental cheese samples [57]. The reflectance method was also used to measure milk fat color differences due to various parameters like feed of cattle [73].

2.5.4.2   Colorimeter

A colorimeter is a device of fairly simple design based upon the visual concepts of color [49]. The sample is illuminated with a single type of light (such as incandescent or pulsed xenon) at a 45° angle relative to the perpendicular line to the plane of the mounted sample. The reflected light is measured directly perpendicular to the sample through a series of three and sometimes four colored filters which represent the relative amounts of red, green and blue light reflected from the sample. More specifically these filters are designed to ideally simulate the three functions, x, y, z so that the instrument directly measures the three tristimulus values X, Y, Z for the specific illuminant being used [63]. The other essential parts of a tristimulus colorimeter are a white light source, an array of photometers and, nowadays, a computer or an interface to one. The computer is used to collect responses as well as carry out data transformations between CIE and other color scale systems or between different standard white light sources or white diffusers. Data from a colorimeter is given as a three-point output, commonly CIELAB, Hunter Lab or Y, x, y [34, 31]. The reflectance meter was used for color measurement of *basundi* and the color is expressed in terms of percent transparency (T) for brown and yellow color [13].

2.5.4.3   Stimuli Meter

Stimuli meters are based on the principle wherein color measurement is carried out by using a combination of stimuli to match given stimulus, for example, the color of solvent or material being measured is carried out by matching against a given system of known colors. Examples of these are Munsell Systems, the Donaldson instrument and Lovibond tin-tometers [28].

2.5.4.4   The Munsell Scale

In colorimetry, the Munsell color system is a color space that specifies colors based on three color dimensions: hue, value (lightness), and chroma (color purity or colorfulness). Munsell was the first to separate hue, value, and chroma into perceptually uniform and independent dimensions, and was the first to systematically illustrate the colors in three dimensional space [35]. There are 10 hues and each hue is subdivided into 10 shades. The value is evaluated on a scale that ranges from 0 at black to 10 at white. The chroma is expressed on an arbitrary scale of saturation ranging from 0 to 18. The Munsell system assigns numerical values to these three color dimensions. The color produced in the processing of evaporated milk is measured by means of the Munsell system of disc colorimetry [3, 88].

2.5.4.5   Color Charts

A dictionary consisting of 56 charts is classified into seven main groups of hues presented in order of their spectra. Each group, comprised of eight plates, the first plate was painted white with successively darkening shades of gray until the color appeared black. These colors were defined using CIE terms and are convertible to Munsell values [43]. The disadvantage of the color chart was its difficulty to match when the surface was different and possible deterioration of the painted paper-standard through exposure. A color chart for determination of the color of *basundi* was prepared using reflectance meter and had color combination of Red:Green:Blue as 205:150:0 [13].

2.5.5   COLOR MODELS

Three color models are commonly used to define color: RGB (red, green, and blue) model; the CMYK (cyan, magenta, yellow and black) model; and the L*a*b* model. While these color models have been established as useful quantifiers of color, it has also been pointed out that the spectrum of colors seen by the human eye is wider than the gamut (the range of colors that a color system can display or print) available in any color model [63].

2.5.5.1   RGB Model

RGB model is an additive color model that uses transmitted light to display colors in which red, green, and blue light are added together in various combinations of proportions and intensities to reproduce a broad array of colors. RGB is derived from the initials of the three additive primary colors, red (R), green (G) and blue (B). The model is said to relate closely to the way human perceives color on the retina. The model is device dependent, since its range of colors varies with the display device [58].

2.5.5.2   CMYK Model

CMYK refers to the four inks used in most color printing: cyan (C), magenta (M), yellow (Y), and key black (K). In contrast to the RGB model, CMYK is subtractive in nature. Since RGB and CMYK spaces are both device-dependent spaces, inter-conversion of the obtained color values using these models is neither simple nor mathematically established [2].

2.5.5.3   Lab Model

The L*a*b* color is a device independent model, providing consistent color regardless of the input or output device such as digital camera, scanner, monitor and printer [31]. Among the three models discussed herewith, the L*a*b* model has the largest gamut encompassing all colors in the RGB and CMYK gamut [2]. The L*a*b* values is the most widely reported color scale in food research studies. Unlike the RGB and CMYK color models, Lab color is designed to approximate human vision; it is attributed with an aspiration to perceptual uniformity, and its L component closely matches human perception of lightness. The model is also conducive to make accurate color balance corrections by modifying output curves in 'a' and 'b' components, or to adjust the lightness contrast using the L component [2].

The surface color of images acquired by imaging system can be measured using Adobe Photoshop in Lab mode. The L a b values thus obtained can be converted in terms of CIELAB L*, lightness ranging from zero (black) to 100 (white), a* ranging from +60 (red) to −60 (green) and b* ranging from +60 (yellow) to −60 (blue) using Eqs. (2)–(4). From the extracted values, color descriptors like hue, chroma, yellowness index and whiteness index were derived by computation [56], as indicated in Eqs. (5)–(8), respectively.

$$L* = \frac{100 \times L}{255} \tag{2}$$

$$a* = \frac{240 \times a}{255} - 120 \tag{3}$$

$$b* = \frac{240 \times b}{255} - 120 \tag{4}$$

$$C* = \sqrt{a*^2 + b*^2} \tag{5}$$

$$h* = \tan^{-1} \frac{b*}{a*} \tag{6}$$

$$\text{Yellowness Index} = \frac{142.86 \times b*}{L*} \tag{7}$$

$$\text{Whiteness Index} = 100\sqrt{(100 - L*)^2 + a*^2 + b*^2} \tag{8}$$

2.6   COMPUTER VISION SYSTEM

Nowadays the computer is routinely integrated to color measurement systems, by transferring the data collected from either a colorimeter or spectrophotometer to be transformed to XYZ, CIELAB, Hunter Lab or other color system provided by the software. Advancements in digital technology have also enabled use of scanner, camera and software for color measurement purposes. Following its origin in the 1960s, computer vision has experienced growth with its applications expanding in diverse fields not only for color measurements but also for process automations: medical diagnostic imaging; factory automation; remote sensing; forensics, robot guidance, etc. [23]. Computer vision and image analysis, are non-destructive and cost-effective techniques for sorting and grading of agricultural and food products during handling processes and processing operations [61].

2.6.1   APPLICATIONS OF COMPUTER VISION METHODS

Computer vision methods have wide applications in the food industry. Studies have been carried out to analyze visual characteristics of dairy, bakery, meat [79, 80] products, fruits and vegetables [5, 6, 80].

2.6.1.1   Dairy Products

Computer vision is used routinely in the quality assessment of yogurt, cheese, and pizza [19, 41, 74 and 84] and it is easy to analyze functional properties of Cheddar and Mozzarella cheeses with machine vision during cooking [86]. Images for each cheese slice before cooking and at periodic time intervals during cooking were captured and browning factor (BF) is used successfully to describe a reasonable index for cheese browning. Compared with conventional methods using colorimeters, the computer vision method is efficient and provides more information on the color change of cheese by making continuous measurement possible along with the advantage of handling uneven colored surfaces for cooked Mozzarella cheese. A computer vision system was coupled to a cheese vat, to measure color changes in the curd/whey mixture during syneresis [17]. Decreased white/yellow area ratio and increased RGB metric and Lab metric were observed to be caused by three factors intrinsic to curd syneresis: (a) the shrinkage of curd particles during syneresis, (b) the expulsion of whey from the curd during syneresis, and (c) the sinking of curd particles to the bottom of the vat during constant stirring. The study established the feasibility of computer vision and colorimeter measurements for monitoring syneresis. With the objective to determine color changes during storage of set type whole-fat and low-fat yogurts using a machine vision system (MVS) during storage, digitized images of yogurt acquired by digital color video camera, computer, and corresponding software and the obtained data were modeled with an artificial neural network (ANN) for prediction of shelf life [72]. The model was proposed as an alternative method to control the expiration date of yogurt shown in labeling and provide consumers with a safer food supply.

After applying the developed image-processing algorithm to recognize individual cheese shred and automatically measuring the shred length it was found that the algorithm recognized shreds well, even when they were overlapping. It was also reported that the shred length measurement errors were as low as 0.2% with a high of 10% in the worst case [50]. An objective method was developed and evaluated to measure the area occupied by calcium lactate crystals on the surface of naturally smoked Cheddar cheese samples using digital photography and analyzed using the Metamorph offline program. It was concluded that image analysis was well suited for evaluating changes in crystal coverage during cheese aging because of non-destructive type measurements with minimal rupture of cheese and good repeatability [60, 61]. Quantification of *Lactobaccilus spp.* in fermented milks grown in MRS agar was carried out by capturing digital color images through flatbed scanner [4].

Photoshop has been used to analyze the color for food samples, although the software was not originally designed for this purpose. However, it is already one of the most powerful software for color analysis, and the manufacturer and users are regularly making enhancements [21].

By using flatbed scanner and Adobe Photoshop the colors of espresso drink prepared with and without whey as a base ingredient were compared. Fresh coffee samples were scanned at a constant scanning resolution and R, G, B values of the scanned images were measured with Adobe Photoshop. It was found that color of whey coffee was almost similar to color of milk coffee. The suggested method for comparison of colors of product was safe and reliable [48]. The similar method was used for the study of color changes of *gulabjamuns* during frying [44]. Fried *gulabjaimuns* were scanned and using Adobe Photoshop software chroma, hue L a b values were computed. The study established the utility of computer vision system in measuring colors of heterogeneous products like *gulabjamun.*

2.6.1.2   Bakery Products

Color is the most desirable attribute in baking of food materials. The color of microwaved pizza can be effectively measured with simple digital imaging method. A high-resolution digital camera (2 mega-pixel or above) can be employed for capturing the color images of the food sample under proper lighting. Once the color images of the pizza samples were captured, the color can be analyzed qualitatively and quantitatively using Photoshop. The digital imaging method allows measurements and analyzes of the color of food surfaces that are adequate for food engineering research [90].

A computer aided reading system can be engaged for successful evaluation for grading of muffins based on the color. This was reported to improve the standardization and quantification of the inspection processes [1]. Physical features such as size, shape, dough color and fraction of top surface area of chocolate chip cookies can be estimated by using digital images [12].

A simple and new method for the evaluation of baking process on bread quality through the measurement of bread crust thickness, which was done based on the prediction of bread crust thickness by digital imaging and the L a b color system is available [32]. Porosity of white bread was measured by employing flatbed scanning which was reported as fast, easy to use, cheap, robust, and independent of external light conditions and with good accuracy [16].

2.6.1.3   Meat and Related Products

Color is the most important organoleptic characteristics and it influences the acceptability of the food product. The comparison of the instrumental color measurements of a Minolta colorimeter and a digital camera to measure color parameters is available in the literature. The parameters measured by digital camera were found to be more useful in predicting the meat quality [53]. This was due to the fact that the camera took measurements over the entire surface of samples and thus a more representative measurement was taken compared to colorimeter [42]. Computer vision was also used for assessment of fresh pork color [42, 78, 79].

Computer vision has been used to grade salmons replacing human labor [47]. Computer vision system is used to adjudge quality of salmons based on color score derived from RGB values of image taken by camera and L*a*b* color space. In the fish industry, despite the slow uptake, computer vision has been gradually gaining the necessary acceptance for quality evaluation applications [74, 24].

2.6.1.4   Fruits and Vegetables

In fruit processing industry color is the most important parameter hence it is demonstrated that the color of Golden apples can be measured using digitized video images and the apples could be ordered based on the CIE-color information and this was highly related to color ranking by a test panel [68].

Computer vision system involving measurement of peel color in terms of L, a, b values using histogram window method of Adobe Photoshop, standard L*, a*, b* values, the hue and chroma values along with the total color difference could be used to quantify overall changes of peel and pulp of mangoes, thus, enabling customization, standardization and storage studies of various fruits [65].

2.6.1.5   Chocolate

Chocolate blooming can be analyzed by measuring L*, a*, b* values, chroma (C*) values, whitening index (WI), hue (H°) percentage bloom and energy of Fourier using computer vision system and image analysis [5].

2.6.2 APPLICATION OF IMAGE ANALYSIS FOR DETECTION OF ADULTERATION

With increased expectations for food products of high quality and safety standards, the need for accurate, fast and objective quality determination of these characteristics in food products continues to grow.

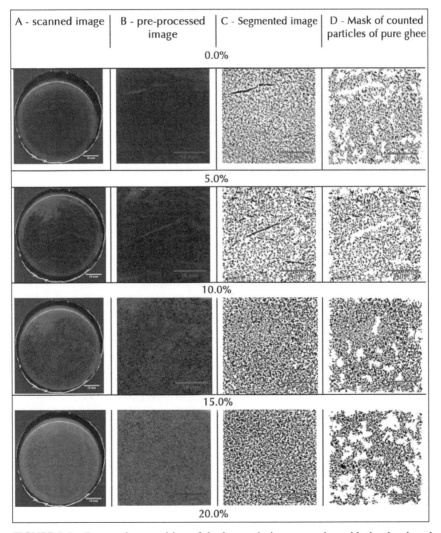

FIGURE 2.8 Progressive transition of the image during processing with the developed protocol.

Computer vision provides one alternative for an automated, non-destructive and cost-effective technique to accomplish these requirements. This inspection approach based on image analysis and processing has found a variety of different applications in the food industry. Considerable research has highlighted its potential for the inspection and grading of fruits and vegetables. Computer vision has been successfully adopted for the quality analysis of meat and fish, pizza, cheese, and bread. Likewise grain quality and characteristics have been examined by this technique [8].

Detection of minced meat fraudulently substituted with pork and vice versa can be detected by using multispectral imaging and multivariate data analysis with 98.48% overall correct classification [66].

Digital image analysis can analyze precisely and quickly a large number of ground coffee powders for detecting the adulteration in roast and ground coffee beans with coffee husk and straw, maize, brown sugar and soybean [67] also the impurities in virgin olive oil can be detected efficiently with computer vision system and pattern recognition [46].

The adulteration of cow ghee with vegetable fat can be detected by application of image analysis protocol. Controlled crystallization is the key factor in detecting the adulteration of ghee with image analysis. Number of parameters such as pixel intensity, particle count, particle diameter, fractal dimension, color parameters can be used as descriptors for detecting the adulteration. The sequence of operations such as conversion to 8-bit image, subtraction of background, enhancement of edges and segmentation may be carried out on the acquired image. Open source software such as ImageJ can be used to carry out image processing operations. Figure 2.8 shows the acquired and processed images of adulterated samples of ghee of various proportions [87]. The combination of derived parameters can be used for detection of adulteration of *ghee* [87].

2.7 CONCLUSIONS

Quality is an important responsibility of the stakeholders of food processing industry and there is need to develop rapid, non-destructive objective tools to evaluate and monitor the same during both online and offline modes. Image processing has been steadily gaining acceptance as a relevant methodology in quality control and assurance and has the potential to form an important component of this activity in the future.

2.8   SUMMARY

The major aim of this chapter is to provide an introduction to basic concepts and methodologies for digital image processing and to develop a confidence in readers that this tool can be implemented for studying the morphological, textural and color characteristics of the food material. This chapter includes information on: What is image? What is image processing? Purpose and scope of image processing; How to acquire images? Image acquisition systems; Image processing software; How to calibrate digital image? What are the basic steps in image processing? How to measure area, porosity, particle count, particle diameter, angle, intensity, etc.? What is texture? How it is different from traditional definition? Classification of image texture; How to measure texture? Importance of color in foods and how to measure it? These concepts are illustrated with practical examples.

KEYWORDS

- adulteration
- angular second moment
- application of computer vision methods
- box counting method
- color chart
- color measurement
- color model
- colorimeter
- computer vision system
- contrast
- correlation

- **CYMK model**
- **detection of adulteration**
- **digital camera**
- **digital image**
- **digital microscopy**
- **entropy**
- **fractal**
- **fractal dimension**
- *ghee*
- **gradient based segmentation**
- **gray level co-occurrence matrix**
- **grey level**
- **hausdorff dimension**
- **illumination**
- **image acquisition devices**
- **image pre-processing**
- **image processing**
- **image segmentation**
- **image texture analysis**
- **inverse difference moment**
- **lab model**
- **model based texture**
- **munsell scale**
- **particle analysis**
- **particle size distribution**
- **pixel**
- **porosity**
- **region based segmentation**
- **RGB model**
- **scanner**
- **spectrophotometer**
- **statistical texture**

- **stimuli meter**
- **structural texture**
- **sum of squares**
- **texture**
- **thresholding based segmentation**
- **transform based texture**
- **void fraction**

REFERENCES

1. Abdullah, M. Z., Aziz, S. A., & Dos-Mohamed, A. M. (2000). Quality inspection of bakery products using a color-based machine vision system. *Journal of Food Quality, 23,* 39–50.
2. Anonymous. (2002). *Adobe Systems.* Adobe Photoshop 7.0 User Guide, Adobe Systems Inc., San Jose, CA.
3. Barrette, A. H., & Peleg, M. (1995). Application of fractal analysis to food structure. *LWT, 28*(6), 553–563.
4. Bell, R. W., & Webb, B. H. (1943). The relationship between high temperature fore-warming and the color and heat stability of evaporated milk of different solids content. *Journal of Dairy Science, 26,* 579–585.
5. Borin, A., Ferrao, M. F., Mello, C., Cordi, L., Pataca, L. C. M., Duran, N., & Poppi, R. J. (2007). Quantification of *lactobaccilus* in fermented milks by multivariate image analysis with least square method vector machines. *Analytical and Bioanalytical Chemistry, 387*(3), 1105–1112.
6. Briones, V., & Aguilera, J. M. (2005). Image Analysis of changes in surface color of chocolate. *Food Research International, 38*(1), 87–94.
7. Brosnan, T., & Sun, D. W. (2002). Inspection and grading of agricultural and food products by computer vision systems: A Review. *Computers and Electronics in Agriculture, 36,* 193–213.
8. Brosnan, T., & Sun, D. W. (2004). Improving quality inspection of food products by computer vision: A Review. *Journal of Food Engineering, 61,* 3–16.
9. Caccamo, M., Melilli, C., Barbano, D. M., Portell, G., Marino, G., & Licitra, G. (2004b). Measurement of holes and mechanical openness in cheese by image analysis. *Journal of Dairy Science, 87,* 739–748.
10. Curtin, D. P. (2007). A short course in sensors, pixels and image sizes. http://www.shortcourses.com/sensors/index.html.
11. Dalen, G. V. (2005). Characterization of rice using flat bed scanning and image analysis, Chapter 6. In: *Food Policy, Control and Research,* edited by Riley, A. P. New York: Nova Biomedical Books, pp. 149–186.

12. Davidson, V. J., Ryks, J., & Chu, T. (2001). Fuzzy models to predict consumer ratings for biscuits based on digital features. *IEEE Transactions on Fuzzy Systems, 9*(1), 62–67.

13. Dharaiya, C. N. (2006). *Quality Evaluation of Basundi Marketed in Bangalore City.* MTech Thesis, National Dairy Research Institute, Bangalore.

14. Du, C. J., & Sun, D. W. (2004). Recent developments in the applications of image processing techniques for food quality evaluation. *Trends in Food Science and Technology, 15*, 230–249.

15. Dziuba, J., Babuchowski, A., Smoczynski, M., & Smietana, Z. (1999). Fractal analysis of caseinate structure. *International Dairy Journal, 9*, 287–292.

16. Esteller, M. S., Zancanaro, O., Palmeira, C. N. S., & Lannes, S. C. (2006). The effect of kefir addition on microstructure parameters and physical properties of porous white bread. *European Food Research and Technology, 222*(1–2), 26–31.

17. Everard, C. D., O'Callaghan, D. J., Fagan, C. C., O'Donnell, C. P., Castillo, M., & Payne, F. A. (2007). Computer vision and color measurement techniques for inline monitoring of cheese curd syneresis. *Journal of Dairy Science, 90*(3), 3162–3170.

18. Francis, F. J., & Clydesdale, F. M. (1975). *Food Colorimetry – Theory and Applications.* AVI Publishing Co. Westport, CT, USA.

19. Gerrard, D. E., Gao, X., & Tan, J. (1996). Beef marbling and color score determination by image processing. *Journal of Food Science, 61*(1), 145–148.

20. Gonzalez, R. C., & Woods, R. E. (2008). *Digital Image Processing,* 3rd edition, Prentice Hall, London, p. 24.

21. González-Tomás, L., & Costell, E. (2006). Relation between consumers' perceptions of color and texture of dairy desserts and instrumental measurements using a generalized procrustes analysis. *Journal of Dairy Science, 89*, 4511–4519.

22. Goodrum, J. W., & Elster, R. T. (1992). Machine vision for crack detection in rotating eggs. *Transactions of the ASAE, 35*, 1323–1328.

23. Gunasekaran, S. (1996). Computer vision technology for food quality assurance. *Trends in Food Science and Technology, 7*, 245–256.

24. Gunnlaugsson, G. A. (1997). Vision technology, intelligent fish processing systems. In: *Seafood from Producer to Consumer, Integrated Approach to Quality,* edited by Luten, J. B., Børresen, T., & Oehlenschläger, J., Elsevier Science, The Netherlands, pp. 351–359.

25. Hareesh, K. S., & Narendra, V. G. (2010). Prospects of computer vision automated grading and sorting systems in agricultural and food products for quality evaluation. *International Journal of Computer Applications, 1*(4), 1–12.

26. Imre, A. R., & Bogaert, J. (2004). The fractal dimension as a measure of the quality habitats. *Acta Biotheoretica, 52*, 41–56.

27. Ingrassia, R., Costa, J. P., Hindalgo, M. E., Canales, M. M., Castellini, H. C., Riquelme, B., & Risso, P. (2013). Application of a digital image procedure to evaluate microstructure of caseinate and soy protein acid gels. *Food Science and Technology, 53*, 120–127.

28. Jacobs, M. B. (1999). *The Chemical Analysis of Food and Food Products.* Third Edition, CSB Publications and Distributors, New Delhi.

29. Jahne, B. (2004). *Practical Handbook on Image Processing for Scientific and Technical Applications,* II edition. CRC Press, p.34.

30. Jia, P., Evans, M. D., & Ghate, S. R. (1996). Catfish feature identification via computer vision. *Transactions of the ASAE, 39*, 1923–1931.

31. Joshi, P. (2004). Color measurement of foods by color reflectance. In: *Color in Food – Improving Quality,* edited by MacDougall, D. B., Woodhead Publishing Limited, England, pp. 81–114.

32. Jusoh, Y. M., China, N. L., Yusof, Y. A., & Rahman, R. A. (2009). Bread crust thickness measurement using digital imaging and Lab color systems. *Journal of Food Engineering, 94*(3), 366–371.

33. Karim, N., Wilson, T. V., & Gurevich, G. (2006). How Digital Camera Works. http://electronics.howstuffworks.com/camerasphotography/digital/digital-camera.htm.

34. Kress-Rogers, E., & Christopher, J. B. B. (2001). *Instrumentation and Sensors for the Food Industry.* Woodhead Publishing Ltd., Cambridge, England.

35. Kuehni, R. G. (2002). The early development of the Munsell system. Second Edition, *Color Research and Application, 27*(1), 20–27.

36. Lawless, H. T., & Heymann, H. (1998). *Sensory Evaluation of Food: Principle and Practices.* Chapman and Hall, New York, pp. 406–429.

37. Leemans, V., Magein, H., & Destain, M. F. (1999). Defect segmentation on 'Jonagold'apples using color vision and a Bayesian classification method. *Computers and Electronics in Agriculture, 23*, 43–53.

38. Li, Q., Xie, Q., Yu, S., & Gao, Q. (2014). Application of digital image analysis method to study the gelatinization process of starch/ sodium chloride solution system. *Food Hydrocollides, 35*, 392–402.

39. Li, Q. Z., & Wang, M. H. (1999). Development and prospect of real time fruit grading technique based on computer vision. *Transactions of the Chinese Society of Agricultural Machinery, 30*(6), 1–7.

40. Liang, B., Sebright, J. L., Shi, Y., Hartel, R. W., & Perepezko, J. H. (2006). Approaches to quantification of microstructure for model lipid systems. *Journal of the American Oil Chemists Society, 83*(5), 389–399.

41. Locht, P., Thomsen, K., & Mikkelsen, P. (1997). Full color image analysis as a tool for quality control and process development in the food industry. ASAE Annual International Meeting, Paper no. 9733006, ASAE, 2950 Niles Road, St. Joseph, Michigan, 49085–9659, USA.

42. Lua, J., Tan, P., Shatadal, P., & Gerrard, D. E. (2000). Evaluation of pork color by using computer vision. *Meat science, 56*(1), 57–60.

43. Maerz, A., & Paul, M. R. (1950). Dictionary of color. Second edition, McGraw-Hill, Inc., New York, USA.

44. Magdaline, E. E. F., Menon, R. R., Heartwin, A. P., Rao, J. K., & Surendranath, B. (2009). Image analysis and kinetics of color changes during frying of *gulabjamuns*. Poster paper presented at XX Indian Convention of Food Scientists and Technologists, 21–23 Dec., Bangalore. Souvenir AD-38, p.9.

45. Marangoni, A. G., (2002). The nature of fractality in fat crystal networks. *Trends in Food Science and Technology, 13*, 37–47.

46. Marchal, P. C., Gila, D. M., Garcia, J. G., & Ortega, G. J. (2013). Expert system based on computer vision to estimate the content of impurities in olive oil samples. *Journal of Food Engineering, 119*, 220–228.

47. Misimi, E., Erikson, U., & Skavhaug, A. (2008). Quality grading of Atlantic salmon (*Salmo salar*) by computer vision. *Journal of Food Science, 73*(5), 211–217.
48. Nawale, P. K., Vyawahare, A. S., Aravindakshan, P., & Rao, J. K. (2009). Dairy based espresso drinks preparation using whey as a base ingredient. Poster paper presented at XX Indian Convention of Food Scientists and Technologists, 21–23 Dec., Bangalore. Souvenir AD-38, 14.
49. Nelson, V. (1948). The spectrophotometric determination of the color of milk. *Journal of Dairy Science, 31*(6), 409–414.
50. Ni, H., & Gunasekaran, S. (1995). A computer vision system for determining quality of cheese shreds. *In: Food Processing Automation IV*, Proceedings of the FPAC Conference, ASAE, 2950 Niles Road, St. Joseph, Michigan 49085–9659, USA.
51. Novini, A. (1990). Fundamentals of machine vision component selection, *In: Food Processing Automation II-* Proceedings of the 1990 conference, Lexinton, KY: ASAE, p.60.
52. Ogava, H., Nasu, H., Takeshige, M., Funabashi, H., Saito, M., & Matsuoka, H. (2012). Noise free accurate count of microbial colonies by time-lapse shadow image analysis. *Journal of Microbial Methods, 91*, 420–428.
53. O'Sullivan, M. G., Byrne, D. V., Martens, H., Gidskehaug, L. H., Andersen, H. J., & Martens, M. (2003). Evaluation of pork color, prediction of visual sensory quality of meat from instrumental and computer vision methods of color analysis. *Meat Science, 65*(2), 909–918.
54. Panigrahi, S., Misra, M. K., Bern, C., & Marley, S. (1995). Background segmentation and dimensional measurement of corn germplasm. *Transactions of the ASAE, 38*, 291–297.
55. Papadakis, S. E., Abdul-Malek, S., Kamden, R. E., & Yam, K. L. (2000). A versatile and inexpensive technique for measuring color of foods. *Food Technology, 54*(12), 48–51.
56. Pathare, P. B., Opara, U. L., & Al-Said, F. A. (2013). Color measurement and analysis in fresh and processed foods, A Review. *Food Bioprocess Technology. 6*, 36–60.
57. Pillonel, L., Dufour, E., Schaller, E., Bosset, J. O., Baerdemaeker, J. D., & Karoui, R. (2007). Prediction of color of European Emmental cheeses by using near infrared spectroscopy, a feasibility study. *European Food Research and Technology, 226*(1–2), 63–69.
58. Poynton, C. A. (2003). Digital video and HDTV, algorithms and interfaces. Morgan Kaufmann. ISBN 1558607927. http://books.google.com/books.
59. Quevedo, R. A., Aguilera, J. M., & Pedreschi, F. (2008). Color of salmon filets by computer vision and sensory panel. Food and Bioprocess Technology, http://www.springerlink.com/content/653w 7054m870rx41/.
60. Quevedo, R. A., Carlos, L. G., Aguilera, J. M., & Cadoche, L. (2002). Description of food surfaces and microstructural changes using fractal image texture analysis. *Journal of Food Engineering, 53*, 361–371.
61. Rajbhandari, P., & Kindstedt, P. S. (2005). Development and application of image analysis to quantify calcium lactate crystals on the surface of smoked Cheddar cheese. *Journal of Dairy Science, 88*(12), 4157–4164.
62. Rajbhandari, P., & Kindstedt, P. S. (2008). Characterization of calcium lactate crystals on cheddar cheese by image analysis. Journal of Dairy Science, *91*, 2190–2195.

63. Randall, D. L. (1997). Instruments for the color management. In: *Color Technology in the Textile Industry*. Second Edition, American Association of Textile Chemist and Colorists, pp. 11–17.

64. Ranganna, S. (2002). *Handbook of Analysis and Quality Control for Fruits and Vegetable Products*. Second Edition, Tata McGraw Hill Publishing Company, New Delhi.

65. Ravindra, M. R., & Goswami, T. K. (2008). Comparative performance of pre-cooling methods for storage of mango. *Journal of Food Process Engineering, 31*(3), 355–371.

66. Ropadi, A. I., Pavlidis, D. E., Mohareb, F., Panagou, E. Z., & Nychas, G. J. E. (2015). Multispectral image analysis approach to detect adulteration of beef and pork in raw meats. *Food Research International, 67*, 12–18.

67. Sano, E. E., Assad, E. D., Cunha, S. A. R., Correa, T. B. S., & Rodrigues, H. R. (2003). Quantifying adulteration in roast coffee powders by digital image processing. *Journal of Food Quality, 26*, 123–134.

68. Schrevens, E., & Raeymaeckers, L. (2005). Color characterization of golden delicious apples using digital image processing. *Acta Horticulturae, 304*, 159–166.

69. Shamsgovara, A. (2012). Analytic and numerical calculations of fractal dimensions. Department of Mathematics, Royal Institute of Technology, KTH, pp. 3–45.

70. Shishkin, Y. L., Dmitrienko, S. G., Medvedeva, O. M., Badakova, S. A., & Pyatkova, L. N. (2004). Use of a scanner and digital image – processing software for the quantification of adsorbed substances. *Journal of Analytical Chemistry, 59*(2), 102–106.

71. So, J. D., & Wheaton, F. W. (1996). Computer vision applied to detection of oyster hinge lines. *Transactions of the ASAE, 39*, 1557–1566.

72. Sofu, A., & Ekinci, F. Y. (2007). Estimation of storage time of yogurt with artificialneural network modeling. *Journal of Dairy Science, 90*, 3118–3125.

73. Solah, V. A., Staines, V., Honda, S., & Limley, H. A. (2007). Measurement of milk color and composition, effect of dietary intervention on Western Australian Holstein-Friesian cow's milk quality. *Journal Food Science, 72*(8), 560–566.

74. Strachan, N. J. C., & Murray, C. K. (1991). Image analysis in the fish and fish industries. In: *Fish Quality Control by Computer Vision*, edited by Pau, L. F., & Olafsson, R. Marcel Dekker, New York, USA, pp. 209–223.

75. Sun, D. W. (2000). Inspecting pizza topping percentage and distribution by a computer vision method. *Journal of Food Engineering, 44*, 245–249.

76. Sun, D. W., & Brosnan, T. (2003). Pizza quality evaluation using computer vision-part I pizza base and sauce spread. *Journal of Food Engineering, 57*, 81–89.

77. Sutheerawattananonda, M., Fulcher, R. G., Martin, F. B., & Bastian, E. D. (1997). Fluorescence image analysis of process cheese manufactured with trisodium citrate and sodium chloride. *Journal of Dairy Science, 80*(4), 620–627.

78. Tan, F. J., Morgan, M. T., Ludas, L. I., Forrest, J. C., & Gerrard, D. E. (2000). Assessment of fresh pork color with color machine vision. *Journal of Animal Science, 78*, 3078–3085.

79. Tan, F. J. (2004). Meat quality evaluation by computer vision. *Journal of Food Engineering, 61*, 27–35.

80. Tao, Y., Heinemann, P. H., Varghese, Z., Morrow, C. T., & Sommer, H. J. (1995). Machine vision for color inspection of potatoes and apples. *Transactions of the ASAE 38*(5), 1555–1561.

81. Tillett, R. D. (1990). Image analysis for agricultural processes. Division Note DN 1585, Silsoe Research Institute.

82. Timmerman (1998). Computer vision system for online sorting of pot plants based on learning techniques. *Acta Horticulturae*, *42*(1), 91–98.

83. Torres, I. C., Ruio, J. M. A., & Ipsen, R. (2012). Using fractal image analysis to characterize microstructure of low-fat stirred yogurt manufactured with microparticulated whey protein. *Journal of Food Engineering*, *109*, 721–729.

84. Tyson, J. (2001). How scanners work. http://HowStuffWorks.com.

85. Wang, H. H., & Sun, D. W. (2002a). Melting characteristics of cheese, analysis of effects of cooking conditions using computer vision technology. *Journal of Food Engineering*, *51*(4), 305–310.

86. Wang, H. H., & Sun, D. W. (2002b). Correlation between cheese meltability determined with a computer vision method and with Arnott and Schreiber tests. *Journal of Food Science*, *67*(2), 745–749.

87. Wasnik, P. G. (2015). *Development of Process Protocol for Image Analysis and Its Application for Detection of Cow Ghee with Vegetable Fat*. PhD thesis, ICAR-National Dairy Research Institute, Karnal, India.

88. Webb, B. H., & Holm, G. E. (1930). Color of evaporated milks. *Journal of Dairy Science, 13*, 25–39.

89. Wu, Y., Lin, Q., Chen, Z., Wu, W., & Xiao, H. (2012). Fractal dimension of the retrogradation of rice starch by digital image processing. *Journal of Food Engineering, 109*, 182–187.

90. Yam, K. L., & Papadakis, S. (2004). A simple digital imaging method for measuring and analyzing color of food surfaces. *Journal of Food Engineering*, *61*(1), 137–142.

91. Yang, Q. (1994). An approach to apple surface feature detection by machine vision. *Computers and Electronics in Agriculture*, *11*, 249–264.

92. Zheng, C. X., Da-Wen Sun and Zheng, L. Y. (2006). Recent Developments and Applications of Image Features for Food Quality Evaluation and Inspection: A Review. *Trends in Food Science and Technology*, *17*(12), 642–655.

93. Zion, B., Chen, P., & McCarthy, M. J. (1995). Detection of bruises in magnetic resonance images of apples. *Computers and Electronics in Agriculture*, *13*, 289–299.

CHAPTER 3

PASSIVATION: A METHOD TO ENSURE QUALITY OF DAIRY AND FOOD PROCESSING EQUIPMENT

GEETESH SINHA, KRISHAN DEWANGAN, and A. K. AGRAWAL

CONTENTS

3.1   INTRODUCTION

We are always concerned about the quality of processed food but it cannot be guaranteed until quality of processing equipment is ensured. The manufacturers and users of modern dairy and food processing equipment demand the use of stainless steel (SS) as the predominant material of construction. SS has become the standard material of construction because of its ability to maintain a high level of performance, while keeping corrosion to a minimum [14].

The metals used for food contact surfaces must be non-toxic, non-tainting, insoluble (in food), highly resistant to corrosion, easy to clean and keep bright, light yet strong, good agent of heat transfer, good in appearance throughout life, low in cost, non-absorbent and durable, etc. The metal SS possesses most of these properties. Therefore, generally food-processing equipment are made of SS. These are certain alloys of iron and chromium, which are highly resistant to corrosion. These alloys sometimes contain nickel and small percentage of molybdenum, tungsten and copper. This complex group of alloys is known as stainless steel. It can be welded, forged, rolled and machined [9]. Although unwantedly, it may be subjected to some contamination/ abnormalities due to mill scales, iron particle impregnation because of hot and cold rolling, pit corrosion, stress corrosion, inter granular corrosion in the *heat affected zone* (HAZ) and ruptured passive layer, etc. The presence of these contaminations/ impurities in SS equipment may adversely affect the metallurgical or sanitary conditions or stability of a surface, or contaminate a process fluid [15].

Passivation is an important surface treatment that helps assure the successful corrosion – resistant performance of SS used for product – contact surface (i.e., tubing/piping, tanks and machined parts used in pumps, valves, homogenizers, de-aerators, process-monitoring instruments, flow meters, ingredient feeders, blenders mixers, dryer pasteurizers, heat exchangers, conveyors and foreign body detectors) [6, 14]. It is performed when free iron, oxide scale, rust, iron particles, metal chips or other nonvolatile deposits might adversely affect the metallurgical or sanitary conditions or stability of the surface, the mechanical operation of a part, component or system, or contaminate the process fluid [15].

This chapter describes the passivation process to enhance the quality of dairy products.

3.2 DAIRY AND FOOD PROCESSING EQUIPMENT: USE OF STAINLESS STEEL IN MANUFACTURING

The SS is the steel that is correctly heat treated and finished, resists oxidation and corrosive attack from corrosive media. The SS has important application in food processing plant due to its marked resistance to corrosion by the atmosphere and a range of acid and alkaline solutions. There are fairly different distinct types of stainless steels having large percentage of chromium (4–22%) and sometimes of nickel (0–26%). The stainless steels are broadly grouped into three groups according to their microstructure namely: (i) martensite; (ii) ferrite; and (iii) austenite steels. The austenite SS is probably most important group of SS alloys. The main characteristics of this type are high ductility, work hardening ability, good corrosion resistance and high tensile properties. These alloys are highly resistant to many acids and cold nitric acid. These steels are used in manufacture of vats and pipes that are used in the dairy and food plants. The food processing equipment like kettle, tank, household items like cooking utensils and dairy utensils like milk can are some applications of this type of steels [3, 5].

The SS possesses favorable engineering properties like hardness, malleability and ductility. It can be fabricated easily to any size and shape. It can be easily welded although cannot be soldered. The equipment made with SS can be imparted high degree of surface finish (mirror polish) and therefore attractive in appearance. The SS equipment last longer on continuous service. They are actually non-toxic, non-tainting and have no influence on product quality in regard to flavor. The metallic surface of these equipment must be insoluble that comes in contact with both products and cleaning solutions. The SS equipment surfaces particularly product contact surfaces are easy to clean and sterilize [5].

3.3 NEED OF PASSIVATION OF FOOD PROCESSING EQUIPMENT MADE OF STAINLESS STEEL

Particles of iron or tool steel or abrasive particles may be embedded in or smeared on or into the surfaces of SS components during: handling and processing such as rolling, forming, machining, pressing, tumbling and lapping. If allowed to remain, these particles may corrode and produce rust spots on the SS, due to the formation of a galvanic couple between two dissimilar metals that can promote a corrosive reaction. To dissolve the embedded or smeared iron and prevent this condition, as well as restore the original corrosion-resistant surface, semi-finished or finished parts need to be given a *passivation* treatment [1].

Food Scientists/engineers alike are somewhat lack of knowledge when it comes to the relationship between corrosion resistant (stainless) steel and chemical passivation. Some believe it is effective because it is a cleaning process. Others credit the enhanced corrosion resistant properties to the thin, transparent oxide film resulting from passivation. Verification tests, including copper sulfate immersion, and accelerated corrosion tests, such as salt spray, high humidity, and water immersion, undisputedly confirm the effectiveness of passivation. Advanced material engineers in aerospace, electronics, medical, and similar high-tech industries have utilized passivation technology for many years. Their applications demand the maximum performance from components manufactured from stainless steels, and they realize that passivation is one of the most effective methods of achieving the desired results [4].

3.4 PICKLING

Pickling is typically performed to remove tightly adherent oxide films resulting from hot- forming, heat-treating, welding and other high temperature operations. Welding or heat treatment often produces complex oxides that can vary in color. All these oxides are generally referred to as "scale" and must be removed. Pickling is pre-cursor step to passivation. Prior to pickling, the heavy surface soils such as oil, grease, buffing compounds, drawing compounds, some scale, heavy rust, dye and paint markings, tape, adhesive residue and other foreign substances must be removed. This step may be accomplished by the use of alkaline cleaners, solvent cleaning, vapor degreasing, ultrasonic cleaning, steam cleaning, water-jetting, or other mechanical cleaning. Pre-cleaning is not required if oxide or scale is the only soil on the surface [2].

Where applicable, alternative mechanical methods such as blasting, shot peening, tumbling, and wheel abrading may also be performed. Abrasives containing iron should not be used. In many cases, pickling of stainless steels is performed in two steps: one for softening the scale and one for final scale removal. Over-pickling, under-pickling and pitting usually are direct results of lack of control over process variables including acid concentrations, solution temperature and contact time. The processes of pickling and passivation must be clearly distinguished. The process specified is to "*pickle and passivate* (a two-step method)" in order to create a surface on SS that would be resistant to corrosion. *Pickling* (or chemical de-scaling) is done to remove scale while *passivation* is done to make the surface more passive and corrosion resistant. Passivation is performed on clean stainless steel, provided the surface has been thoroughly cleaned or de-scaled [13].

3.5   WHAT IS PASSIVATION?

According to ASTM-A-380, passivation is "*the removal of exogenous iron or iron compounds from the surface of a stainless steel by means of a chemical dissolution, most typically by a treatment with an acid solution that will remove the surface contamination but will not significantly affect the stainless steel itself.*" In addition, it also describes passivation as "*the chemical treatment of a stainless steel with a mild oxidant, such as a nitric acid solution, for the purpose of enhancing the spontaneous formation of the protective passive film*" [1].

In layman's terms, the passivation process removes "free iron" contamination left behind on the surface of the SS as a result of machining and fabricating processes. These contaminants are potential corrosion sites, which if not removed, result in premature corrosion and ultimately result in deterioration of the component. In addition, the passivation process facilitates the formation of a very thin, transparent oxide film, which protects the SS from "selective" oxidation (corrosion). *Passivation* is the formation of a hard non-reactive surface film that inhibits further corrosion. This layer is usually an oxide or nitride that is a few atoms thick [1, 10].

3.6 EFFECTS OF PASSIVATION

Passivation is actually a comprehensive term denoted to describe a range of processes for achieving the desired effect on stainless steel. Therefore, it is necessary to encompass the various effects covered precisely by passivation [6, 7].

1. Passivation is the process by which SS will spontaneously form a chemically inactive surface when exposed to air or other oxygen-containing environments. Steels containing more than 11% Chromium are capable of forming an invisible, inert or passive, self-repairing oxide film on their surface. It is this passive layer that gives stainless steels their corrosion resistance. If a SS surface is scratched, then more chromium is exposed which reacts with oxygen allowing the passive layer to reform.
2. Passivation is the chemical treatment of a SS surface with a mild oxidant such as citric acid passivation solution. This process is to accelerate the process noted above in step 1.
3. Passivation is the removal of exogenous iron or iron compounds from the surface of SS by means of a chemical dissolution, most typically by a treatment with a citric acid passivation solution that will remove the surface contamination but will not significantly affect the SS itself.
4. Passivation is also accomplished by electropolishing, which is an electrochemical process that is a super passivator of SS and results in a more passive surface than the other methods mentioned above.

3.7 STEPS IN PASSIVATION

3.7.1 SURFACE CLEANING, CHEMICAL CLEANING, AND DEGREASING

This stage allows pickling and passivation chemicals to come in contact with metal surface which has to passivate.

3.7.2 PICKLING PROCESS

This stage allows to itch metal surface uniformly, which is controlled itching. In this stage metal surface gets activated uniformly.

3.7.3 PASSIVATION PROCESS

This stage eliminates corrosion susceptibility of surface of the metals creating uniform passive layer. Chemical surface conversion takes place. Surface becomes passive for further air-oxidation.

3.7.4 CONFIRMATION OF PASSIVATION

In this stage, the passivation effectiveness must be confirmed by one or more specified test practices to qualify for certification.

3.8 HOW IS THE PASSIVATION PROCESS PERFORMED?

The process typically begins with a thorough cleaning cycle. It is intended to remove oils, greases, forming compounds, lubricants, coolants, cutting fluids, and other undesirable organic and metallic residue left behind as a result of fabrication and machining processes (Figure 3.1). General degreasing and cleaning can be accomplished by a variety of commonly accepted methods, including vapor degreasing, solvent cleaning, and alkaline soaking. After removal of the organic and metallic residues, the parts are placed into the appropriate passivation solution. Although there are many variations of passivating solutions, yet the overwhelming choice is still the nitric acid based solutions. Recently, there has been substantial research performed to develop alternative processes and solutions that are more "environmentally friendly," yet equally effective. Although alternative solutions containing citric acid and other types of proprietary chemistry are available, they have not been as widely accepted commercially as nitric acid based solutions [11].

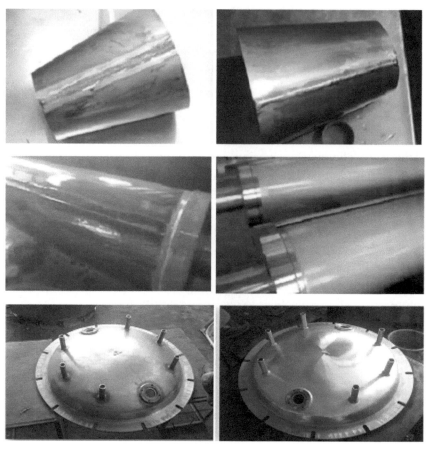

FIGURE 3.1   Utensils or parts of equipment before (left) and after (right) passivation.

The three major variables that must be considered and controlled for the passivation process selection are time, temperature, and concentration. Typical immersion times are between 20 and 120 minutes. Typical bath temperatures range between room temperature and 70°C. Nitric acid concentrations in the 20–50% by volume range are generally specified. Many specifications include the use of sodium dichromate in the passivation solution, or as a post passivation rinse, to aid in the formation of a chromic oxide film. Careful solution control, including water purity, ppm (parts per million) of metallic impurities, and chemical maintenance, are crucial for success. The type of SS being processed is the determining factor when selecting the most effective passivation process. Bath selection (time, temperature, and concentration) is a function of the type of alloy being processed. A thorough knowledge of the material types and passivation processes is paramount to achieving the desired results. Conversely, improper bath and process selection and/or process control will produce unacceptable results; and in extreme cases, can lead to catastrophic failure, including extreme pitting, etching and/or total dissolution of the entire component [7, 10, 14, 15].

3.9 APPLICATION METHOD: USER FRIENDLY APPROACH

The pickling and passivation chemical is available in ready to use formulation for application on equipment by any one of the following methods:

1. by dipping;
2. by spraying;
3. by swabbing with lint free cloth; and
4. by immersion.

The solution should be used at room temperature.

3.10 PROCESS OF PASSIVATION

The process calls for the following sequence:

1. Degreasing the product/plant (swabbing or spray).
2. Removal of welding seams.
3. Spray/brush/swab/immerse the equipment by pickling passivation chemical.
4. During application of the product, let the solution react with the surface for about 15–20 minutes.
5. Spray wash with normal water-water rinsing (120 bars).
6. Drying.

3.11 PRECAUTIONS DURING PASSIVATION

Passivation should only be performed by trained, experienced technicians familiar with the potential hazards associated with the science.

Safety practices must be fully understood when handling passivation chemicals. Special boots, gloves, aprons and other safety equipment must be utilized. Tanks, heaters and ventilation, as well as baskets and racks, must be appropriately engineered to perform the process. Iron or steel parts or equipment must never be introduced to the process, or the results can be devastating [11].

3.12   SPECIFICATIONS AND VERIFICATION TESTING OF PASSIVATION

There are few generally accepted industry specifications available for reference when choosing a passivation process. They offer time, temperature, and concentration information, and subsequent testing requirements to validate the effectiveness of the process. Many large corporations have developed internal specifications to control their unique requirements regarding passivation and verification testing. Regardless of the situation, it is usually prudent to reference a proven procedure when requesting passivation [1]. The most commonly referenced industry specifications are ASTMA-967 and ASTM-A-380. Both standards are well written, well-defined documents, which provide guidance on the entire process, from manufacturing to final testing requirements.

One of the most commonly specified verification tests is the copper sulfate test. Passivated parts are immersed in a copper sulfate solution for 6 minutes, rinsed, and visually examined. Any copper (pink) color indicates the presence of free iron, and the test is considered unacceptable. Other validation tests include a 2 hour salt spray or 24 hour high-humidity test. These tests are performed by placing passivated parts in a highly controlled chamber, which creates an accelerated corrosive environment. After subjecting the test pieces to the corrosive atmosphere for the prescribed exposure periods, the parts are removed and evaluated [1, 8, 12].

3.13   SUCCESSFUL PASSIVATION: RECOMMENDATIONS TO FOOD PROCESSING EQUIPMENT MANUFACTURERS

It is seen that most of the equipment manufacturers use the passivation solution for only the weld pool line or the heat dissipation areas. But that does not ensure from risk of corrosion since the pitting on the remaining surface remain untreated as well as the corrosion susceptible gray zone remains forever. The simple spatters of the grinding also get indented into the surface and it causes the seeding for the corrosion. The following practices will reduce gross contamination during manufacturing and increase the chances of successful passivation and test results:

- Never use grinding wheels, sanding materials, or wire brushes made of iron, iron oxide, steel, zinc, or other undesirable materials that may cause contamination of the SS surface.
- Grinding wheels, sanding wheels, and wire brushes that have been previously used on other metals should not be used on stainless steel.
- The use of carbide or other non-metallic tooling is recommended whenever possible.
- Use only clean, unused abrasives such as glass beads or iron-free silica or alumina sand for abrasive blasting. Never use steel shot or grit, or abrasives which have been used to blast other materials.
- Thorough cleaning prior to any thermal processing is critical. Stress relieving, annealing, drawing, or other hot-forming processes can actually draw surface contaminants deeper into the substrate, making them almost impossible to remove during passivation.
- Care should be taken during all thermal processes to avoid the formation of discoloration (oxides). Passivation is not designed to remove discoloration, and will not penetrate heavy oxide layers. In extreme situations, additional pickling and descaling operations are required prior to passivation to remove the discoloration. Controlled atmosphere ovens are highly recommended for all thermal processes to reduce airborne contamination and prevent oxides from developing [14].

3.14   CONCLUSIONS

During equipment manufacturing, machining, fabricating and heat-treating practices can substantially affect the corrosion resistance of the component. The passivation will enhance the corrosion resistance of stainless steels, but to realize the maximum performance from these high-tech alloys, all parties involved with manufacturing must understand their responsibility in maintaining the integrity of the material throughout the process.

Pickling or chemical descaling is but one of several pretreatment steps available for preparing an article for further processing such as passivation or electro-polishing, or to perform a superior cleaning operation of welded structures. Passivation process is both an art and a science. Passivation is a process performed to make a surface passive, for example, a surface film is created that causes the surface to lose its chemical reactivity. Passivation unipotentializes the SS with the oxygen absorbed by the metal surface, creating a monomolecular oxide film. Passivation can result in the very much-desired low corrosion rate of the metal.

3.15   SUMMARY

Pickling and passivation is used to reduce contamination and corrosion on stainless steel. It is the innovation technology to change the surface area of SS for strong, better and long serving. Normally dairy and food equipment manufacturing industry uses SS of grades of AISI 304, 304L, 316, 316L, etc., as material of construction. The raw pieces of SS materials are cut, rolled, welded and brazed, in order to give final shape to the food processing equipment. During fabrication, the materials develop several metallurgical changes which may be responsible for contamination of the surface of final equipment. It is therefore absolutely essential to ensure pickling and passivation of SS plants and equipment. Otherwise contamination will come in contact with the metal and ultimately result in the contamination of food product.

KEYWORDS

- dairy processing equipment
- electro polishing
- food processing equipment
- galvanization
- passivation
- pickling
- stainless steel

REFERENCES

1. American Society for Testing and Materials (ASTM), (1996). *Cleaning and Descaling Stainless Steel Parts, Equipment and System (ASTM-A-380).* ASTM, West Conshohocken, PA, USA.
2. Anonymous (1986). *Pickling Handbook Surface Treatment of Stainless Steels.* Voestalpine Bohler Welding GmbH, p. 12.
3. Anonymous, (1988). *Cleaning and Descaling Stainless Steels.* A Designers Handbook Series by Nickel Development Institute.
4. Bornmyr, A., Toesch, J., & Winkler, F. (2009). *Pickling Handbook: Surface Treatment of Stainless Steels.* Böhler Welding GmbH.
5. Holah, J. (2000). *Food Processing Equipment Design and Cleanability.* The National Food Centre, Ireland.
6. http://en.wikipedia.org/wiki/Passivation.
7. http://en.wikipedia.org/wiki/Passivation (chemistry).
8. http://www.astm.org/Standards/A380.htm.
9. http://www.astropak.com/ultra-pass-passivation.php.
10. http://www.delstar.com/passivating.htm.
11. http://www.electrohio.com/Finishing/Passivation/Passivation.htm.
12. http://www.passivationindia.com.
13. http://www.picklingandpassivation.com/index.html.
14. Maller, R. R. (1998). Passivation of stainless steel. *Trends in Food Science and Technology, 9,* 28–32.
15. Tuthill. H. (2002). *Stainless Steels and Specialty Alloys for Modern Pulp and Paper Mills.* Reference Book Series No 11-025 by Nickel Development Institute, p. 139.

CHAPTER 4

TECHNOLOGY OF PROTEIN RICH VEGETABLE BASED FORMULATED FOODS

DINESH CHANDRA RAI and ASHOK KUMAR YADAV

CONTENTS

4.1 INTRODUCTION

Formulated foods are foods that are prepared or manufactured according to plan from individual components, to yield products having specified physical, chemical and functional properties. These are foods that have been taken apart and put together in a new form. Designed, engineered or formulated from ingredients, they may or may not include additives, vitamins and minerals. Formulated foods are engineered foods and are same as fabricated foods. There are mainly two types of formulated foods:

- Those designed to simulate natural counterparts and are referred as "analogs" such as meat and dairy analogs.
- Those no prior counterparts and are manufactured for specific functional properties by mixing two or more food products.

The food processing industry is giving increased emphasis to the production and utilization of alternate protein isolate products as functional and nutritional ingredients in an expanding number of formulated food products. Alternate protein sources such as soy and other vegetable proteins offer additional flexibility in formulating foods due to their economics, availability, functionality and nutritional properties. It also needs for developing soy and vegetable protein isolates with improved flavor, color and functionality for producing simulated dairy foods. It also considers alternative technologies for incorporating soy and vegetable proteins into the formulation so that they may function properly for forming stable solutions, emulsions, foams and gels that resemble those in their natural dairy food counterparts. Many types of formulated foods are now available and selected categories are shown Table 4.1.

Consumers are now more concerned on nutritious, healthy and natural foods [4]. Fruit and vegetable blends present numerous advantages, such as flavor and nutritional added-value [24]. Blended beverage has appeared as a good option of nutritionally improved product with good sensory properties [5, 11, 17]. Some of the Indian traditional formulated foods are listed below:

- Curd based formulations;
- Halwa;
- Kheer;
- Other sweat meats (such as channa roll, sweet curd, laddoo, etc.);

TABLE 4.1 Types of Formulated Foods

Baby foods	Foods for infants, pureed or strained fruits and vegetables with cereals, mashed fruits in yogurt
Convenience foods	Snack packs, T.V. dinners, etc.
Dairy analogs	Formulated non dairy products, soy based products
Geriatric foods	Foods formulated for old aged people as banana split oatmeal
Low calorie foods	Diet of less than 800 kilocalories per day, for example, Tofu, dried fruits
Meat analogs	Texturized vegetable protein foods
Novelty foods	Imitation Cavier, French fried molded onion rings
Snack foods	Extruded products such as roasted chickpeas, herbed cashews, baked kale chips
Soft moist foods	Intermediate moisture foods
Special purpose dietary foods	Low cholesterol, low fat, sugar free, low sodium, etc.

- Paneer and Channa formulations;
- Using ayurvedic traditions – Ksheer pak (e.g., arjuna chhal is cooked in milk);
- Yoghurt based formulations.

This chapter discusses needs for developing fruits, vegetables and soy proteins isolates with improved flavor, color and functionality for producing formulated foods. The chapter also considers alternative technologies for incorporating fruits, vegetables and soy proteins into the formulation so that they may function properly for forming stable solutions, emulsions, foams and gels that resemble those in their natural counterparts.

4.2 PRINCIPLES OF FOOD FORMULATIONS

Formulated foods are manufactured by combining the three basic building blocks of the food (protein, fat and carbohydrate) in the way to provide best texture, flavor and other desirable characteristics. This involves manipulation of these three components along with water, vitamins, preservation, etc. to design product with predictable composition, flavor, texture and storage properties. In the design of formulated food, it is important to emphasize on solid-liquid phase equilibria, multi component solubility behavior, polymorphism, nucleation, crystal growth, wetting, emulsification, stability of disperse phase and mechanical properties of crystal assemblies.

Protein fortification in food and beverages is a key imperative to meet global challenges in nutritional deficiencies. Animal proteins typically provide complete protein, but soy is one of the best plant sources that also provide a complete protein. There has been a notable shift toward plant-derived proteins that offer similar or superior functional properties. Sensory properties are important to successful penetration of plant protein ingredients into applications dominated by animal proteins, as well as for development of new applications.

4.3   NEED FOR FORMULATION AND PROTEIN CHEMISTRY

Formulation of foods is required due to:

- Lower cost: The cost of certain formulated foods may be lower while providing about the same nutrition and satiety, for example, margarine.
- The nutritional properties of the food may be enhanced: Fruit juices fortified with Vitamin C, margaine with Vitamin A, cereals with iron and other nutrition.
- Food can be made more convenient.
- Food can be formulated to fill special needs.
- Formulated foods can fill other meal advantages. They can improve the balance of the meal.
- They can increase palatability and finally can provide satiety and other advantages.

Proteins are added to food for nutritional reasons and for their functionality. Functional abilities include viscosity enhancement and water binding, gelation, aeration and foaming and emulsification with contributions to a food's flavor, texture and color.

When formulating a food or beverage, it is advisable to first consider why a protein ingredient would be used. For example, if it is just for viscosity or emulsification, alternatives such as starch or a lipid-based emulsifier are available, because proteins are the most expensive macro-ingredient as compared to carbohydrates and fats. Beyond functionality, other factors that influence which protein is chosen for a product's formulation includes the percent of protein within the ingredient, digestibility, allergenicity, label simplicity, desired label claims, animal welfare, and amino acid profile and score. A wide range of food components can contribute proteins at a wide range of costs. Soy proteins are one of the most economical. They offer good nutritional quality and the soy industry provides soy proteins in numerous forms for specific needs.

There are a number of dairy food systems, for example, fluid milk, infant formula, coffee whitener, sweet and sour cream, margarine, cheese, frozen dessert and whipped topping, which offer opportunities for utilizing vegetable proteins in place of milk protein products [7]. Significant progress has been made in developing technologies for utilizing vegetable proteins in a number of simulated dairy food systems; however, several important problem areas must be solved before vegetable protein can be universally used in such dairy food systems. First, it will be necessary to produce "second generation" vegetable protein isolates [26] with improved flavor, color, and functionality compared to those presently available. Second, it will be necessary also to devise new and improved technologies for formulating the simulated dairy food products that will promote the vegetable protein's ability to provide the necessary functional attribute. Initially, it would be advantageous to gain a thorough understanding of the fundamental properties of the vegetable protein system which control and limit their functionality in each of the application areas. A similar degree of understanding of the basic properties of the vegetable protein system will be helpful in attempting to overcome technical problems associated with utilizing them in simulated dairy food systems. It is felt that concentrating on the basics offers more promise for solving the serious problem areas than merely attempting to develop a special protein isolate or a special technology for each particular food application.

Pea protein is non-genetically modified (GMO), is not a known allergen and contains no gluten. In agriculture, the low-water usage and nitrogen fixation properties of pulse crops makes them more sustainable than some other protein crops. Peas are low in cysteine and methionine, but high in lysine, resulting in high Protein Digestibility Corrected Amino Acid Score (PDCAAS).

Pea protein properties and applications depend on the method used for isolation, which result in different albumin, vicilin and legumin ratios. This likely explains the differences in behavior between various pea protein products on the market. Generally, pea proteins show good water binding, gelation and emulsification but lesser foaming properties.

Other types of protein that are under development or that have recently entered the market include potato, rice, canola-rapeseed (mustard), quinoa, oat, flax, hemp, algae and leaf material. To utilize low-cost proteins, the physiochemical properties need to be understood and tailored to the intended use. The development and understanding of soy protein as an ingredient for the food industry can, therefore, serve as a model for the utilization of other plant-based proteins (Figure 4.1).

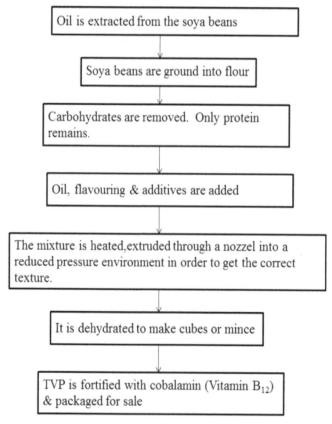

FIGURE 4.1   Flow diagram for manufacturing of texturized vegetable protein.

Meat analogs containing fibrous, vegetable proteinaceous materials (often referred to as textured vegetable proteins) are used as meat substitutes. Meat analogs may be prepared by combining the textured vegetable protein with an edible binder, fats, flavoring agents, etc. and fabricated so as to resemble natural meat cuts. Hydrated, textured vegetable proteins are also used as proteinaceous diluents or extenders in comminuted meat products.

Pliable vegetable proteins suitable for use as extenders or textured vegetable proteins in meat analogs are prepared by forming a homogeneous aqueous dispersion of water-soluble vegetable protein and edible plasticizers, drying and heat-denaturing the dispersion so as to form a mass and hydrating the mass with an aqueous acid to convert the hydrate to a pliable protein product. Flow diagram for manufacturing of texturized vegetable protein is shown in Figure 4.1. Illustrative product formulating ingredients include soy protein concentrates, polyols, triglycerides and lactic acid as an acidulant.

Nowadays, food industry is marked by the high volume of waste produced. According to the recent research conducted by FAO, about 1.3 billion tons of food is wasted worldwide per year, which represents one-third of the total food industry production [14]. The largest amount of loss is found by fruits and vegetables, representing 0.5 billion tons. In developing countries, fruit and vegetable losses are severe at the agricultural stage but are mainly explained by the processing step, which accounts for 25% of losses [14].

Fruits and vegetables are extensively processed and the residues are often discarded. However, due to their rich composition, they could be used to minimize food waste. Development of food products based on the solid residue generated from the manufacture of an isotonic beverage could be accepted (Figure 4.2). These types of beverages could be produced based on integral exploitation of several fruits and vegetables: orange, passion fruit, watermelon, lettuce, courgette, carrot, spinach, mint, taro and cucumber. The remaining residue could be processed into flour. The fruit and vegetable residue (FVR) flour could be incorporated with different levels into biscuits and cereal bars (Figure 4.2).

High fiber, protein and mineral contents and also the water holding capacity (WHC) and oxygen holding capacity (OHC) of the FVR flour

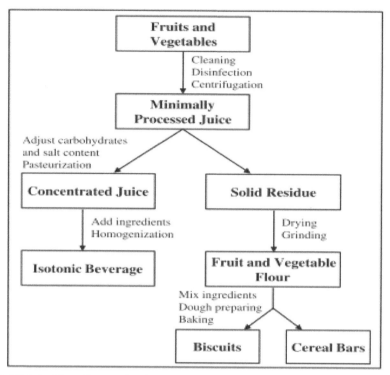

FIGURE 4.2 Flow diagram for the isotonic beverage, fruit and vegetable residue flour, biscuit and cereal bars production.

are potentially suitable for use in food applications as a new low-calorie and functional raw material. The designed products have high fiber content, reasonable consumer acceptance and were microbiologically stable. Such research promotes the reducing food waste since whole plant tissues have been used leading to the maximum exploitation of food raw materials.

4.4 PRODUCTS MADE FROM BLEND OF MILK AND COCONUT MILK

Coconut milk contains carbohydrates (mainly sucrose and some starch), lipid and minerals like P, Ca, & K. Coconut protein is rich in lysine, methionine and tryptophan [21]. Yoghurt had been made from mixtures of cow milk and coconut milk in different combinations or even from blend of coconut milk and SMP [9]. This practice would be an interesting option in high coconut producing regions and countries.

4.5 MILK SOLIDS WITH VEGETABLES

It has been reported that the reduction by 50% in risk of developing some forms of cancer through intake of 400 to 600 g of fruit and vegetable per day [15]. Broccoli is one of the richest sources of health promoting gluco-sinolates, antioxidants and essential nutrients viz., fiber, Ca, Mg, Se, Zn, ascorbic acid, folate, β-carotene, protein, etc. Cheese has health promoting characteristics associated with certain immune functions in the body due to bioactive peptides and conjugated linoleic acid (CLA). Cheese is rich in minerals like Ca, Mg, P and fat-soluble vitamins and possesses anti-caries function. A blend of Blue cheese powder containing up to 10.0 to 20.0% of freeze dried broccoli sprouts powder has been acceptable as a novel 'health-promoting' food. Such cheese powder had total antioxidant capacity (ascorbic acid equivalent), total polyphenols (gallic acid equivalent), total chlorophyll and total carotenoids of 53.5, 4504.2, 147.1 and 59.8 µg/g dry weight basis (DWB), respectively [22].

Carrot is rich in β-carotene (concentration of 39.6 and 23.9 mg/100 g in fresh and powdered carrot respectively), the precursor of vitamin A and contains an appreciable amount of vitamin B_1, B_2, B_6 and anthocyanin pigment. The sweetness of carrot is due to presence of sucrose, maltose and glucose. Carrot intake may enhance immune system, and protect against high blood pressure, osteoporosis, cataracts, arthritis, bronchial asthma and urinary tract infection [3]. Buffalo or cow skim milk has been blended with carrot juice (7.0% TS) and added with 8.0% sugar and 0.2% gelatin to prepare a nutritious beverage; and levels above 20.0% carrot juice incorporation resulted in sedimentation [23].

A sterilized carrot *kheer* (sweetened concentrated milk) has been developed using milk added with 8.0% sugar, dry fruit and 30.0% of shredded carrot that was previously cooked in *ghee*. The technology of preparing carrot *burfi* (*Khoa* based sweet meat) is very popular in India whose method has been standardized [18]; the product had 8.2% fat and 33.3%

TS [20]. A *kheer* mix has been formulated based on dehydrated carrot, skim milk, sugar, corn flour, cashew and cardamom [16]. A healthy bottle gourd *halwa* (similar to *burfi*) has been successfully standardized [8]. Cheese powders with additional flavor profiles such as cheese with bacon, onion or tomato, are products that would be suitable for quiches, pizzas and pasta meals [1].

4.6 MILK SOLIDS WITH FRUIT

Fruit are invariably used for flavoring of several dairy products. However, due to the presence of phytochemicals in most of the fruits, its involvement has increased looking at the 'wellnesses' of the product containing it. Fruits are rich sources of various important phytonutrients namely, vitamins, minerals, antioxidants and dietary fibers. Incorporation of fruits in milk products not only aids in 'value-addition' and 'product diversification,' but also helps in checking the post harvest losses.

Merging of dairy products and fruit beverage markets with the introduction of 'juice-ceuticals' like fruit yogurt beverages that are typical examples of hybrid dairy products offering health, flavor and convenience. Typical examples related to inclusion of fruit solids in dairy based products include ice cream and frozen desserts, stirred yogurt, fat spreads, etc.

4.7 MILK SOLIDS WITH VEGETABLE OILS: FILLED DAIRY PRODUCTS, CHEESE ANALOGUES AND FAT SPREADS

4.7.1 FILLED PRODUCTS

Filled products are the one in which milk fat is replaced completely by vegetable oils and fats. This is mainly done to have a dietetic product without cholesterol and with low levels of saturated fats. Typical examples include filled milk, filled whipped cream, filled ice cream, etc. Corn, olive, and groundnut oil were found suitable for preparing 'filled strawberry yogurt' containing 1.5% vegetable oil. The filled yogurt had higher proportion of PUFA and mono unsaturated fatty acids [2].

4.7.2 NON-DAIRY COFFEE WHITENER

The formulation for a non-dairy coffee whitener/creamer comprised of (on dry matter basis) 60.0 to 65.0% corn syrup solids and maltodextrin, 20.0 to 32.0% vegetable oil having 35 to 40°C melting point, 2.0 to 5.0% Na-caseinate, 1.0 to 3.0% disodium phosphate, 1.0 to 3.0% emulsifier, stabilizer and cream flavor [13].

4.7.3 CHEESE ANALOGUES

Analogue cheese, cheese substitute of imitation cheese are synonyms for diverse type of cheese materials that use vegetable oils instead of butter fat and contain casein or caseinates as the protein source. The protein source can also be from vegetable (soybean, groundnut) sources. The texture and flavor profile of such cheese analogs is governed by the type of oil, protein, starch, hydrocolloid and emulsifying salts used in the formulation. The vegetable oil is required to be partly hydrogenated or suitably modified in order to elevate its melting point to a level near to that of milk fat [6].

4.7.4 MIXED FAT SPREADS

A good example of mixed fat spread is 'Bregott,' developed and marketed by the Swedish Dairies Board. The product was made using ripened cream (35.0% fat, pH 4.6 to 4.7) to which refined soy bean oil was added (at levels of 20.0% of total fat) and after aging churned to obtain fat spread that had superior spreadability at refrigeration temperature [25]. Safflower seed based fat spread has been prepared from *chakka* (partly dehydrated *dahi*) obtained from a blend of buffalo milk and safflower milk (1:1); the additives were 1.0% tri-sodium citrate, 0.1% potassium sorbate, 1.5% common salt and 10.0% of cheese as flavoring. The low-fat spread had low cholesterol and was rich in PUFA (Polyunsaturated fatty acid) compared to product made from cow/buffalo milks; the cost and calorie was also lower for the former product [10]. Spreads with increased content of ω-3 fatty acids have been developed using olive oil [19].

4.8   PRODUCTION OF SIMULATED DAIRY FOOD SYSTEMS

Simulated dairy food products require soy or vegetable protein isolates with significantly improved flavor, color, and functional properties. The need for such high quality protein products stems from the fact that milk and most dairy food products have an extremely mild and delicate flavor, low color level and exhibit a characteristic texture that will not accommodate even a minor alteration. Even though certain fermented dairy foods and cheese products possess a highly viscous or gelled structure and have a characteristic flavor and color, soy and vegetable protein products must not detectably alter these properties or they will be readily identified as inferior products by the consumer. Figure 4.3 depicts some of the examples of the soy-based foods.

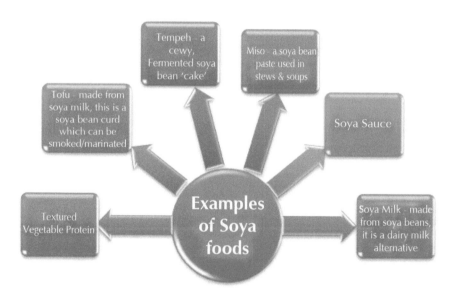

FIGURE 4.3   Examples of soy based foods.

4.9   FORMULATION OF VEGETABLE PROTEIN AND FIBER: CHALLENGES AND SOLUTIONS

Protein and fiber are added to food systems for many reasons, both functional and nutritional. However, with their addition comes the need for ingredient and processing adjustments, depending on the final food and its desired characteristics.

Strategies to overcome these issues include use of multiple sources of protein and fiber. In addition to protein powders, nuggets or crisps can be high in protein and also contain fiber. Coatings can be protein or fiber fortified. Cereal pieces, like oats, wheat flakes, nuts, pulse flour, or pieces and seeds, are other sources of protein and fiber. Protein hydrolysates are helpful to overcome such problems.

Processing technology innovations will also be needed to successfully produce simulated dairy food products using soy and vegetable proteins. Properly established procedures for solubilizing, blending and incorporating soy and vegetable proteins into the formulation, which will presumably replicate the composition and overall properties of milk or the other natural dairy food system, will be essential for successfully utilizing these proteins in simulated dairy food and other system.

4.10   CONCLUSIONS

Vegetable proteins have source of adequate protein for the formulation of new food products. It has been proved from many researches that attention should be paid to the mixture of vegetable proteins to meet the adequate nutritional quality needed for the promotion of growth of infants, adults and aged people. The results of vegetable based formulated foods expatiate in terms of nitrogen retention in various tissues of the body. Food consumption, nitrogen retention, average protein level and optimum growth rate were found to be inter-related. From the economic and health benefit point of view, newly formulated diets could be produced in a much cheaper rate than the commercial available products. Investigation of such products will be useful to industries and common people in generally to alleviate protein energy malnutrition in developing countries.

4.11   SUMMARY

At present, food processing industry is giving increased emphasis on the production and utilization of alternate protein isolate products as functional and nutritional ingredients in an expanding number of formulated food products. Alternate protein sources such as fruits, vegetables and soy proteins offer additional flexibility in formulating foods due to their economics, availability, functionality and nutritional properties.

This chapter discusses needs for developing fruits, vegetables and soy proteins isolates with improved flavor, color and functionality for producing formulated foods. It also considers alternative technologies for incorporating fruits, vegetables and soy proteins into the formulation so that they may function properly for forming stable solutions, emulsions, foams and gels that resemble those in their natural counterparts.

Properly established procedures for solubilizing, blending and incorporating fruits, vegetables and soy proteins into the formulation, which will presumably replicate the composition and overall properties of milk or the other natural food system, will be essential for successfully utilizing these proteins in simulated food products. Food manufacturers throughout the world have developed many dairy-related products that contain fruits, vegetables and soy proteins and sodium caseinate. Also, dietary, religious, and ethnic constraints have become increasingly important considerations in the formulation and market of these products.

Any food product made out of milk will obviously be highly nutritious. However, milk has certain limitations like allerginicity, lactose intolerance, cholesterol, saturated fat content, etc. With advancement in technology, several dairy derived ingredients are being produced that when used in conjunction with other food ingredients in food manufacture can yield 'value-added' products that have balanced nutrition, and exhibiting superior functionalities as well as 'wellness' in food application. This implicates the synergy of dairy with other food ingredients/products; such combination exhibits effects that are probably not obtained when either ingredient is used singly. Dairy along with fruit and vegetable ingredients may get transformed into 'formulated foods.'

Combining the neutraceutical components of dairy and non-dairy based food item, a new 'value added' product can emerge that can fulfill the ever increasing demand for 'wellness food' by the 'health conscious' consumers. With further research on dairy ingredients and their interaction with other food constituents and ingredients, more and more new 'novel' products would be developed for 'enjoyment' and 'wellness' of the consumers.

KEYWORDS

- cholesterol
- conjugated linoleic acid
- fabricated foods
- formulated foods
- genetically modified
- malnutrition
- Maltodextrin
- miso
- mozzarella
- neutraceutical
- oxygen holding capacity
- phytochemicals
- polyunsaturated fatty acid
- protein digestibility corrected amino acid score
- protein hydrolysates
- tempeh
- texturized vegetable protein
- tofu
- value-added
- water holding capacity
- β-carotene
- ω-3 fatty acids

REFERENCES

1. Anon. (1994). Powder performance. *Dairy Ind. Intl.*, *59*, 42–44.
2. Barrantes, E., Tamime, A. Y., & Sword, A. (1994). Oils versus milk fat. *Dairy Ind. Intl.*, *59*, 25–30.
3. Beom, J., Yong, S., & Myung, H. (1998). Antioxidant activity of vegetables and blends in iron catalyzed model system. *J. Food Sci. Nutr.*, *3*, 309–314.
4. Betoret, E., Betoret, N., Vidal, D., & Fito, P. (2011). Functional foods development: Trends and technologies. *Trends Food Sci Technol.*, *22*, 498–508.
5. Bhardwaj, R. L., & Pandey, S. (2011). Juice blends—a way of utilization of under-utilized fruits, vegetables, and spices: a review. *Crit Rev Food Sci Nutr.*, *51*, 563–570.
6. Chavan, R. S., & Jana, A. (2007). Cheese substitutes: An alternative to natural cheese: a review. *Intl. J. Food Sci. Technol. Nutr.*, *2*, 25–39.
7. Circle, S. J. (1974). JAOCS 51, 198A.
8. Dalal, T. (2008). Healthy halwas – Bottle gourd halwa. In: *Mithai*. Chapter 3. Sanjay and Co. Pub., Mumbai, pp. 53–54.
9. Davide, C. L., Peralta, C. N., Sarmago, I. G., & Sarmago, L. J. (1990). Yoghurt production from a dairy blend of coconut milk and skimmed milk powder. *Philippine J. Coconut Stud.*, *11*, 51–58.
10. Deshmukh, M. S., Patil, G. R., Sontakke, A. T., Kalyankar, S. D., & Padghan, P. V. (2003). Storage study of low-fat spread from safflower milk blended with buffalo milk. *J. Dairying Foods and Home Sci.*, *22*, 63–66.
11. Dominguez-Perles, R., Moreno, D. A., Carvajal, M., & Garcia-Viguera, C. (2011). Composition and antioxidant capacity of a novel beverage produced with green tea and minimally-processed byproducts of broccoli. *Innov Food Sci Emerg Technol.*, *12*, 361–368.
12. Ferreira, M. S. L., Santos, M. C. P., Moro, T. M. A., Basto, G. J., Andrade, R. M. S., & Gonçalves, E. C. B. A. (2015). Formulation and characterization of functional foods based on fruit and vegetable residue flour. *J Food Sci Technol.*, *52*(2), 822–830.
13. Gardiner, D. S. (1977). *Non-Dairy Creamer Compositions*. United States Patent No. 4,046,926.
14. Gustavsson, J., Cederberg, C., Sonesson, U., & Van Otterdijk, R. A. M. (2011). *Global Food Losses and Food Waste: Extent, Causes and Prevention*. Food and Agriculture Organization of the United Nations (FAO), Rome.
15. Heber, D. (2004). Vegetables, fruits and phytestrogens in the prevention of diseases. *J. Postgrad. Med.*, *50*, 145–149.
16. Manjunatha, S. S., Kumar, B. L., Mohan, G., & Das, D. K. (2003). Development and evaluation of carrot kheer mix. *J. Food Sci. Technol.*, *40*, 310–312.
17. Martins, R. C., Chiapetta, S. C., Paula, F. D., & Goncalves, E. C. B. A. (2011). Evaluation isotonic drink fruit and vegetables shelf life in 30 days. *Braz J Food Nutr.*, *22*, 623–629.
18. Mathur, S. (2008). Carrot burfi. In: *Indian Sweets*. Ocean Book Ltd. Pub., pp. 35–36.
19. Mortensen, B. K. (2009). Production of yellow fats and spreads. In: *Dairy Fats and Related Products*, edited by Tamime, A. Y., Blackwell Pub. Ltd., pp. 167–194.
20. Qureshi, M. A., Goel, B. K., & Uprit, S. (2007). Development and standardization of sterilized carrot kheer. Egypt. *J. Dairy Sci.*, *35*, 195–197.

21. Seow, C. C., & Gwee, C. N. (1997). Coconut milk: chemistry and technology. *Int. J. Food Sci. Technol.*, *32*, 189–201.

22. Sharma, K. D., Stahler, K., Smith, B., & Melton, L. (2011). Antioxidant capacity, polyphenolics and pigments of broccoli cheese powder blends. *J. Food Sci. Technol.*, *48*, 510–514.

23. Singh, C., Grewal, K. S., & Sharma, H. K. (2005). Preparation and properties of carrot flavored milk beverage. *J. Dairying Foods Home Sci.*, *24*, 184–189.

24. Sun-Waterhouse, D. (2011). The development of fruit-based functional foods targeting the health and wellness market: a review. *Int J Food Sci Technol.*, *46*, 899–920.

25. Wilbey, R. A. (1986). Production of butter and dairy-based spreads. In: *Modern Dairy Technology, Advances in Milk Processing*, edited by Robinson, R. K. Elsevier Appl. Sci. Pub., London, *1*(3), 93–129.

26. Wolf, W. J., & Cowan, J. C. (1975). *Soybeans as a Food Source*. CRC Press, Cleveland, OH.

PART II

PROCESSING METHODS AND THEIR APPLICATIONS IN THE DAIRY INDUSTRY

APPLICATION OF SCRAPED SURFACE HEAT EXCHANGER IN MANUFACTURING OF DAIRY PRODUCTS: A REVIEW

A. V. DHOTRE and A. G. BHADANIA

CONTENTS

5.1   INTRODUCTION

For processing milk to milk products, number of unit operations are employed, namely: filtration, clarification, fractionation, chilling, freezing, preheating, homogenization, pasteurization, sterilization, concentration, drying, curing, etc. Majority of these operations involve heat transfer, which is commonly done by employing indirect type of heat exchangers, wherein the heat transfer takes place through a metallic heat transfer surface. The inherent problems associated with the use of indirect heat exchangers include fouling of heat transfer surface, low degree of turbulence over heat transfer surface on both sides due to boundary layer formation, difficulty in cleaning and sanitization, failure to handle viscous and/ or particulated products.

These shortcomings are greatly mitigated in a scraped surface heat exchanger (SSHE). The scrapers in SSHE continuously scrape the product off the surface, thus eliminate fouling, increase turbulence, aid in controlling product residence time, mix the product, avoid localized treatment and also facilitate the cleaning operation. These features have greatly increased the applications of SSHE for manufacturing dairy products especially traditional dairy products (TDP) over last three decades. SSHE has also been used for the manufacture of several western dairy products [5, 30]. The SSHE has been employed for many processing milk into many dairy products majority of which are the indigenous milk products of India.

This chapter is an attempt to discuss these applications in the dairy industry.

5.2   SCRAPED SURFACE HEAT EXCHANGER

The SSHE also-called as swiped surface heat exchanger is a special type of tubular heat exchanger. In its simplest form, a SSHE consists of: (1) jacketed stationery cylinder called as product tube or barrel; (2) a rotating cylinder or shat carrying scraper blades; (3) inlet and outlet ports; and (4) accessories for sensing, controlling and safety purpose. The material to be processed flows axially through the annular section between the stationary outer cylinder and rotor. The moving blades scrape the heat transfer surface periodically to prevent the deposition of material and promote mixing [28].

The mechanical action of scraper blades on the heating surface enhances heat transfer, prevents fouling and to some extent regulates the residence time distribution. Extensive information has been presented [5, 30] on flow patterns, mixing effects, residence time distribution, heat transfer, power requirements of SSHE and their applications in food processing. The inherent features of SSHE make it suitable to handle the products with high viscosity and particulate structure, which occurs in most of the dairy product especially in traditional Indian dairy products (TIDP). Moreover, the operating parameters (such as residence time, temperature, pressure, speed of rotation, etc.) can be easily controlled as compared to the kettle method. Therefore, the SSHEs have replaced many kettle based units that were used in the manufacture of dairy products [7]. The numerous applications of SSHEs are discussed product wise in this chapter.

5.3 MILK CONCENTRATION FOR MAKING KHOA AND ALLIED PRODUCTS

Khoa is the product obtained by partial desiccation or dehydration through boiling of whole milk under atmospheric conditions. Requirement of constant stirring and scraping while heating is the specific feature of the khoa making process [29]. The concentration of milk solids is almost increased by 5 times during the manufacture of khoa.

In order to overcome the drawbacks of traditional method of khoa making, such as limited capacity, lot of time and labor requirement and necessity to clean pan between batches, Banerjee, et al. [8] designed and developed equipment for continuous khoa making (Figure 5.1). It consists of a steam jacketed drum heater, open steam jacketed pans and power driven scrapers. In this equipment, milk at the rate of 50 liter per hour is gradually concentrated by heating it in a steam jacketed drum heater operated at 3 kg/cm^2 (294.3 kPa) steam pressure followed by further heating and concentrating it in open steam jacketed pans. Since the unit was made of mild steel, the product was prone to oxidation and discoloration [49], the plant did not work effectively owing to the lack of controls for regulated supply of milk [14]. Therefore, the method for production of khoa with this machine was specially standardized [18] with suggestion to improve some constructional features of the machine. The improvements were incorporated and new design was fabricated using stainless steel [35, 47], which performed satisfactorily on small and medium scale production. A khoa making equipment consisting of a hemispherical open pan weld mounted on a cylindrical water jacket has also been reported for village level operations [6].

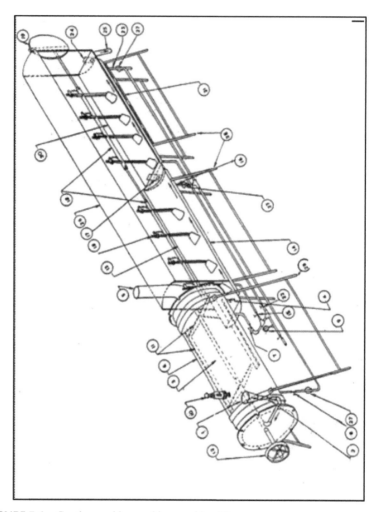

FIGURE 5.1   Continuous khoa making machine [8].

Alfa Laval has introduced Convap: an adaptation of the basic Contherm unit that allows the unit to operate as a SSHE evaporator [7]. Contherm heat exchanger can easily be converted into Convap by adding a vapor dome, instead of a top head assembly and with the addition of auxiliaries like vapor separator, condenser and a vacuum pump. It has been adopted at Chitle farm, Pune for khoa making [22].

Aiming for better thermal efficiency, a Conical Process Vat working on SSHE principle has been developed at NDRI, Karnal. It consists of a steam jacketed conical vat with cone angle of 60°. The jacket is partitioned into four segments to provide variable heating area for efficient use of thermal energy. The rotary scrapers were designed to offer a uniform centrifugal force of scraping at all points on heat transfer surfaces. The scraping assembly is coupled to a variable speed drive to control rpm at different stages of khoa preparation. There is provision for recirculation of the milk during concentration so that the heat transfer area of the cone can be reduced to get better thermal performance [22], with decrease in the bulk. Sometimes due to recirculation, quality of the product deceases during conversion to khoa. However, it has been successfully used for preparation of basundi [48] and has been integrated with SSHE for preparation of rabri [15].

A SSHE based batch equipment for hygienic manufacture of khoa was also developed [36] consisting of a stationary-jacketed drum with steam inlet, condensate outlet through steam trap and pressure gage. Power operated spring loaded scraper enhanced heat transfer and mixing of the product. The steam temperature of 121°C and scraper speed of 28 rpm were reported as optimum operating parameters. A prototype khoa making machine based on horizontal SSHE was developed using mild steel and subsequently modified using stainless steel [16] with feed capacity of 50 kg of milk per hour per batch. The unit consists of a steam jacket divided into three compartments for better control of heating process as the content reduces during later part of khoa making. The scraper blades are spring loaded on the assembly, which is arranged in such a way that the whole surface is efficiently scraped [22]. The perforations on upper side of drum facilitate milk feeding, exhaust of vapor and supervision of concentration process. The steam jacket can be fed with cold water for cooling the finished product. Upon evaluation for the batch-wise production of khoa and allied products, the machine was found suitable for preparing 32 kg of carrot-milk halwa in 2.5 h, 30 kg of bottle guard halwa in 3 h, 35 kg of rice-milk kheer in 45 minutes, 25.5 kg of basundi in 1 h and 18 kg tomato ketchup in 2 h by making relevant changes in the operating variables of the machine [51]. However, for production of large quantities for industrial operation, the need of continuous khoa producing unit was felt. The National Dairy Development Board at Anand [43] has developed continuous khoa making machine using liquid full type Inclined Scraped Surface Heat Exchanger (ISSHE). In this machine, the milk is fed by positive displacement pump at the lower end of inner cylinder of ISSHE, where it forms a pool of boiling milk. The jacket has three – compartments. The rotor is screw shaped with scraper blades fitted in a staggering manner. Thus, while rotating the scraper, blades perform the basic function and the screw like portion conveys the milk towards the upper end of ISSHE. The milk flows back to the pool and the concentrated milk (solid mass conveyed to the top) is discharged as khoa. The limitation with the machine was that it requires pre-concentrated milk to around 45% TS [23].

Beside liquid full SSHE, the thin film SSHE was also applied in milk dairy operations [3]. A continuous khoa making unit, operating on the principle of thin film scraped surface heat exchanger (TFSSHE) [27], produced acceptable quality khoa when the equipment was operated under standardized conditions. But higher rotor speed had adverse effect on flavor and texture of khoa. Therefore, two TFSSHEs were used in cascade manner. First SSHE has four adjustable clearance scraper blades rotating at 3.3 rps and second SSHE has two adjustable clearance scraper blades and two helical blades to push the product forward, rotating at 2.5 rps. Performance of this system was tested on large scale and was found satisfactory; and khoa prepared was better than that prepared with batchwise operation. The unit was further modified as continuous burfi making machine by adding a third stage [33] and its performance was evaluated for burfi making [22].

During conversion of milk to khoa, its TS concentration increases to 60–67%, for example, almost 5 times, radically changing its physicochemical properties. Therefore a single unit converting milk to khoa cannot be compatible at all stages throughout the concentration process and that also with engineering parameter as well as technological and quality aspects. One set of design parameters, which is efficient during initial stages of concentration, will fail in the final stage and vice versa. Similarly, the unit operating at higher efficiency may give technologically inferior product and vice versa. Taking this into consideration, Christie and Shah [17] designed and developed a three-stage unit for continuous manufacture of khoa at GAU, Anand (Now, AAU-Anand). As shown in Figure 5.2, the machine comprised of three jacketed cylinders placed in a cascade arrangement with some slope to facilitate easy drainage. Each of the three SSHEs was designed according to the requirement of heat transfer area,

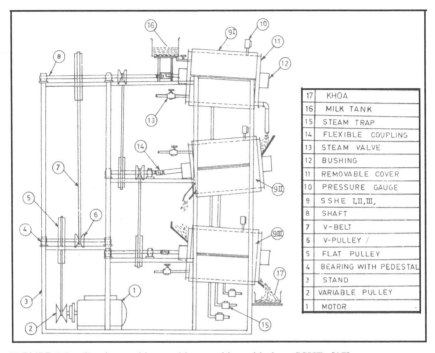

17	KHOA
16	MILK TANK
15	STEAM TRAP
14	FLEXIBLE COUPLING
13	STEAM VALVE
12	BUSHING
11	REMOVABLE COVER
10	PRESSURE GAUGE
9	SSHE I,II,III,
8	SHAFT
7	V-BELT
6	V-PULLEY /
5	FLAT PULLEY
4	BEARING WITH PEDESTAL
3	STAND
2	VARIABLE PULLEY
1	MOTOR

FIGURE 5.2   Continuous khoa making machine with three SSHEs [17].

temperature gradients and scraping speeds with respect to the TS concentration of milk at that particular stage. The design was further improved and tested for its heat transfer behavior [10–13].

As subsequent work for industrial application, a three stage continuous khoa making machine has been developed at NDRI, Karnal [23]. The machine in Figure 5.3 consists of three SSHE of identical length, diameter and effective heating length. All three SSHEs are provided with vapor outlets, spring loaded safety valves on jackets and glass wool insulations. The rotor assembly of first two stages consists of four variable clearance blades with spring support while that of third has only two of such blades and remaining two are skewed blades. The third stage is also provided with sugar dosing device consisting of hopper and VFD driven auger. The milk is fed geometrically to the top most SSHE by a progressive cavity pump. Concentrate from first to second and second to third flows down under gravity. A magnetic flow meter is provided in the feed line to the

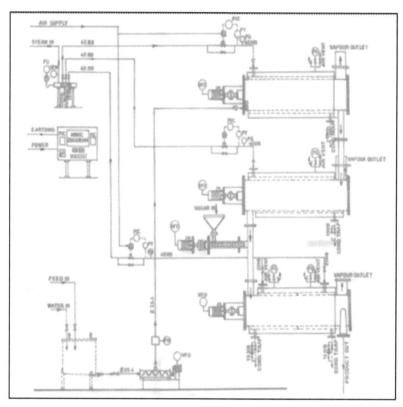

FIGURE 5.3   Modified khoa making machine [23].

first stage. The steam supply to each jacket is regulated by electrically controlled pneumatic valves. Under industrial trials, the machine has been reported to produce 50 kg/h of khoa when fed with raw milk and 120 kg/h of khoa when fed with concentrated milk (30% TS).

The successful (traditional) method of khoa manufacture clearly demarcates two unit operations. First is the fast evaporation to concentrate the milk to 45% TS by intensive heat transfer to have a desirable flavor and second is the slow heating and vigorous scraping to achieve final concentration and desirable textural attributes. The concentration with flavor development can be achieved in SSHE whereas the texture development can be accomplished in the Conical Process Vat [34]. For designing a specialized heat exchanger, it is essential to know the parameters governing heat transfer in the particular process. Following correlation can be used to calculate overall heat transfer coefficient during evaporation of milk up to 45% TS:

$$U_0 = 620.22E - 32.3S + 585.6V_c + 133.3B - 11.7 - 0.23E^2 + 0.48S^2$$
$$-73.6V_c^2 - 6.3B^2 - 0.17T^2 - 0.25ES + 0.64EV_c - 0.03EB + 0.71Et$$
$$+2.24SV_c + 0.76SB - 0.21ST - 12.0V_cB - 0.46V_cT - 0.302BT$$

(1)

where, S is TS in milk, Vc is the circumferential velocity, E is percent vapor fraction, B is the number of solids and T is the temperature difference.

The scraped film heat transfer coefficient during evaporation of milk to high concentration can be predicted by the following correlation [27].

$$N_u = \left[6615.1(N_{re})_G \right]\left[0.133(P_r)_G^{0.08} \right](\Delta T / T_S)^{0.28}$$ (2)

where, $(Nre)_G$ is generalized Nusselt number, $(Pr)_G$ is the generalized Prandtl number, ΔT is the temperature difference, and Ts is steam temperature.

5.3.1 BASUNDI

Basundi is a traditional sweetened concentrated milk but it is characterized by its peculiar body texture consisting of very minute flakes and cooked flavor. Most of the SSHE based units developed for manufacture of khoa have been studied for the preparation of basundi and some were found satisfactory. The exclusive attempt to mechanize basundi making was made at AAU – Anand. Three pilot models (viz. cylindrical type, conical type, and Karahi type) were developed for Basundi making on the principle of SSHE (SSHE). All the models were tested for their heat transfer behavior under different operating conditions. Heat transfer and energy consumption were estimated for design optimization [46]. The heat utilization in these heat exchangers was evaluated with and without induced draft on milk surface [40]. Mechanization of manufacture of Basundi was tried using batch type stainless steel version of SSHE developed at SMC College of Dairy Science, Anand Agricultural University, Anand. The process parameters are optimized for and the product was compared favorably with products made by conventional method in the sensory and rheological profile, with better score and color.

Manufacture of Basundi was tried at NDRI, Karnal, using conical process vat and two-stage thin film SSHE with standardized buffalo milk. Basundi prepared in conical process vat, was good in body, texture, appearance and overall acceptability for processing time between 80 to 100 min [21, 48].

A Continuous Basundi Making Machine (CBM) has been developed at SMC College of Dairy Science – AAU-Anand, based on the principle of Scrap Surface Heat Exchanger (SSHE). It consists of concentration unit of three SSHEs and chilling units of two SSHEs with specially designed scrapers, variable frequency drive (VFD) to facilitate variation of speed of scrapers, resistance temperature detector (RTD) sensors and other controls to optimize processing parameters, which resulted into better quality product in terms of sensory and rheological attributes [40]. There is a forced draft-cooling tower to condense the vapors.

5.3.2   UHT PROCESSING OF MILK

The UHT processing includes heating the milk at 121°C for 2 sec or other suitable time temperature combination. The temperature, being more than the normal boiling point of milk, requires heating under high pressure to avoid evaporation and concentration. A cascaded three thin film SSHEs unit employing high pressure pump was developed for UHT processing of milk [26]. Each SSHE cylinder was made air-tight, the provision for heat regeneration and cooling required were also satisfactorily achieved in the three SSHEs.

Based on the design of contherm SSHE, Tetra Pack developed a continuous aseptic processing module (Tetratherm) for indirect UHT treatment. But that design was more suitable for tomato products, soups, sauces, desserts and viscous products even with particles than milk.

5.4 MECHANIZATION OF GHEE MAKING

Several alterations in the basic design of SSHE have been tried in order to exploit the SSHE in dairy processing. The SSHE with vertically placed product tube have been used where the vapor removal is either not required or not required essentially in the product tube itself. Ghee, the most valuable indigenous milk product, contains more than 99.5% fat and less than 0.5% moisture.

The first prototype of continuous ghee making equipment consisted of two SSHEs and two vapor separators [44]. The butter is first molten in jacketed vat and is passed to first SSHE, where it is heated to 112°C. The superheated liquid is then flashed in the vapor separator removing water. The concentrated liquid is passed to second SSHE, where it is heated to 115°C and flashed in the second vapor separator to get the final product, ghee. Due to use of SSHE, the high overall heat transfer coefficient is obtained and the temperatures 112°C and 115°C, which are essential to develop typical ghee flavor, can be quickly achieved. The operating pressure in both the SSHE is above atmosphere. Therefore, the moisture removal is possible on the principle of flash evaporation. The plant has a reported capacity of 100 kg/h.

Another continuous ghee-making plant was designed with two vertical falling film SSHE [1]. The machine is characterized by high heat transfer coefficients and better control over the quality of product, but there was no significant saving in steam consumption.

A study was undertaken to prepare ghee from cream and butter by using a vertical SSHE (120 mm diameter) having two scraper blades. Different flow rates ranging from 48 kg/h to 90 kg/h and scraper speeds from 200 rpm to 500 rpm were studied and it was found that ghee of acceptable quality could be prepared with an energy consumption of 99–187 kCal/kg of butter with that set up against the jacketed kettle, which consumed 293 kCal/kg. The film heat transfer coefficient was found to be directly correlated with the speed of the scraper and independent of the product flow rate [32].

When ghee is made from butter, the butter is first melted and the water content in it, which generally accounts for 16%, needs to be removed. It is accomplished using stratification followed by evaporation or the evaporation alone. A continuous ghee making system of capacity 500 kg of ghee per hour from butter was developed. The butter was melted employing a continuous butter melting unit and final water evaporation with a horizontal SSHE. The unit consumed 35 kg/h of steam against that of 68 kg/h required in jacketed ghee kettle of the same capacity [2, 4]. The ghee prepared in the unit was also reported to possess better shelf life [9]. The Panchmahal Dairy at Godhra (Gujarat) has also developed an industrial method for ghee making with an aim to reduce fat and SNF losses by inclusion of a serum separator and sipro-heater. The method may employ the SSHE [40].

5.5   MECHANIZATION OF MANUFACTURING OF SANDESH

Sandesh is a popular begali sweet made from chhana, a heat-acid coagulation product of milk. The process of making sandesh involves making channa from preferably cow milk, mixing sugar @ 30–35%, kneading vigorously and cooking at 70–75°C for 10 minutes. The prototype equipment for the manufacture of chhana with a capacity of 40 kg/h has been developed at National Dairy Research Institute – Karnal [45] and at Indian Institute of Technology (IIT) – Kharagpur [40]. The IIT Kharagpur demonstrated the production of sandesh in a continuous manner. Screw extruder is modified into a vented type extruder to release the vapors formed during intense heating [6].

A pilot scale SSHE has been designed and developed at AAU-Anand to mechanize sandesh making process. The SSHE works on the principle of votator type scraper assembly having four spring loaded scrapers and half round jacket. The other half of the cylinder was left unjacketed and was made perforated for the escape of vapors that generate during the process of making sandesh from chhana [37, 38]. During the batchwise manufacture of sandesh using the machine under optimum operating conditions, the U-values ranged from 131 to 639 W/m^2K, the outer convective heat transfer coefficient, ho of 12890 W/m^2K and the inner convective heat transfer coefficient, hi of 442.32 W/m^2K. The sensory attributes of such produced sandesh were at par with that prepared by the traditional method.

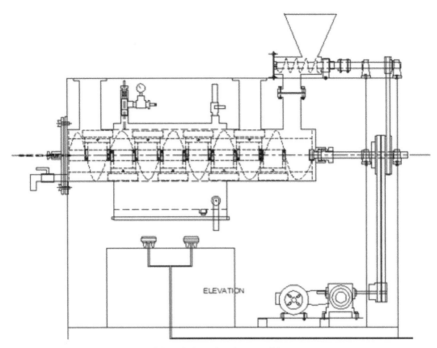

FIGURE 5.4 Continuous shrikhand thermization machine.

5.6 SHRIKHAND THERMIZATION

Shrikhand is a cultured milk product made by partially dewheying curd and blending it with sugar and/or other additives like salts, flavorings, color, fruit pulp, spices. It is popular throughout India as well as in Sri Lanka, Nepal and Pakistan. As recommended by the FSSA, shrikhand should contain minimum of 58% solids and upto 72.5% sugar (dry basis). Shrikhand contains plethora of desirable and undesirable microorganisms that bring out various biochemical changes during storage putting the safety of the stored product in question. Therefore, in spite of so high sugar and less moisture content, shrikhand is stored below 0°C and cold chain needs to be maintained from production to consumption. A post-production heat treatment known as thermization has been recommended for enhancing the shelf life of shrikhand [39, 42]. The Sugam dairy at Baroda have adopted thermization of shrikhand in a steam jacketed vat.

Suitably designed equipment for thermization of the Shrikhand is essential to achieve desirable attributes in the product. Considering the viscous nature of shrikhand, a SSHE based *Continuous Shrikhand Thermization Machine* has been designed and developed [20]. The machine in Figure 5.4 has a jacketed product tube, specially designed spring loaded scraper assembly, feeder assembly, VFD for regulating scraper speed and sensors to monitor the operating parameters of the machine. The machine uses LPG as source of thermal energy to produce steam in the jacket. While thermizing shrikhand at 70°C, 75°C and 80°C, at different scraper speeds of 20 rpm, 35 rpm and 50 rpm, an acceptable quality product was obtained. The values of overall heat transfer coefficient (U) ranged from 267.77 to 487.67 W/m²K and the optimum capacity was 160.71 kg/h to 124.07 kg/h depending upon the TS content of chakka used for Shrikhand making. The product thermized in the machine was reported to have better sensory attributes and shelf life under refrigerated as well as ambient storage [19].

5.7   MISCELLANEOUS APPLICATIONS

A batch type Halwasan Making Machine (BHM) has also been reported by AAU – Anand. The machine works on the principle of scrape surface kettle and Halwasan prepared by using BHM is very good in quality and rheological attributes, the keeping quality of the product is reported to be 22 days at room temperature compared to the keeping quality of 8–10 days of the Halwasan made by conventional method [40].

An integrated plant for mechanized production of TIDP has been reported [40] with a milk handling capacity of 250 kg/h. The plant consists of three basic units: plate heat exchanger (PHE), twin cylinder film scraped surface heat exchanger (Twin SSHE) and Batch type Steam Jacketed Kettle. This integrated plant is suitable for mechanized value added Products like Basundi, Kheer, Peda, Thabdi, Burfi, Gajar Halwa, Halwasan, etc.

Gerstenberg Schroder, an equipment manufacturing company in Germany has developed an SSHE named as Consister, for thermal processing with simultaneous mixing of viscous fermented products. It includes a vertical product tube within which the scarper assembly rotates. The drive to the assembly has been located at bottom. The speed of the product is higher than the rotational speed achieved by the product in product tube, therefore the relative motion between the product and scraper assembly facilitate mixing. It has been used for the thermization of quark (200 kg/h, 70–72°C for 30–40 sec), a western fermented milk product resembling *Shrikhand*. Besides this, the SSHE has proven its applicability in cooling molasses, concentration of whey solids, aseptic processing of fruits for yogurt and ice cream, concentration of juices that tend to crystalize and, etc.

5.8 CONCLUSIONS

Efficient heating or cooling of complex fluids is an important requirement for several operations in dairy and food industry. SSHE is more suitable for heat transfer to viscous and heat sensitive products. It is easier to control time, temperature, speed and pressure in SSHE as compared to other type of heat exchangers. This makes it possible to achieve more even and improved quality of products. Manufacture of various dairy and food products using scraped surface processing plant increases their shelf-life. Flow patterns, mixing effects, residence time distribution, heat transfer and power requirements are major parameters for design and performance evaluation of SSHE. The choice and design of heat exchanger depends on a large number of factors such as flow rate of both streams, heat sensitivity and corrosivity of the fluids, terminal temperatures, properties of both streams, operating pressure, allowable pressure drop, fouling tendencies, inspection-cleaning and extension possibilities, type and phase of fluids, material of construction, heat recovery, flow arrangement and cost of heat exchanger. These heat exchangers can be classified according to transfer processes, degree of heat transfer, surface compactness, construction features, flow arrangements, number of fluids and heat transfer mechanism, scraper arrangement, blade clearance, design of product tube, etc. The scraping action of the blades rotating within the heat transfer tube continuously removes the product film from the heat exchanger surface and prevents burning-on, scaling or crystal build up on the wall. During operation, the product is brought in contact with a heat transfer surface that is rapidly and continuously scraped there by exposing the surface to the passage of untreated product.

The major advantages of SSHE include short residence time in heated zone so heat sensitive product can be conveniently processed, high heat transfer coefficient, narrow residence time distribution, wide viscosity range, minimum surface fouling, liquids with foaming tendency can be easily handled, low product inventory, waste heat recovery is possible for energy conservation, efficient use of thermal energy, better process control and optimization. SSHE basically consists of a cylindrical rotating shaft (the "rotor") within a concentric hollow stationary cylinder (the "stator") so as to form an annular region along which the process fluid is pumped. For maximum efficiency the product tube is manufactured from a material with a high heat-transfer coefficient. The blades are usually made from stainless steel as well as Teflon material depending on the application. Two or four blades located around the rotor throughout the product tube, though sometimes they are axially staggered in order to improve mixing. The rotating scraper helps in improving efficiency, proper mixing and imparting desirable consistency and texture to the product. The scraper may be grouped into fixed clearance scraper and variable clearance scraper. The product velocity at the heat transfer surface depends on the speed of the scraper and the effective number of the rows of the blades and does not depend on the product flow rate. Liquid-full operation and thin-film operations are commonly used in SSHE.

The overall heat transfer coefficient (U) represents the intensity of heat transfer from one fluid to another through a wall separating them. The U-value is dependent on steam side film heat transfer coefficient (h_o), product side film heat transfer coefficient (h_i) and wall heat transfer coefficient (h_w). The value of h_w is constant as it is fixed with the design of the SSHE. Heat and mass transfer equations are used in order to calculate the U-values of SSHEs. The U value is a function of blade rotational speed and mass flow rate, number of blades, temperature, feed rate, steam pressure, etc. The product side film heat transfer coefficient depends on various operating conditions of the SSHE such as scraper speed, concentration of the milk, design of the scraper, etc.

All the metallic parts that come in contact with the product should be made of SS and the nonmetallic parts coming in contact with product should be nontoxic, non-corrosive to product as well as cleaning solutions. More advanced designs of SSHE include features such as holes in the blades (which reduce power consumption), oval stators (which reduce "channeling" in which fluid passes through the exchanger relatively unprocessed), and non-centrally mounted shafts (which enhance mixing and prevent material from building up on the underside of the blades). Forward and backward tapering and Step arrangement of SSHE cylinder is also an advance design of SSHE. Rotating disc or cone type (Screw type) with advanced instrumentation plays an important role in process control and automation.

5.9 SUMMARY

By virtue of its distinguishing features such as fouling prevention, no localized heating/cooling, high heat transfer coefficient, better control on product movement, etc., the SSHEs have proven most suitable to handle viscous and/or particulate products. The operating parameters such as residence time, temperature, pressure, speed of rotation can be easily controlled in SSHE. Consequently, it has been applied widely to mechanize the production of dairy products especially the traditional Indian dairy products. Majority of Indian dairy products, which are prepared by concentrating milk leading to semi solid viscous mass have been manufactured using SSHEs. Preparation of good quality khoa and its derived products like peda, burfi, halwa and its variants are well known applications of SSHE. Apart from the milk concentration process, the SSHE have been applied in sensible heating/cooling treatments including UHT processing of milk and Thermization of Shrikhand. There is still ample scope for other applications of SSHE in manufacture of other dairy products too with some thoughtful alterations in the basic design of SSHE.

KEYWORDS

- basundi
- burfi
- CPV
- ghee
- halwa
- halwasan
- khoa making
- mechanization

- **peda**
- **sandesh**
- **shrikhand**
- **SSHE**
- **thermization**

REFERENCES

1. Abichandani, H., Agrawala, S. P., Verma, R. D., & Bector, B. S. (1978). Advances in continuous ghee-making technique, *Indian Dairyman, 30*(11), 769–771.
2. Abichandani, H., Bector, B. S., & Sarma, S. C. (1995). Continuous ghee making system – design, operation and performance, *Indian Journal of Dairy Science, 48*(11), 646–650.
3. Abichandani, H., Dodeja, A. K., & Sarma, S. C. (1989). Applications of thin film scraped surface heat exchangers, *Indian Dairyman, 41*(1), 21–24.
4. Abichandani, H., Dodeja, A. K., & Sarma, S. C. (1996). A unique system for continuous manufacture of ghee and khoa, *Indian Dairyman, 48*(10), 11–12.
5. Abichandani, H., Sarma, S. C., & Heldman, D. R. (1987). Hydrodynamics and heat transfer in thin film SSHE: A Review. *Journal of Food Process Engineering, USA, 9*(2), 143–172.
6. Agrawal, S. P. (2003). Development of equipment for manufacture of Indian milk products and future strategies. *Indian Dairyman, 55*(8), 33–37.
7. Alam, A. A. (1988). Process equipment manufacture. *Indian Dairyman, 40*, 285–290.
8. Banerjee, A. K., Verma, I. S., & Bagchi, B. (1968). Pilot plant for continuous manufacture of khoa. *Indian Dairyman, 20*, 81–84.
9. Bector, B. S., Abichandani, H., & Sarma, S. C. (1996). Shelf life of ghee manufactured in continuous ghee making system. *Indian Journal of Dairy Science, 49*(6), 398–405.
10. Bhadania, A. G., Patel, S., Shah, B. P., & Shah, U. S. (2004a). Determination of film heat transfer coefficients of Scraped Surface Heat Exchanger during manufacture of Khoa. *Indian Food Industry, 23*(3), 44–47.
11. Bhadania, A. G., Patel, S. M., Shah, B. P., & Shah, U. S. (2004b). Effect of steam pressure on overall heat transfer of surface heat exchanger during manufacture of Khoa. *Beverage and Food World, 3*, 33–35.
12. Bhadania, A. G., Shah, B. P., & Shah, U. S. (2004c). Effect of heat transfer on physico chemical and sensory quality of khoa manufactured using Scraped Surface Heat Exchanger. *Journal of Food Science and Technology, 41*(6), 656–660.
13. Bhadania, A. G., Shah, B. P., & Shah, U. S. (2005). Energy requirement of Scraped Surface Heat Exchanger (SSHE) during manufacture of Khoa. *Indian Food Industry, 86*, 13–17.

14. Boghra, V. R., & Rajorhia, G. S. (1982). Utilization of pre concentrated and dried milk for khoa making. *Asian Journal of Dairy Science, 1*, 6.
15. Chopde, S., B. Kumar, P. S. Minz, & P. Sawale (2013). Feasibility study for mechanized production of rabri. *Asian Journal of Dairy & Food Research, 32*(1), 30–34.
16. Christie, I. S., & Shah, U. S. (1990) development of khoa making machine. *Indian Dairyman, 42*, 249–252.
17. Christie, I. S., & Shah, U. S. (1992). Development of a three stage continuous khoa making machine. *Indian Dairyman, 44*(1), 1–4.
18. De, S., & Singh, B. P. (1970). Continuous production of khoa. *Indian Dairyman, 22*, 294.
19. Dhotre, A. V., Bhadania, A. G., & Shah, B. P. (2009). Effect of mechanized thermization on quality of shrikhand. *International Journal of Food Science, Technology & Nutrition, 3*(1–2), 11–23.
20. Dhotre, A. V. (2006). Development and performance evaluation of scraped surface heat exchanger for continuous thermization of shrikhand. MSc Thesis submitted to Anand Agricultural University, Anand.
21. Dodeja, A. K., & Abichandani, H. (2004). Development of unique system for continuous production of ghee and khoa. *Beverage and Food World, 31*(8), 37–38.
22. Dodeja, A. K., & Agrawala, S. P. (2005). Mechanization for large scale production of indigenous milk products: a review. *Indian Journal of Animal Sciences, 75*(9), 1118–1125.
23. Dodeja, A. K., & N. Kishor (2011). Continuous khoa making machine – Operational features and performance of new version. *Indian J of Dairy Science, 64*(4), 283–289.
24. Dodeja, A. K., Abichandani, H., Sarma, S. C., Dharam Pal, & Pal, D. (1992). Continuous khoa making system-design, operation and performance. *Indian Journal of Dairy Science, 45*(12), 671–674.
25. Dodeja, A. K., Sarma, S. C., & Abhichandani, H. (1989a). Thin film scraped surface heat exchanger and plate heat exchanger – A comparative study. *Indian Journal of Dairy Science, 42*, 757.
26. Dodeja, A. K., Sarma, S. C., & Abichandani, H. (1989b). Development of thin film scraped surface heat exchanger for uht processing of milk. *Indian Journal of Dairy Science, 42*(4), 760–764.
27. Dodeja, A. K., Sarma, S. C., & Abichandani, H. (1990). Heat transfer during evaporation of milk to high solids in thin film scraped surface heat exchanger. *Journal of Food Process Engineering, 12*(3), 211–225.
28. Gandhi, N., & Parikh, P. (2015). Thermal analysis of scraped surface heat exchanger used in food industries. *International J Innovative Science, Engineering & Technology, 2*(5), 622–627.
29. Gupta, S. K., Agrawala, S. P., Patel, A. A., & Sawhney, I. K. (1987). Development of equipments for indigenous dairy products, *Indian Dairyman, 39*(9), 419–425.
30. Harrod, M. (1986). Scraped surface heat exchangers: a review. *Journal of Food Process Engineering, 9*, 1–62.
31. Kohli, R. K., & Sarma, S. C. (1983). Application of scraped surface heat exchanger for making ghee. *Transactions of the ASAE American Society of Agricultural Engineers, 26*(4), 1271–1274.
32. Kohli, R. K., & Sarma, S. C. (1990). Mechanization in ghee. *Indian Journal of Dairy Science, 43*(2), 181–184.

33. Kumar, B. K., & Dodeja, A. K. (2003). Development of continuous burfi making machine. *Indian Journal of Dairy Science, 56*(5), 274–277.

34. Mahesh, K., B. Kumar, & P. S. Minz (2012). Optimization of process parameters for manufacture of khoa using Response Surface Methodology. *Indian Journal of Dairy Science, 65*(2), 107–114.

35. More, G. R. (1983). Development of khoa making equipment. *Indian Dairyman, 35,* 275.

36. More, G. R. (1987). Development of semi-mechanized khoa making equipment. *Indian Journal of Dairy Science, 40,* 246.

37. Patel, J. S., & Bhadania, A. G. (2005a). Applications of SSHE in mechanization of Sandesh making process. *Beverage & Food World,* 46–50.

38. Patel, J. S., & Bhadania, A. G. (2005b). Heat transfer behavior of scraped surface heat exchanger during manufacturing sandesh. *Processed Food Industry,* 21–26.

39. Patel, R. S., & Abd-El-Salam (1986). Shrikhand – an Indian analog of Western quarg. *Cultured Dairy Products Journal, 21*(1), 6–7.

40. Patel, S. (2013). Mechanized manufacture of traditional milk products. Proceedings of National Seminar on Mechanized Production of Indian Dairy Products held at AAU-Anand, Gujarat during 2–3 Sept, pp. 44–51.

41. Patel, S. M. (1990). Study on heat transfer performance of SSHE during Khoa making, MSc Thesis, Gujarat Agricultural University, S. K. Nagar.

42. Prajapati, J. P, Upadhyay, K. G., & Desai, H. K. (1993). Quality appraisal of heated shrikhand stored at refrigerated temperature. *Cultured Dairy Products Journal, 28*(2), 14–17.

43. Punjarath, J. S., Veeranjaneyulu, B., Mathunni, M. S., Samal, S. K., & Aneja, R. P. (1990). Inclined SSHE for continuous khoa making. *Indian Journal of Dairy Science, 43*(2), 225–230.

44. Punjrath, J. S. (1974). New development in ghee making. *Indian Dairyman, 26,* 275–287.

45. Punjrath, J. S. (1991). Indigenous milk products of India: The related research and technological requirements in process equipment, *Indian dairyman, 43*(2), 75–87.

46. Rajasekhar, T., Shah, B. P., Bhadania, A. G., & Prajapati, P. S. (2001). Application of scraped surface heat exchangers in dairy and food industry: A Review. *Indian Journal of Dairy & Biosciences, 12,* 8–12.

47. Rajorhia, G. S., & Srinivasan, M. R. (1975). *Annual Report.* National Dairy Research Institute, Karnal, India.

48. Ranjeet, K. (2003). Studies on the manufacture of basundi using conical process vat. M. Sc. Thesis, NDRI, Karnal.

49. Rizvi, S. S. H., Mann, R. S., & Ali, S. I. (1987). A case study of appropriate technology transfer: Development of an automated continuous khoa powder manufacturing process. *Indian Dairyman, 39,* 63.

50. Singh, A. K., & Dodeja, A. K. (2012). Manufacture of basundi using three-stage sshe. *Indian Journal of Dairy Science, 65*(3), 197–207.

51. Upadhyay, J. B., Bhadania, A. G., Christic, I. S., & Shah, U. S. (1993). Manufacture of khoa based sweets and other food products on SSHE: an encouraging experience. *Indian Dairyman, 45,* 224.

CHAPTER 6

APPLICATION OF HIGH PRESSURE PROCESSING IN THE DAIRY INDUSTRY: A REVIEW

SANTOSH S. CHOPDE, BHAVESH B. CHAVHAN, and MADHAV R. PATIL

CONTENTS

6.1 INTRODUCTION

Consumer's demand for minimally processed, fresh, additive-free and shelf-stable products have prompted food scientists to explore other processing technologies as alternative to traditional treatments that rely on heating or cooling operations. Traditional treatments ensure high level of food safety but they lead to degradation of various food quality and nutritious attributes. Considerable efforts have been made by the scientists to minimize the undesirable effects on food quality. As a result, numerous non-thermal processing technologies, particularly high pressure (HP), power ultrasonic, pulsed electric field, ultra violet decontamination, pulsed white light, ionizing radiation, etc., have been evolved. Among these, high pressure processing (HPP) is the most promising technology for food applications. It is also recognized as high-hydrostatic pressure processing (HHHP) or ultrahigh-pressure processing (UHPP). Regardless of its nomenclature, the technology has been cited as one of the best innovations in food processing during the past 50 years [21].

In HPP, food is subjected to very HP in the range of 100 to 1000 MPa at processing temperature from 0°C to 100°C with exposure time of few seconds to 20 minutes. In 1991, the first commercial HP processed product appeared in the Japan, where this technology is now being used for processing products such as jams, sauces, fruit juices, rice cakes and desserts. HPP of food results in:

- Inactivation of microorganisms and enzymes at ambient temperature or even lower temperatures.
- Modification of biopolymers.
- Retention of sensorial attributes of foods such as color, taste, flavor, etc.
- Alteration in product functionality.

Beside microbial destruction, the HP affects protein structure and mineral equilibrium, suggesting its different applications in dairy products. The pioneering research into the application of HP to milk dates back to the end of the 19th century [31]. Comprehensive knowledge about the mechanism and kinetics of pressure induced degradation, denaturation, inactivation of dairy compounds like nutrients, proteins, microorganisms, enzymes, etc. is required for executing this new technology in the dairy industry.

This chapter reviews the mechanism of HPP, basic components of a HPP system, the effects of HP on the physico-chemical and other properties of milk, and applications of HPP in the dairy industry.

6.2   MECHANISM OF HIGH PRESSURE PROCESSING

Application of pressure over the food system results in various physical and chemical changes. During pressure treatment, physical compression of the system occur which results in a volume reduction and an increase in temperature and energy. Following three principles govern the behavior of foods under pressure.

1. **Le Chatelier's principle:** It states that any phenomenon (phase transition, change in molecular configuration, chemical reaction) accompanied by a decrease in volume is enhanced by pressure.
2. **Principle of microscopic ordering**: It indicates that at constant temperature, an increase in pressure increases the degree of ordering of molecules of a given substance. Therefore, pressure and temperature exert antagonistic forces on molecular structure and chemical reactions [6].
3. **Isostatic principle:** It states that pressure is transmitted quasi-instantaneously and uniformly distributed throughout the entire sample, whether in direct contact with the pressurizing medium or insulated from it in a flexible container. Thus, the process time is independent of sample size and geometry, assuming uniform thermal distribution within the sample.

In view of these, at a relatively low temperature (0–40°C) covalent bonds are almost unaffected by HP while the tertiary and quaternary structures of molecules which are maintained chiefly by hydrophobic and ionic interactions are altered by HP >200 MPa [30]. HPP is usually carried out with water as a hydraulic fluid to facilitate the operation and compatibility with food materials [22].

6.3   HIGH PRESSURE PROCESSING SYSTEM FOR FOODS: BASIC COMPONENTS

The non-availability of appropriate equipment hampered early application of HP. However, recent development in equipment design has ensured global acknowledgement of the potential for such a technology in food processing. The advances accomplished in ceramics and metallurgical industry during the 1970s and 1980s has led to the possibility of treating food by this method at industrial level. Similarly, developments in mechanical engineering have permitted large high-pressure vessels to be constructed at reasonable cost with sufficient durability to withstand thousands of pressure cycles without failure.

A typical HPP unit consists of a pressure vessel and pressure-generating device. Food packages are loaded into the vessel and top is closed. The pressure medium, usually water containing a small amount of soluble oil, is pumped into the vessel from the bottom. Once the desired pressure is reached, it can be maintained without further need for energy input. The process is isostatic, so pressure is transmitted rapidly and uniformly throughout the pressure medium and food with little or no heating. Pressure is exerted equally from all sides so that there is no 'squashing' effect and product is not affected.

6.4   MILK SYSTEM

Studies on application of HPP for preservation of milk started in 1899 when Hite et al., demonstrated the extension of shelf life of HP treated milk [31]. HPP of milk results in inhibition and destruction of microorganism, modifications in physicochemical/technological properties and changes in RCT [5]. Further, it leads to disintegration casein micelles into casein particles of smaller diameter, with a decrease in milk turbidity and lightness, and an increase of viscosity of the milk [54]. This section covers the application of HPP in inhibition and destruction of microorganisms in milk and modification in physico-chemical/technological properties will be studied.

6.4.1   APPLICATION OF HPP IN INHIBITION AND DESTRUCTION OF MICROORGANISMS IN MILK

Milk is often regarded as being nature's most complete food. It plays an important role for human nutrition and is one of the most frequently sold food worldwide. Milk is a perishable commodity and spoils very easily. Its low acidity, high water activity and high nutrient content make it the perfect breeding ground for both vegetative and spore forming bacteria, including those which cause food poisoning (pathogens). Spore forming bacteria are thermo-resistant and important for milk deterioration. Heat treatments such as pasteurization, sterilization, ultra-high temperature (UHT) are the main treatment applied to microbial stabilization of milk, since they are simple. The heat processing of milk affects vitamins (losses around 10% of folic acid and 15% of B-complex vitamins) and denatures proteins, resulting in the release of sulfured compounds [29].

Many studies conducted on the inactivation of pathogenic and spoilage microorganisms by HPP have demonstrated that it is possible to obtain 'raw' milk with treatment of pressure about 400 to 600 MPa of a microbiological quality comparable to that of pasteurized (72°C, 15 s) milk depending on the microbiological quality of milk but not sterilized milk due to HP resistant spores [10]. A pressure treatment of 400 MPa for 15 min or 600 MPa for 3 min at 20°C to milk gives shelf life of 10 days at a storage temperature of 10°C [51].

The factors that affect the resistance of microorganism to the applied pressure in food include:

- HP processing conditions (pressure, time, temperature, cycles, etc.);
- Food constituents and the properties; and
- The physiological state of the microorganism.

The bacterial spores are always more resistant than vegetative cells and they can survive at pressure of 1000 MPa. Bacterial spores, however, can often be stimulated to germinate by pressures between 50–300 MPa. Germinated spores can then be killed by heat or mild pressure treatments. Gram-positive microorganisms tend to be more resistant to pressure than gram-negative microorganisms. Gram-positive microorganisms need the application of 500–600 MPa at 25°C for 10 min to achieve inactivation, while gram-negative microorganisms are inactivated with treatments of 300 400 MPa at 25°C for 10 min. Vegetative forms of yeasts and molds are the most pressure sensitive [56].

Studies on the effect of variation in pressure resistance on growth stage (exponential and stationary phase) with respect to growth temperature (8° and 30°C) between two strains of *Listeria monocytogenes*, *Bacillus cereus*, and *Pseudomonas fluorescens* reported that exponential-phase cells were significantly less resistant to pressure than stationary-phase cells for all of the three species studied. Growth temperature had a significant effect at the two growth stages studied. Pressure treatment at 8°C induced significantly less spore germination than at 30°C [45].

On the other hand, studies on Ewe's milk containing 6% fat inoculated with *Listeria innocua* 910 CECT at a concentration of 10^7 CFU/ml showed that low-temperature (2°C) pressurizations produced higher *Listeria innocua* inactivation than treatments at room temperatures (25°C). The kinetics of destruction of *L. innocua* were first order with *D*-values of 3.12 min at 2°C and 400 MPa, and 4 min at 25°C and 400 MPa [27].

In general, the HP induced inactivation was greater on *P. fluorescens* > *E. coli* > *L. innocua* > *L. helveticus* > *S. aureus*. The temperature effect in addition to the HP on microorganisms was different: *P. fluorescens, L. innocua,* and *L. helveticus* showed higher resistance to HP at room temperature (25°C) than at low temperature (4°C), whereas *E. coli,* and *S. aureus* showed less resistance to HP at room temperature than at low temperature.

6.4.2 APPLICATION OF HPP IN MODIFICATION OF PHYSICO-CHEMICAL PROPERTIES OF MILK

HPP of milk results in significant effects on many constituents of milk. The structure of casein micelles is disrupted and the whey proteins, α-lactalbumin and β-lactoglobulin, are denatured. The α-lactalbumin is more resistant to pressure than the β-lactoglobulin. HPP shifts the mineral balance in milk and moderately HPs in the range of 100–400 MPa causes crystallization of milk fat. However, milk enzymes seem to be quite resistant to pressure. Pressure-induced effects on individual milk constituents, alters many properties of milk [14]. However, to fully understand the effects of HP treatment on milk and to evaluate the full potential of this process in dairy technology, further research is required in several focus areas, including the reversibility of pressure-induced changes in milk and the physical stability of HP-treated milk.

6.4.3 EFFECTS OF HPP ON APPEARANCE AND COLOR

The earliest and most prominent difference between heat and pressure treatment of skim milk is the appearance of the milk directly after treatment. On heating, skim milk becomes whiter while pressure treated skim milk appears translucent or semi-translucent with slightly yellow hue. On storage, heated milk remains white regardless of storage temperature. Pressure treated milk retains semi-transparent appearances if stored at 5°C while storage at room temperature makes it more turbid [17].

HP treated ewe's milk showed decrease in lightness ($L*$) and an increase in greenness ($-a*$) and yellowness ($+b*$). This decrease in $L*$ value is due to disintegration of casein micelle [28]. HPP of skim milk at 600 MPa for 15 min resulted in significant changes in the $L*$, $b*$ and $a*$ values, that could be also perceived visually [48]. Warming of samples from 4° to 43°C derived back color values of HP milk towards the values of untreated milk, although not to the same initial point. Treatment at 200 MPa at 20°C had only a slight effect on the $L*$ values, but treatment at 250–450 MPa significantly decreased the $L*$ value of pasteurized or reconstituted skim milk. Treatment at >450 MPa had little further effect on the $L*$ values [36]. Cow's milk showed less sensitivity to pressure with respect to color change [47].

6.4.4   EFFECTS OF HPP ON PH OF MILK

HP treatment affects the mineral distribution, chiefly calcium and phosphate and level of ionized minerals. HP treatment raises the concentration of ionic calcium in milk [43]. Increase in the concentration of phosphate in the milk serum, increases the milk pH [65]. Pressure up to 600 MPa had no significant effect on the pH of skim bovine milk. Several succeeding studies have observed increases, of varying magnitude, in the pH of HP treated caprine milk or raw, pasteurized or UHT treated bovine milk. The degree of pH shift depends on magnitude of pressure and temperature. The degree of pH shift was more for higher pressure and lower temperature. Further, the pH shift is more in UHT treated milk than pasteurized or raw milk. Increase in pH is rapidly reversible on subsequent storage at 20°C, but virtually irreversible at 5°C [36, 65].

6.4.5   EFFECTS OF HPP ON PARTICLE SIZE CHANGES

HPP of skim milk at about 300 MPa substantially decreases the size of the particles. The average decrease in size is from about 200 to 100 nm regardless of temperature at pressurization. Pressure treatments in the range of

200 to 300 MPa causes the particle size to increase which is affected by pH of milk, temperature, pressure and duration of pressure processing. Early reports suggested that the denaturation of the whey proteins may be responsible for this aggregation phenomenon [32], but successive studies carried on whey protein depleted systems point out that the aggregation is owed entirely to the casein micelles [2]. Preheat treated milk at 90°C for 10 min has shown that particle size changes similar to those in unheated milk over the entire pressure range, demonstrating that the pre-denaturation of the whey proteins has little effect on the aggregation or disaggregation phenomenon in pressure-treated milk [32]. The changes in particle size of HP treated milk was accompanied by a considerable increase in the level of serum phase casein, and the changes in size and serum phase casein appeared to be correlated [1]. The aggregation of the casein micelles increases as the temperature and the duration of pressure treatment were increased. Pressure treatment of about 250 MPa and at temperatures above 20°C resulted in the formation two distinct populations: one with the particle size larger than the original casein micelles and other with smaller than original size of casein micelle [26]. When the pH of the milk was reduced from the natural pH (6.7) to 6.5, the particle size was decreased with the increase in pressure, with a marked decrease in size above 200 MPa. Study on ewe's milk reported that HP up to 500 MPa produces some modifications on size and distribution of milk fat globules of ewe's milk. HP treatments at 50°C showed a tendency to increase the number of small globules in the range 1–2 μm, whereas at 4°C the tendency was the reverse [28].

6.4.6 EFFECTS OF HPP ON DESTRUCTION OF VOLATILE COMPOUNDS

HPP of milk at room and mild temperatures only disorders moderately weak chemical bonds such as hydrogen bonds, hydrophobic bonds, ionic bonds. Thus, it had no effect on small molecules such as vitamins, amino acids, simple sugars and flavor compounds [13]. HPP of milk at 400 MPa resulted in no significant loss of vitamins B1 and B6 [55]. HP treatments at 400 MPa for 15 min at 40–60°C reduced the proteolytic activity and at

25–60°C maintained or improved the organoleptic properties of milk. This suggests that these combined treatments could be used to produce milk of good sensory properties with an increased shelf-life [25].

6.5   APPLICATION OF HIGH PRESSURE PROCESSING IN CHEESE PRODUCTION

In cheese manufacturing, many research groups have studied the application of HP in making cheese from pressure treated milk, microorganism inactivation in cheese, acceleration of cheese ripening and increased cheese yield. This section covers the research findings of studies conducted on application of HPP in cheese processing.

6.5.1   CHEESE PRODUCTION FROM PRESSURE TREATED MILK

The physicochemical and sensory properties of cheese are the most valued. Therefore, ensuring that the processing technologies applied to them do not affect these unique attributes in a negative fashion is of utmost importance. Milk pasteurization is recognized to adversely affect the development of many sensory characteristics of cheese, leading to alteration in texture and often delayed maturation. HPP did not alter the composition of fresh cheese particularly its total solid, ash, fat and soluble nitrogen contents. However, non-protein nitrogen of HP cheese remained lower than the non-treated fresh cheese. HP treated milk cheese contained higher moisture, salt and total free amino acid contents than pasteurized milk cheese (control) [60]. Table 6.1 summarizes the results of studies conducted on cheese with regard to the impact HP treatments on physicochemical, rheological and sensory properties.

HP treatment (500 MPa) improves sensory characteristics of cheese [40]. Changes in organoleptic characteristics of the cheese are not perceived by sensory judges if it is HP treated within pressure range of 200 to 500 MPa for 15 min [57]. HP treated cheese shows a similar level of lipolysis to cheese made from raw milk; whereas the level of lipolysis in cheese made from pasteurized milk was lower [11].

TABLE 6.1 Effects of HPP Treatments on Quality Parameters of Cheese

Parameter Evaluated	Cheese Variety	Instant of application	HP treatment applied P (MPa)/T (Min)/T (°C)	HP induced modification	Ref.
Color	Mato	1 day	500/5, 10, or 15/10	$L*$ and $a*$ decreased, whereas $b*$ increased compared to control cheese.	[12]
	Cheddar, Turkish white-brined	1 day	50–400/5, 10, or 15/22–25	Increasing pressure intensity and holding time did not affect $L*$, but $a*$ decreased and $b*$ increased compared to control cheese.	[53]
Rheological properties	Cheddar	1 and 4 months	200–800/5/25	Pressures up to 300 MPa applied to 1- month old cheese had no significant effect. At 800 MPa, cheese had similar fracture stress and Young's modulus as control cheese. Pressure applied to 4-month old cheese increased fracture work.	[62]
	Cheddar	1 day	400/10/25	Increased fracture strain and fracture stress values, lower fluidity, flowability, and stretchability increased up to 21 day, but to a lesser extent than in control cheese.	[53]
	Low-moisture mozzarella	1 and 5 Days	400/20/25	Reduced time required to attain satisfactory cooking performance (by 15 day). Increased fluidity, flowability, stretchability, and reduced melting time on heating at 280–°C.	[50]
	Gouda	3 day	50, 225, or 400/1 h/14	Less rigid and solid-like, more viscoelastic, and had less resistance to flow at longer times.	[46]
Sensory properties	Hispanico	15 days	400/5/10	Treatments applied to immature cheese limit the formation of volatile compounds. However, differences become less significant during ripening.	[4]
	Ewes' milk cheese	1 or 15 Days	200 or 500/10/12		[38]
	Raw goat milk cheese	1, 3, or 50 day	400 or 600/7/10	Treatments applied at more advanced stages do not cause significant differences compared to control cheese.	[16]

This behavior was explained by heat-sensitive but partial pressure-resistant characteristics of the indigenous milk lipase. Further, hard surface which was apparent in a HP processed milk cheese (500 MPa in the range of 30 ± 5 min at 25°C) was subjected to sensory analysis and sensory judges showed preference for processed cheese over the non-processed cheese regardless of texture attributes [59].

6.5.2 EFFECTS OF HPP ON RENNET COAGULATION TIME (RCT)

HP treatment at <150 MPa had no effect on the RCT, whereas pressure treatment at 200 to 600 MPa significantly reduced RCT [17]. However, RCT was reduced markedly after the treatment at 200–600 MPa. In general, rennet coagulation properties of milk subjected to 200 MPa for 30 min were enhanced [42]. The pressure-induced disintegration of the casein micelles appears to reduce the RCT and the cutting time, whereas the denaturation of the whey proteins increases the RCT and the cutting time, in a similar fashion to that for the heated milk. RCT decreases with increasing pressure and treatment time [34]. Temperature of treatment and pH of the milk has considerable effect on the RCT of HP treated milk. Treatment at 50–60°C (200 MPa) delayed the rennet coagulation of milk [43]. Acidification of milk (pH-5.5) before HP treatment decreased its RCT, whereas alkalization (pH-7.0) had the opposite effect [3]. The strength of the rennet-induced coagulum from heated milk treated at 250–600 MPa for 30 min or 400 or 600 MPa for 0 min was considerably higher than that of unheated unpressurized milk.

6.5.3 EFFECTS OF HPP ON MICROORGANISM INACTIVATION IN CHEESE

Food scientists have employed HP technology in cheese processing to inactivate toxigenic and infectious pathogens such as *Escherichia coli*, *Staphylococcus aureus*, *Listeria monocytogenes*, *Aeromonas hydrophila*, *Salmonella enterica*, and *Yersinia enterocolitica*, as well as spoilage microorganisms such as *Staphylococcus carnosus, Enterococcus* spp., coliforms, yeasts, and molds, and also microbial spores from *Bacillus subtilis, Bacillus cereus*, and *Penicillium roqueforti*. Many studies have revealed that HP treatments cause structural and functional alterations in vegetative cells and spores leading to cell injury or death. These include cell membrane disruption or increased permeability, ribosomal destruction, collapse of intracellular vacuoles, denaturation of membrane-bound proteins, damage to the proton efflux system, inactivation of key enzymes, including those involved in DNA replication and transcription, release of dipicolinic acid and small acid-soluble spore proteins, and hydrolysis of spore core and cortex [7].

The level of microbial inactivation obtained is a function of applied pressure and temperature [58]. The Gram positive *S. aureus* species was more resistant to pressure than the Gram negative *E. coli* species with the latter showing increasing sensitivity to pressure on going from 10°C to 30°C. At lower pressures (<300 MPa), the mold species was more resistant than the bacteria, and was much more sensitive at higher pressures. In the case of *E. coli* and the molds, similar trends and degree of inactivation by HPP were observed for strains within species. Log phase cells are more sensitive to pressure than stationary phase cells. Pressure treatment of $\geq$ 300 MPa for 5 min to fresh lactic curd cheese effectively controlled the outgrowth of yeasts and extends product shelf-life from 3 to 6–8 weeks [15].

Microbial resistance to HP depends not only on the intrinsic resistance of the microorganisms, but also on the physiological state [44]. The cells of *S. aureus* ATCC 6538 and *E. coli* K-12 in the exponential phase of growth were more sensitive to HPP treatments in Cheddar cheese slurry than cells in the stationary phase [49]. The higher pressure conditions (345 and 550 MPa) and longer exposure times (10 and 30 min) achieved a greater reduction in numbers of undesirable bacteria in the natural microflora of Swiss cheese slurries (coliforms, presumptive coagulase-positive *Staphylococcus*, yeasts, and molds) and in starter LAB added to milk for acid production and flavor development [18]. A greater antimicrobial impact can be achieved with moderate pressure treatments and shorter pressure holding times when combining high temperatures with HPP treatments. However, the use of high temperatures could lead to undesirable effects in certain cheese quality parameters. Treatments at 50°C caused high whey losses and unacceptable textural characteristics.

6.5.4   EFFECTS OF HPP ON YIELD OF CHEESE

In cheese making process, it is very essential to obtain the maximum achievable recovery of substance from milk, because the higher the recovered percentage of solids, the greater the amount of cheese obtained and therefore gain in economic terms. Cheese yield is influenced by various factors: perhaps the single most important factor is the composition of the milk, which itself depends on the species and breed of animal, the stage of lactation and the somatic cell count of the milk. Furthermore, pre-treatment of cheese-milk can also increase the yield of cheese.

Heat treatment of the milk intended for cheese making can increase its yield. Through incorporation of whey proteins in the cheese curd, this treatment is very common in the manufacture of fresh cheese. However, for the manufacture of semi-hard and hard cheeses, heat treatment adversely affects other cheese making properties of milk: It significantly increases the RCT of milk and reduces the firmness of the rennet-induced coagulum. High pressuring of milk could be a good solution over this. HP treated milk gives higher yields [19]. Arias et al. [3] reported that treatment of milk at 200 MPa had no effect on wet curd yield, although denaturation of β-lactoglobulin was observed at 200 MPa whereas at 300–400 MPa wet curd yield was significantly increased upto 20% and reduced both the loss of protein in whey and the volume of whey. Increased treatment time, up to 60 min at 400 MPa increased wet curd yield and reduced protein loss in whey; and the changes were greatest during the first 20 min of treatment [42]. Authors also reported that increased cheese yield is primarily due to greater moisture retention, secondly due to incorporation of some denatured β-lactoglobulin. Also, the casein micelles and fat globules in HP-treated milk may not aggregate as closely as in untreated milk, therefore allowing more moisture to be entrapped in the cheese [19].

6.5.5   EFFECTS OF HPP ON RIPENING OF CHEESE

Cheese ripening is a time-consuming and expensive process due to high storage costs. Its acceleration is highly desirable. Pioneering research on this topic was performed by Yokoyama et al. [64], who reported significant reduction inthe ripening times of Japanese Cheddar and Parmesan-type

TABLE 6.2 Effects of HP Treatments on the Ripening Process of Different Cheese Varieties

Cheese variety	Instant of application	Treatment applied P (MPa)/t (min or h)/T (°C)	HP induced changes	Ref.
Proteolysis				
Cheddar	After salting	50/72 h/25	Similar taste and FAA content of a 6 month-old commercial cheese obtained in 3 d (Cheddar: 26.5 mg/g, Parmensan: 76.7 mg/g).	[64]
Cheddar	2, 7, 14, or 21 days	50/72 h/25	Faster αs1-casein hydrolysis and accumulation of αs1-I-casein. Increased pH 4.6 SN/TN below 150 MPa and FAA levels. Total FAA decreased as pressure increased.	[50]
Blue-veined	42 days	400–600/10/20	Accelerated breakdown of β- and αs2-casein and increased levels of PTAh SN/TN.	[63]
Gouda	After brining, 5 or 10 days	50 or 500/20–100/14	No changes in pH 4.6 SN, PTA SN/TN, FAA contentand SDS-PAGE profiles.	[40]
Lipolysis				
Full-fat Cheddar	1 day	400/10/25	Lipolysis was not significantly different from control over 180 d	[53]
Ewes' milk cheese	1 or 15 days	200–500/10/12	Lowest concentration of total FFA at pressure treatments of 400 to 500 MPa applied on d 15 after 60 d of ripening compared to other treatments. Highest levels of FFAs were obtained at 300 MPa applied on day 1 compared to other treatments.	[38]
Blue-veined	42 day	400–600/10/20	Reduced lipolytic activity of *P. roqueforti*.	[63]
Glycolysis				
Full-fat Cheddar	1 day	400/10/25	Concentration of total lactate in HHP-treated cheese was significantly lower compared to the control after 180 d of ripening	[53]

Note: h = time in hours; FAA = free amino acids; SN = soluble nitrogen; TN = total nitrogen; PTA = phosphotungstic acid; FFA = free fatty acids.

cheese without affecting sensory attributes [64]. Many food scientists have assessed the application of HP treatments to accelerate the ripening of cheese (Table 6.2).

HP treatments are able to accelerate cheese ripening by altering the enzyme structure, conforming changes in the casein matrix making it more prone to the action of proteases and bacterial lysis promoting the release of microbial enzymes that promote biochemical reactions [63]. HP treatments also increase pH (0.1 to 0.7 units) and modify water distribution of certain cheese varieties, promoting conditions for enzymatic activity.

The initial hydrolysis of caseins in milk is carried out mainly from the action of plasmin, chymosin, and to a lesser extent by pepsin. HP treatment at 800 MPa for 60 min at 8°C did not inactivate plasmin in 14-day-old Cheddar cheese, while at 20°C and 30°C its activity was reduced by 15% and 50%, respectively compared to controls [33].

Free amino acid (FAA) levels were 16.2, 20.3, 26.5, and 25.3 mg/g after HP treatments at 5, 15, 50, and 200 MPa, respectively, while the FAA level was 21.3 mg/g in the control cheese. There was non-significant difference between the taste of Cheddar cheese HP-treated at 50 MPa and commercial control cheese. With increase of pressure from 100 to 400 MPa, production of total FAA was decreased. Conversely, increasing processing time up to 60 h, total FAA levels were increased. These research studies on Cheddar cheese ripening clearly demonstrate that HPP treatments enhanced proteolysis. HP enhanced proteolysis did not altered the pathways of proteolysis, thus flavor and texture development were very similar to traditional commercial Cheddar cheese. Low to moderate HP treated (50 to 150 MPa) young Cheddar cheese showed accelerated proteolysis, whereas higher HP treated conditions (≥400 MPa) may help to arrest the ripening process at a desired stage, thus maintaining optimum "commercial attributes" for a longer period of time [50].

6.6   EFFECTS OF HIGH PRESSURE PROCESSING ON YOGHURT AND FERMENTED DAIRY PRODUCTS

Yogurt, **yogurt,** or yogurt is a custard-like food with a tart flavor, prepared from milk curdled by bacteria, especially *Lactobacillus bulgaricus* and *Streptococcus thermophilus,* and often sweetened or flavored.

Among the processes involved in fermented milk and yogurt process-ing, the most important are homogenization, pasteurization and fermen-tation. Two strategies have been used to improve fermented milk and yogurt quality and preservation by means of HP: making fermented milk and yogurt from HP treated milk and pressurization to inactivate micro-biota.

Acid set gels prepared from HP treated milk showed improved texture (rigidity and resistance to breaking) and syneresis resistance [37]. Most of the viscosity improvement is achieved after pressurization for 15 min at 400 MPa and for 5 min at 600 MPa with slight further increase up to 60 min. Yoghurt firmness increases as pressure increases, and treatments of 350 MPa at 25°C and 500 MPa at 55°C showed no differences in whey syneresis compared with pasteurized milk [24]. During storage at 4°C for 20 days, yogurts made from HP treated milk showed a good stability in terms of firmness and water retention compared with yogurt made from the untreated milk.

Milk treatment with pressures of 400–600 MPa for 10 min at 25°C can achieve similar results as low temperature pasteurization in terms of pathogenic and spoilage microorganisms inactivation [61]. Research stud-ies have described the disintegration of the casein micelles into smaller particles and the simultaneous increase in the amount of caseins and cal-cium phosphate in the serum phase which improved the water holding capacity of yogurt [35]. Further, the combination of HP and thermal treat-ment increased the yogurt viscosity and lowered gelation times compared to HP treated samples [24].

6.7   EFFECTS OF HIGH PRESSURE ON CREAM, BUTTER AND ICE CREAM

HPP induces fat crystallization, shortens the time required to achieve a desirable solid fat content and thereby reduces the aging time of ice-cream mix and also enhances the physical ripening of cream for butter making [8]. Studies on the freeze fracture and transmission electron microscopy of cream (30 and 43% fat) subjected to HPP of 100 to 150 MPa at 23°C for pasteurization indicated that pressurization induced fat crystallization within the small emulsion droplets mainly at the globule periphery [9].

Fat crystallization was increased with the length of pressure treatment and was maximal after processing at 300–500 MPa. Further, the crystallization proceeded during later storage, after pressure release. Pressurization treatment improved whipping ability of cream when treated for 2 min at 600 MPa and is possibly due to better crystallization properties of milk fat [23]. Excessive pressurization leads to excess denaturation of whey proteins, which results in longer whipping time and destabilization of whipped cream. Studies on the effects of added whey protein concentrate on the over-run and foam stability of ice cream mix had confirmed the effects of HPP on foaming properties of whey proteins in a complex system [41].

HPP (450 MPa at 25°C for 10 to 30 min.) of dairy cream (35% fat) reduced the microbial load significantly [52]. Pressure-assisted freezing may be of special interest to avoid coarse ice crystallization and obtain a smooth texture in various types of ice creams (including low fat) or sherbets. The Unilever Company–India has patented combinations of HP processing and freezing for improved consistency and smoothness, and slower melting of ice cream [39].

Studies on the effects of HPP on pasteurized dairy creams (35% fat) at 450 MPa and 25°C for 15 or 30 min, or at 10 or 40°C for 30 min indicated that pressurization at 450 MPa at 10 or 25°C did not modify fat globule size distribution or flow behavior [20].

6.8 SUMMARY

The need to meet the demand of consumer for safe, nutritious, natural, economic, convenient and delicious food necessitates the processors to look beyond the conventional thermal food processing technologies. Among various non-thermal technologies, HPP is the promising technology. Comprehensive knowledge of the mechanism and kinetics of pressure induced degradation, denaturation, inactivation of dairy compounds like nutrients, proteins, microorganisms, enzymes, etc., this technology can be effectively applied in rennet or acid coagulation of milk, ripening of cheese, syneresis and firmness in fermented milk product, aging of ice cream mix, ripening and fat crystallization of dairy cream and many other dairy products.

KEYWORDS

- appearance of milk
- butter
- cheese
- cheese production
- cheese ripening
- cheese yield
- color of milk
- cream
- fermented milk
- high pressure processing, HPP
- HPP system
- ice cream
- inactivation of microorganisms
- isostatic principle
- le Chatelier's principle
- mechanism of HPP
- milk
- milk processing
- particle size
- pH
- physicochemical properties of milk
- Principle of microscopic ordering
- volatile compound
- yogurt

REFERENCES

1. Anema, S. G., Lowe, E. K., & Stockman, R. (2005). Particle size changes and casein solubilization in high-pressure-treated skim milk. *Food Hydrocolloids, 19,* 257–267.
2. Anema, S. G., Stockman, R., & Lowe, E. K. (2005b). Denaturation of β-lactoglobulin in pressure treated skim milk. *Journal of Agricultural and Food Chemistry, 53,* 7783–7791.

3. Arias, M., Lopez-Fandino, R., & Olano, A. (2000). Influence of pH on the effects of high pressure on milk. *Milchwiss*, *55*(4), 191–194.

4. Avila, M., Garde, S., Fernandez-Garcia, E., Medina, M., & Nunez, M. (2006). Effect of high pressure treatment and a bacteriocin-producing lactic culture on the odor and aroma of Hispanico cheese: Correlation of volatile compounds and sensory analysis. *Journal of Agricultural and Food Chemistry*. *54*, 382–389.

5. Balci, A. T., & Wilbey, R. A. (1999). High pressure processing of milk—the first 100 years in the development of new technology. *International Journal of Dairy Technology*, *52*, 149–155.

6. Balny, C., & Masson, P. (1993). Effects of high pressure on proteins. *Food Rev. Int.*, *9*(4), 611–628.

7. Black, E. P., Stewart, C. M., & Hoover, D. G. (2011). Non-thermal processing technologies for food. In: *Non-Thermal Processing Technologies for Food*, edited by Zhang, H. Q., Barbosa-Canovas, G. V., Balasubramanian, V. M., Dunne, C. P., Farkas, D. F., & Yuan, J. T. C. Oxford: Wiley-Blackwell, pp. 51–71.

8. Buchheim, W., Schrader, K., Morr, C. V., Frede, E., & Schutt, M. (1996). Effects of high pressure on protein, lipid and mineral phase of milk heat treatments and alternative methods. Brussels: *International Dairy Federation*, *9602*, 202–213.

9. Buchheim, W., & Abou El Nour, A. M. (1992). Induction of milk fat crystallization in the emulsified state by high hydrostatic pressure. *Fett Wissenschaft Technologie*, *94*(10), 369–373.

10. Buffa, M., Trujillo, A. J., & Guamis, B. (2001). Rennet coagulation properties of raw, pasteurized and high pressure-treated goat milk. *Milchwissenschaft*, *56*(5), 243–246.

11. Buffa, M., Guamis, B., Pavia, M., & Trujillo, A. J. (2001). Lipolysis in cheese made from raw, pasteurized or high-pressure-treated goats milk. *International Dairy Journal*, *11*(3), 175–179.

12. Capellas, M., Mor-Mur, M., Sendra, E., & Guamis, B. (2001). Effect of high-pressure processing on physico-chemical characteristics of fresh goat's milk cheese (Mato). *International Dairy Journal*, *11*(3), 165–173.

13. Cheftel, J. C. (1992). Effects of high hydrostatic pressure on food constituents: an overview. In: *High Pressure and Biotechnology*, edited by Balny, C., Hayashi, R., Heremans, K., & Masson, P. London, UK: *Colloque INSER My John Libbey Eurotext, Ltd.*, pp. 195–209.

14. Considine, T., Patel, H. A., Anema, S. G., Singh, H., & Creamer, L. K. (2007). Interactions of milk proteins during heat and high hydrostatic pressure treatments — A Review. *Innovative Food Science and Emerging Technologies*, *8*, 1–23.

15. Daryaei, H., Coventry, M. J., Versteeg, C., & Sherkat, F. (2008). Effect of high pressure treatment on starter bacteria and spoilage yeasts in fresh lactic curd cheese of bovine milk. *Innovations in Food Science and Emerging Technologies*, *9*, 201–205.

16. Delgado, F. J., Gonzalez-Crespo, J., Cava, R., & Ramirez, R. (2011). Changes in the volatile profile of a raw goat milk cheese treated by hydrostatic high pressure at different stages of maturation. *Intl Dairy J*, *21*, 135–141.

17. Desorby-Banon, S., Richard, F., & Hardy, J. (1994). Study of acid and rennet coagulation of high pressurized milk. *Journal of Dairy Science*, *77*, 3267–3274.

18. Ding, Y. T., Sang, W. G., Jin, Z., & Harper, J. W. (2001). High pressure treatment of Swiss cheese slurries inactivation of selected microorganisms after treatment and during accelerated ripening. *J Zhejiang University*, *2*, 204–208.

19. Drake, M. A., Harrison, S. L., Asplund, M., Barbosa, C. G., & Swanson, B. G. (1997). High pressure treatment of milk and effects on microbiological and sensory quality of Cheddar cheese. *Journal of Food Science*, *62*(4), 843–845.

20. Dumay, E., Lambert, C., Funtenberger, S., & Cheftel, J. C. (1996). Effects of high pressure on the physico-chemical characteristics of dairy creams and model oilywater emulsions. *Lebensmittel Wissenschaft und Technologie*, *29*(7), 606–625.

21. Dunne, C. P. (2005). High pressure keeps food fresher. Available at http://www.natick.army.mil/about/pao/05/05-22.htm.

22. Earnshaw, R. (1996). High pressure food processing. *Nutri Food Sci*, *2*, 8–11.

23. Eberhard, P., Strahm, W., & Eyer, H. (1999). High pressure treatment of whipped cream. *Agrarforschung*, *6*(9), 352–354.

24. Ferragut, V., Martinez, V. M., Trujillo, A. J., & Guamis, B. (2000). Properties of yogurts made from whole ewe's milk treated by high hydrostatic pressure. *Milchwissenschaft*, *55*(5), 267–269.

25. Garcia Risco, M. R., Olano, A., Ramos, M., & Lopez-Fandino, R. (2000). Micellar changes induced by high pressure. Influence in the proteolytic activity and organoleptic properties of milk. *Journal of Dairy Science*, *83*(10), 2184–2198.

26. Gaucheron, F., Famelart, M. H., Mariette, F., Raulot, K., Michel, F., & Le, Graet, Y. (1997). Combined effects of temperature and high pressure treatments on physicochemical characteristics of skim milk. *Food Chem.*, *59*, 439–447.

27. Gervilla, R., Capellas, M., Ferragut, V., & Guamis, B. (1997). Effect of high hydrostatic pressure on Listeria innocua 910 CECT. *Journal of Food Protection*, *60*(1), 33–37.

28. Gervilla R, Ferragut, V., & Guamis, B. (2001). High hydrostatic pressure effects on color and milk-fat globule of ewe's milk. *Journal of Food Science*, *66*(6), 880–885.

29. Gomes, M. I. F. V. (1995). Contribuiçãoaoestudo da atividade proteolítica residual sobre a estabilidade proteica do leite esterilizado "longa vida." Thesis (PhD in Food Technology) – Faculty of Food Engineering, State University of Campinas.

30. Hendrickx, M. L., Van den, B. I., & Weemaes, C. (1998). Effects of high pressure on enzymes related to food quality. *Trends Food Sci Technol.*, *9*, 197–203.

31. Hite, B. H. (1899). The effect of pressure in the preservation of milk. *West Virginia Agricultural Experimental Station Bulletin*, *58*, 15–35.

32. Huppertz, T., Fox, P. F., & Kelly, A. L. (2004). High pressure treatment of bovine milk: Effects on casein micelles and whey proteins. *Journal of Dairy Research*, *71*, 97–106.

33. Huppertz, T., Fox, P. F., & Kelly, A. L. (2004). Properties of casein micelles in highpressure treated bovine milk. *Food Chemistry*, *87*, 103–110.

34. Huppertz, T., Hinza, K., Zobrista, M. R., Uniacke, T., Kellya, A. L., & Fox, P. F. (2005). Effects of high pressure treatment on the rennet coagulation and cheesemaking properties of heated milk. *Innovative Food Science and Emerging Technologies*, *6*(3), 279–285.

35. Johnston, D. E., Austin, B. A., & Murphy, P. M. (1992). Effects of high hydrostatic pressure on milk. *Milchwiss*, *47*, 760–763.

36. Johnston, D. E., Austin, B. A., & Murphy, R. J. (1992). The effects of high pressure treatment of skim milk. In: *High Pressure and Biotechnology*, edited by Balny, C., Hayashi, R., Heremans, K., Masson, P. Colloque INSERM. London: John Libbey Euro Text Ltd. *224*, 243–247.

37. Johnston, D. E., Murphy, R. J., & Birks, A. W. (1994). Stirred-style yogurt-type product prepared from pressure treated skim-milk. *High Pressure Research, 12*, 215–219.
38. Juan, B., Trujillo, A. J., Guamis, V., Buffa, M., & Ferragut, V. (2007). Rheological, textural and sensory characteristics of high-pressure treated semi-hard ewes' milk cheese. *International Dairy Journal, 17*, 248–254.
39. Keenan, R. D., Wix, L., & Young, D. (1998). *Method for the Preparation of a Foodstuff*. Patent WO 98y18350.
40. Kolakowski, P., Reps, A., & Babuchowski, A. (1998). Characteristics of pressurized ripened cheeses. *Polish Journal of Food and Nutrition Sciences, 7*(3), 473–482.
41. Lim, S. Y., Swanson, B. G., Ross, C. F., & Clark, S. (2007). High hydrostatic pressure modification of whey protein concentrate for improved body and texture of low fat ice-cream. *J Dairy Sci., 91*, 1308–1316.
42. Lopez Fandino, R., Carrascosa, A. V., & Olano, A. (1996). The effects of high pressure on whey protein denaturation and cheese-making properties of raw milk. *Journal of Dairy Science, 79*(6), 929–1126.
43. Lopez Fandino, R., De la Fuente, M. A., Ramos, M., & Olano, A. (1998). Distribution of minerals and proteins between the soluble and colloidal phases of pressurized milks from different species. *Journal of Dairy Research, 65*, 69–78.
44. Manas, P., & Pagan, R. (2005). A review: microbial inactivation by new technologies of food preservation. *Journal of Applied Microbiology, 98*, 1387–1399.
45. McClements, J. M. J., Patterson, M. F., & Linton, M. (2001). The effect of growth stage and growth temperature in high hydrostatic pressure inactivation of some psychrotrophic bacteria in milk. *Journal of Food Protection, 64*, 514–522.
46. Messens, W., Foubert, I., Dewettinck, K., & Huyghebaert, A. (2000). Proteolysis of a high-pressure-treated smear-ripened cheese. *Milchwissenschaf, 55*(6), 328–332.
47. Mussa, D. M., & Ramaswamy, H. (1997). Ultra-high pressure pasteurization of milk: kinetics of microbial destruction and changes in physico-chemical characteristics. *Lebensmittel Wissenschaft and Technologies, 30*(6), 551–557.
48. Needs, E. C., Capellas, M., Bland, A. P., Manoj, P., Macdougal, D. B., & Paul, G. (2000). Comparison of heat and pressure treatments of skim milk, fortified with whey protein concentrate, for set yogurt preparation: effects on milk proteins and gel structure. *Journal of Dairy Research, 67*(3), 329–348.
49. O'Reilly, C., O'Connor, P. M., Kelly, A. L., Beresford, T. P., & Murphy, P. M. (2000). Use of hydrostatic pressure for inactivation of microbial contaminants in cheese. *Applied and Environmental Microbiology. 66*(11), 4890–4896.
50. O'Reilly, C. E., O'Connor, P. M., Murphy, P. M., Kelly, A. L., & Beresford, T. P. (2000). The effect of exposure to pressure of 50 MPa on Cheddar cheese ripening. *Innovative Food Science and Emerging Technologies, 1*(2), 109–117.
51. Rademacher, B., & Kessler, H. G. (1997). High pressure inactivation of microorganisms and enzymes in milk and milk products. In: *High Pressure Bio-Science and Biotechnology*, edited by Heremans, K. Leuven, Belgium: Leuven University Press. pp. 291–293.
52. Raffalli, J., Rosec, J. P., Carlez, A., Dumay, E., Richard, N., & Cheftel, J. C. (1994). High pressure stress and inactivation of *Listeria innocua* in inoculated dairy cream. *Sciences des Aliments, 14*(3), 349–358.

53. Rynne, N. M., Beresford, T. P., Guinee, T. P., Sheehan, E., Delahunty, C. M., & Kelly, A. L. (2008). Effect of high-pressure treatment of 1 day-old full-fat Cheedar cheese on subsequent quality and ripening. *Innovative Food Science and Emerging Technologies*, *9*, 429–440.

54. Schmidt, D. G., & Koops, J. (1977). Properties of artificial casein micelles, Part 2. *Netherlands Milk and Dairy Journal*, *31*, 342–357.

55. Sierra, I., Vidal Valverde, C., & Lopez-Fandino, R. (2000). Effect of high pressure on the vitamin B1 and B6 content in milk. *Milchwissenschaft*, *55*(7), 365–367.

56. Smelt, J. M. (1998). Recent advances in the microbiology of high pressure processing. *Trends in Food Science & Technology*, *9*, 152–158.

57. Szczawinski, J., Szczawinska, M., Stanczack, B., Fonberg-Broczek, M., & Arabas, J. (1997). Effect of high pressure on survival of Listeria monocytogenes in ripened, sliced cheese at ambient temperature. In: *High Pressure Research in Biosciences and Biotechnology*, edited by Heremans, K. Leuven, Belgium: Leuven, pp. 295–298.

58. Tom, B., & Cait, L. (1999). High pressure processing of dairy products. *Dairy Products Research Center*, Fermoy, Co. Cork, Ireland. DPRC No. 22, pp. 1–17.

59. Trujillo, A. J., Capellas, M., Buffa, M., Royo, C., Gervilla, R., Felipe, X., Sendra, E., Saldo, J., Ferragut, V., & Guamis, B. (2000). Application of high pressure treatment for cheese production. *Food Research International*, *33*, 311–316.

60. Trujillo, A. J., Royo, C., Ferragut, V., & Guamis, B. (1999). Ripening profiles of goat cheese produced from milk treated with high pressure. *Journal of Food Science*, *64*(5), 833–837.

61. Trujillo, A. J., Capellas, M., Saldo, J., Gervilla, R., & Guamis, B. (2002). Applications of high-hydrostatic pressure on milk and dairy products: a review. *Innov. Food Sci. Emerg. Technol.*, *4*, 295–307.

62. Wick, C., Nienaber, U., Anggraeni, O., Shellhammer, T. H., & Courtney, P. D. (2004). Texture, proteolysis and viable lactic acid bacteria in commercial Cheddar cheeses treated with high pressure. *Journal of Dairy Research*, *71*, 107–115.

63. Voigt, D. D., Chevalier, F., Qian, M. C., & Kelly, A. L. (2010). Effect of high-pressure treatment on microbiology, proteolysis, lipolysis and levels of flavor compounds in mature blue-veined cheese. *Innovative Food Science and Emerging Technologies*, *11*, 68–77.

64. Yokoyama, H., Sawamura, N., & Motobayashi, N. (1993). *Method for Accelerating Cheese Ripening*. Fuji Oil Company, Assignee. U.S. patent 5,180,596.

65. Zobrist, M. R., Huppertz, T., Uniacke, T., Fox, P. F., & Kelly, A. L. (2005). High pressure induced changes in rennet-coagulation properties of bovine milk. *International Dairy Journal*, *15*, 655–662.

CHAPTER 7

EMERGING MILK PROCESS TECHNOLOGIES: HIGH HYDROSTATIC PRESSURE

ADARSH M. KALLA and DEVARAJU RAJANNA

CONTENTS

7.1   INTRODUCTION

Most of food materials are seasonal and perishable in nature. In the growing season, there may be a local glut, because of insufficient transport facilities, lack of good roads and poor availability of packing materials. In general, the surplus cannot be taken quickly enough to the natural markets in urban areas; and cannot be stored for sale in the off-season because of inadequate local cold storage facilities. Thus the growers do not get a good price for their produce because of the glut and some of it is spoiled resulting in a complete loss. Two approaches are possible for solving this problem. One is the creation/expansion of cold storage facilities in the raw material surplus regions themselves, as also in the major urban consumption centers, to ensure supply of fresh product throughout the year. Another approach is to process the raw material into various products that could be preserved for a long time, and this adds to the value of the product. A goal of food manufacturer is to develop and employ processing technologies that retain or create desirable sensory qualities or reduce undesirable changes in food due to processing. Food processing involves the application of scientific principles to slow down the natural process of food decay caused by micro-organisms, enzymes in the food or environmental factors such as: heat, moisture and sunlight. Much of this knowledge is known traditionally and put into practice by experience and information is handed down through the generations.

The increasing urbanization, rise in middleclass purchasing power and change in food habits have left the consumer demand, driven towards better tasting and additive-free foods with a longer shelf-life. The new information about the links between diet and health has contributed to new demands for foods. At the same time, technological changes in production, processing and distribution, structural change and growth in large-scale retailing, and expansion of trade worldwide have contributed to a rapidly changing market for food products.

The traditional food preservation methods like heating, freezing, dehydration and some of chemical treatments like use of preservatives are most widely used methods. The thermal treatment is one of the most generally used food preservation method. However, thermal treatment reduces possible health hazards arising from pathogenic microorganisms associated with food, but the product quality (taste and flavor) is affected by the heat-induced reactions. The fouling of equipment by deposit formation on walls is governed by specific reactions of food components. These typical undesired reactions reduce the heat transfer coefficient, and increase product losses, resulting in higher operating costs [44].

The increasing popularity of minimally processed foods has resulted in greater health benefits. Furthermore, the ongoing trend has been to eat out and to consume ready-to-eat foods [1]. With this increasing demand for ready-to-eat, fresh, minimally processed foods, including processed fruits and vegetables preserved by relatively mild techniques, and thus ensure product safety and convenience: a high pressure processing (HPP) approach appears to be the best method. The HPP utilizes less destructive method to meet the demand for minimally processed fresh foods, which also meet the ever-increasing food safety standards enacted by FDA and USDA regulations.

This chapter reviews technology of high hydrostatic pressure processing (HHP) with special emphasis on its applications in dairy products (milk and milk products).

7.2 DEFINITION OF HIGH PRESSURE PROCESSING

HPP is a non-thermal method of food processing, where food is subjected to elevated pressures (600 MPa) [39], with or without the addition of heat, to achieve microbial inactivation or to alter the food attributes in order to achieve qualities that are desirable to the consumers. Pressure inactivates most vegetative bacteria at pressures above 600 MPa. HPP retains food quality, maintains natural freshness, and extends microbiological shelf life. The process is also known as HHP or ultra-high pressure processing (UHP).

7.3 HIGH PRESSURE PROCESSING: A BRIEF HISTORY

High pressure technology, to kill microorganisms and preserve food, was discovered in 1899 and has been used with success in chemical, ceramic, carbon allotropy, steel/alloy, composite materials and plastic industries for decades. HPP is similar in concept to cold isostatic pressing of metals and ceramics, except that it demands much higher pressures, faster cycling, high capacity, and sanitation [59, 105]. At the end of 18th century, experiments were carried out by Bert Hite [39] and Royer [83] using high pressure technology for microbial inactivation of milk, fruit juice and vegetables juice [104]. At the Agriculture Research Station in Morgans Town, West Virginia, USA, a high-pressure unit was designed and constructed to pasteurize milk and other food products [39]. This unit could reach pressures in excess of about 700 MPa. The potential use of HP processing was examined for a wide range of foods and beverages, including the pressure inactivation of viruses.

The systemic research was carried out about macroscopic physical behaviors in UHP, such as the compressibility of solid, the phenomenon of melting, the properties of mechanical, the changing of state, and so on [10]. The phenomenon of protein about coagulating at <500 MPa and turning into hard gelatinous body at <700 MPa were investigated. These achievements laid the foundation of UHP applications in food processing [10]. It was pointed out explicitly that the UHP can be used in the commercial processing of fruit juice [20]. The UHP technique was used by Japanese scholars to solve tedious problems in the food processing, which could not solved by heating treatment. The Jam under HPP was produced for the first time [75]. Afterwards, a fresh orange juice "spot squeeze" was sold commercially a compressed food in France. High-pressured food aroused a bigger interest by its unique effect of sterilization, enzyme inactivity and the superiority, which the heat sterilization does not have.

The HPP techniques have also gained momentum in areas of food preservation outside of sterilization and pasteurization. The range of possibilities offered by combining HP with Low Temperatures (HPLT) has allowed the basis of a new field of HP food applications, such as: pressure-supported freezing, thawing and subzero storage. Studies have been conducted to develop and optimize the HPLT processes. New findings regarding the phase transitions of water, with consequential benefits for the food Industry, have recently been revealed [96].

7.4 PRINCIPLES OF HIGH PRESSURE PROCESSING

HPP treatment of food is carried out using batch or semi continuous process. HPP work has been extended to rice products, fish, poultry products and ready to eat meats. HPP treatment can provide fresh like taste, minimal processing and high quality convenient products with an extended shelf life.

HPP has emerged as the most innovative non-thermal food processing technique during past few decades. The most important work involving microbial inactivation in food science by using high pressure is based on principle of isostatic distribution and Le-Chatelier principle.

7.4.1 ISOSTATIC PRINCIPLE

The isostatic principle indicates that pressure transmittance occurs in a uniform and quasi instantaneous manner (Figure 7.1). The food products are compressed by uniform pressure from every direction and then

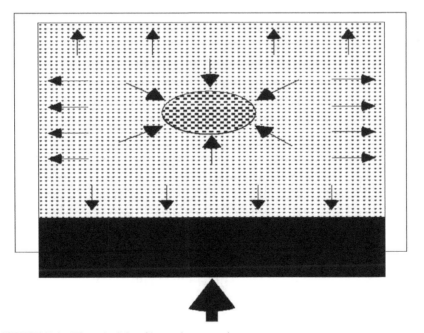

FIGURE 7.1 The principle of isostatic processing.

returned to their original shape when the pressure is released [69]. The pressurization process time is independent of the sample volume. Hence products are compressed independently of product size and geometry. When an aqueous medium is compressed, the compression energy E (in Joule) is equal to:

$$E = \frac{2}{5} \times C \times P \times V_0 \tag{1}$$

where, P = pressure (Pa), C = compressibility of solution, and V_0 = initial volume (m^3).

Therefore, energy required for compression of 1 liter water is 19.2 kJ at 400 MPa as compared to 20.9 kJ for heating one liter of water from 20 to 25°C. The covalent bonds of food constituents are less affected than weak interactions due to low energy levels involved in pressure processing [17, 28].

7.4.2 *LE CHATELIER PRINCIPLE*

The Le Chatelier principle describes the effect of pressure on the basis of absolute reaction rate theory. It states that, if a system at equilibrium experiences a change in concentration, temperature, volume, or partial pressure, then the equilibrium shifts to counteract the imposed change and a new equilibrium is established. This principle is named after Henry Louis Le Chatelier and sometimes Karl Ferdinand Braun, who discovered it independently. The principle has analogs throughout the entire physical world.

Most biochemical reactions cause change in volume. Therefore, biochemical processes are influenced by pressure application. Overall volume change favors the disruption of hydrophobic bonds and dissociation of ionic interactions. Hydrogen bond formation is favored, while covalent bonds are not disrupted by high pressure [52].

7.5 DESCRIPTION OF THE HIGH PRESSURE PROCESSING

In a HPP, the food product is placed in a pressure vessel capable of sustaining the required pressure; the product is submerged in a liquid, which acts as the pressure-transmitting medium. Water may be used as the pressure-transmitting medium, but media containing castor oil, silicone oil, sodium benzoate, ethanol or glycol are also used. The ability of the pressure-transmitting fluid to protect the inner vessel surface from corrosion, the specific HP system being used, the process temperature range and the viscosity of the fluid under pressure are some of the factors involved in selecting the medium. The pressure vessel is the most important component of high hydrostatic- pressure equipment. It is necessary to design the high-pressure vessel to be dimensionally stable to avoid failure. If it fails, it should fail causing leakage before fracture [19].

Industrial HP treatment is currently a batch or semi-continuous process. The selection of equipment depends on the kind of food product to be processed. Solid food products or foods with large solid particles can only be treated in a batch mode. Liquids, slurries and other pump able products have the additional option of semi-continuous production [93].

Currently, most HP machines in industrial use for food processing are batch systems, whereby the product is placed in a high pressure chamber and the vessel is closed, filled with pressure-transmitting medium and pressurized either by pumping medium into the vessel, where the packages of food, surrounded by the pressure-transmitting fluid, are subjected to the same pressure as exists in the vessel itself (Figure 7.2), or

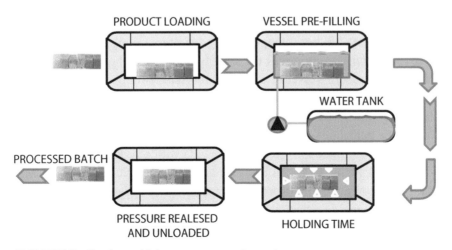

FIGURE 7.2 Batch type high pressure processing system.

by reducing the volume of the pressure chamber, for example by using a piston. If water is used as the pressurizing medium, its compressibility must be accounted for; water is compressed by up to 15% of volume at pressures above 600 MPa. Once the desired pressure is reached, the pump or piston is stopped, the valves are closed and the pressure is maintained without further energy input. After the required hold time has elapsed, the system is depressurized, the vessel is opened and the product unloaded. The system is then reloaded with product, either by operators or machines, depending on the degree of automation possible [93].

The total time for pressurization, holding and depressurization is referred to as the 'cycle time.' The cycle time and the loading factor (i.e., the percentage of the vessel volume actually used for holding packaged product, primarily a factor of package shape) determine the throughput of the system. In a commercial situation and with a batch process, a short holding time under pressure is desirable in order to maximize throughput of product. If the product is able to be pumped, it may be advantageous to pump it into and out of the processing vessel through special high-pressure transfer valves and isolators. To package the product after treatment, additional systems, such as an aseptic filling station, are then required [93].

Current semi-continuous systems use a pressure vessel with a free piston to compress liquid foods. A low-pressure food pump is used to fill the pressure vessel and, as the vessel is filled, the free piston is displaced. When filled, the inlet port is closed and water at high-pressure process is introduced behind the free piston to compress the liquid food. After an appropriate holding time, releasing the pressure on the high-pressure process water decompresses the system. The treated liquid is discharged from the pressure vessel to a sterile hold tank through a discharge port. A low-pressure water pump is used to move the free piston towards the discharge port. The treated liquid food can be filled aseptically into pre-sterilized containers.

A semi-continuous system (Figure 7.3) with a processing capacity of 600 lph of liquid food and a maximum operating pressure of 400 MPa was used commercially to process grapefruit juice in Japan. Multiple units can be sequenced so that, while one unit is being filled, others are in various stages of operation [71]. For any HP system, the working pressure is a very important parameter, not only because the initial price of the

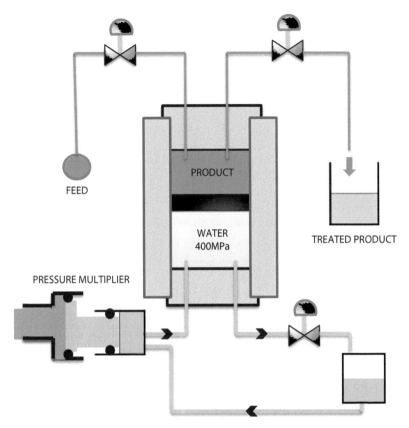

FIGURE 7.3   Semi-continuous high pressure processing system.

equipment increases significantly with its maximum working pressure, but also because a decrease in working pressure can reduce significantly the number of failures, increasing the working life of the equipment [70]. Pressures between 50 and 1000 MPa are commonly used. Keeping the sample under pressure for extended periods of time does not require any additional energy [16]. The work of compression during HP treatment will increase the temperature of foods through adiabatic heating, by approximately 3°C per 100 MPa, depending on the composition of the food [5]. For example, if the food contains a significant amount of fat, such as butter or cream, the temperature rise can be larger. Foods cool down to their original temperature on decompression if no heat is lost to, or gained through, the walls of the pressure vessel during the hold time at pressure.

7.5.1 EFFECTS OF PRESSURE-TEMPERATURE RELATIONSHIPS ON HPP

During the compression phase (t_1–t_2) of pressure treatment, food products experience a decrease in volume as a function of the pressure (Figure 7.4). Both pure water and most moist foods subjected to 600 MPa treatment at ambient temperature will experience about 15% reduction in volume. The product is held under pressure for a certain time (t_2–t_3) before decompression (t_3–t_4). Upon decompression, the product will usually expand back to its initial volume [28]. The compression and decompression can result in a transient temperature change in the product during treatment. The temperature of foods (T_1–T_2) increases as a result of physical compression (P_1–P_2). Product temperature (T_2–T_3) at process pressure (P_2–P_3) is independent of compression rate as long as heat exchange between the product and the surroundings is negligible.

In a perfectly insulated (adiabatic) system, the product will return to its initial temperature upon decompression (P_3–P_4). In practice, however, the product will return to a temperature (T_4) slightly lower than its initial temperature (T_1) as a result of heat loss during the compression (elevated temperature) phase. The rapid heating and cooling resulting from HPP treatment offer a unique way to increase the temperature of the product only during the treatment, and to cool it rapidly thereafter. The temperature increase of food materials under pressure is dependent on factors such as final pressure, product composition, and initial temperature. The temperature of water increases about 3°C for every 100 MPa

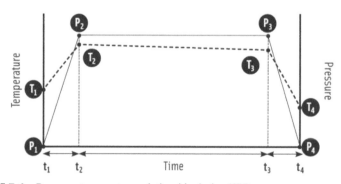

FIGURE 7.4   Pressure–temperature relationship during HPP.

TABLE 7.1 Rate of Temperature Increase (°C per 100 MPa) for Different Foods

Food at 25°C	Temperature increase per 100 MPa, (°C)
2% fat milk	3.0
Beef fat	6.3
Chicken fat	4.5
Mashed potato	3.0
Olive oil	From 8.7 to <6.3
Orange juice	3.0
Salmon	3.2
Soy oil	From 9.1 to <6.2
Tomato salsa	3.0
Water	3.0

pressure increase at room temperature (25°C). On the other hand, fats and oils have a heat of compression value of 8–9°C per 100 MPa, and proteins and carbohydrates have intermediate heat of compression values [72, 78]. Table 7.1 shows the rate of temperature increase (°C per 100 MPa) for different foods.

7.6 OVERVIEW OF TEMPERATURE INCREASE DURING COMPRESSION OF FOODS [92]

7.6.1 EFFECTS OF HIGH PRESSURE ON FOOD CONSTITUENTS

Some of the recent studies have highlighted the effect of HPP on macronutrients and micronutrients of food components. Generally, total protein and total lipid contents are not affected by HPP. Lipids tend to be of greatest interest terms of effect of HPP on the lipid oxidation. Total sugars, sucrose, glucose and fructose are not effect by HPP. Few studies are available that have investigated the effect of HPP. In cabbage, no effect of HPP (up to 500 MPa and 80°C) on total dietary fiber was evident; however soluble fibers were increased, while insoluble fibers decreased at 400 MPa [100]. HPP has shown to enhance glucose retardation/binding in tomato puree [13, 30] suggesting that this technique might be used to develop diabetic foods.

7.6.1.1   Water

Water is a major constituent in most foods and HPP markedly affects water. Water cannot be compressed at normal pressure, but it is partially compressible at high pressure. The water can be compressed upto 4% at 100 MPa and 15% at 600 MPa at 22°C. Foods with high water ratio will show similar trends of compressibility to water. Many physicochemical properties of water are reversibly modified under pressure. Compression of water causes increase in temperature of water @ 2–3°C per 100 MPa [9].

Pressure can increase the ionic product [H+ OH–] of water. It can increase from 10–100 folds with application of 100 MPa pressure. The positive and negative charges are separated under pressure by a driving force called electrostriction. Water molecules rearrange in more compact manner with smaller total volume around electric charges, due to H-bonding and dipole-dipole interactions. The reaction $H_2O = H^+ + OH^-$ causes volume decrease of 21.3 mL per mole at 25°C [45]. Thus, pH of water, weak acids and several buffers decrease by 0.2–0.5 pH units per 100 MPa.

7.6.1.2   Proteins

Proteins are large organic compounds made of amino acids arranged in a linear chain and joined together by peptide bonds between the carboxyl and amino groups of adjacent amino acid residues. High pressure causes denaturation of proteins depending on protein type, processing conditions and applied pressure level (Figure 7.5). Proteins can dissolve or precipitate on application of high pressure. These changes are reversible when pressure applied is in range of 100–300 MPa and irreversible when pressure level applied is higher than 300 MPa.

• The destruction of hydrophobic and ion pair bonds, and unfolding of molecules is called denaturation. At high pressure, oligomeric proteins tend to undergo proteolysis. Monomeric proteins do not show any vulnerability to proteolysis with increase in pressure [90].
• High pressure causes rupturing of non-covalent interactions within protein molecules and further cause's reformation of inter and intra

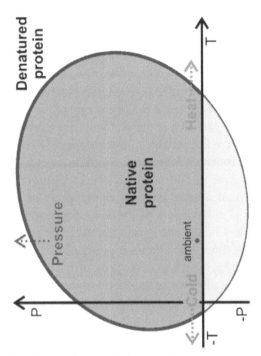

FIGURE 7.5   Elliptical phase diagram of denatured proteins.

molecular bonds. Different types of interactions are responsible for secondary, tertiary and quaternary structure of proteins.

- The quaternary structure is mainly held by hydrophobic interactions that are sensitive to pressure. The tertiary and secondary structures of proteins can be significantly modified at pressures above 200 MPa.
- Final changes in conformation after HPP denaturation can cause full or partial unfolding of polypeptide structure which eventually results in exposure of peptides that can enhance antioxidant activity [60].
- Denaturation is complex process involving intermediate forms leading to multiple denatured products. Secondary structure can even show irreversible denaturation at very high pressure above 700 MPa, leading to irreversible denaturation [6].
- When pressure increases to 100 MPa, temperature of denaturation of protein increases, whereas at higher pressure, temperature of denaturation usually decreases causing elliptical phase diagram of denatured proteins.

- At high pressure, proteins denature usually at room temperature than at higher temperatures.
- Pressure and temperature show antagonistic behavior at molecular level by following principle of microscopic ordering, which says that increase in pressure at constant temperature leads to an ordering of molecules or a decrease in the entropy of the system.
- High pressure has advantage for inducing protein denaturation as these effects are irreversible at pressure level <200 MPa and are not expected to reoccur. Temperature variations can lead to changes in both the volume and the thermal energy of protein but in contrast, at constant temperature under high pressure, the internal energy of the system is independent of pressure, and internal interactions are affected solely by the changes in the volumes of water structure and protein molecules. Denaturation is simply a two stated thermodynamic transition between two states of a protein. Interpretation of denaturation is difficult as the thermodynamic parameters are influenced by binding of the denatured molecules to multiple sites on a protein and this can change the binding of denaturant molecule to multiple sites on a protein and this binding changes the chemical potential of the protein.
- Denaturation induced by pressure means that the volume occupied by the compact folded native conformation is larger than that of unfolded part. Protein unfolding is characterized by a negative molar volume of denaturation. The size of the protein hydration shell increases by attraction of new water molecules by the newly exposed surface amino acid residues but this increase is more than that compensated by the negative contribution from the disruption of electrostatic and hydrophobic interactions and disappearance of voids in the protein not accessible to solvent molecules [63].

7.6.1.3 Enzymes

HPP is most commonly used to inactivate deleterious enzymes, thereby ensuring high quality characteristics of food to be maintained [80]. Effects of HPP on enzymes can be divided in two classes: In first class, we can take pressure which is used to activate some enzymes in food to improve food quality [43]. On other hand, undesirable enzymes in food can be inactivated using high pressure level. In regard to pressure inactivation, there are four distinguished groups of enzymes, based on recovery and loss of activity [62]:

- completely and irreversibly inactivated;
- completely and reversibly inactivated;
- incompletely and irreversibly inactivated;
- incompletely and reversibly inactivated.

Enzymes activity is influenced by pressure induced de-compartmentalization [33]. In an intact food tissue, enzymes and substrates are separated by compartments and pressure can induce membrane damage resulting in leakage of enzyme and substrate which can cause enzymes to contact the substrate. Pressure can cause enzyme inactivation, but there is a minimum level of pressure below which there is no action on enzyme. This pressure inactivation range is dependent on pH, medium, enzyme type, composition and temperature. This has been attributed to an enzyme portion that is irreversibly converted to the inactive form, while a fraction is converted to a pressure-resistant form. Upon pressure release, the pressure-resistant fraction reverts to the equilibrium state, while the irreversibly inactivated enzyme remains unchanged.

It has been found that high pressure enzyme inactivation can be improved by applying pressure cycles. Successive application of high pressure can result in higher inactivation of many enzymes (trypsin, chymotrypsinogen and pepsin). For trypsin and chymotrypsinogen, it has been reported that successive pressure treatments result in a higher degree of inactivation only when pressures above the minimum level required for their inactivation are applied [54].

7.6.1.4 Vitamins

Food vitamins are inevitably and irreversibly damaged during thermal processing and vitamin contents in foodstuffs are closely interrelated to their nutritional quality [67]. Vitamins are highly sensitive to thermal treatments and other similar technologies which can cause great loss of vitamins by leaching. HPP is a cold treatment as it does not involve thermal cooking and is viable for processing of foods containing high amount of vitamins. Vitamins should be determined depending on fact that the amount present in food should be enough to complete nutritional requirement [98].

The effect of HPP on vitamins is mainly focused on ascorbic acid (vitamin C) content; although thiamine, riboflavin and pyridoxal have also been investigated. The majority of these studies have been on fruits and vegetables. Generally, all water soluble vitamins are stable at HPP, with folate showing higher sensitivity [41, 68, 84, 85]. Ascorbic acid has shown to be stable under HPP, unless it is subjected to high pressure and high temperature (>65°C) conditions, where oxidation reactions are enhanced [68]. In addition to its role as antioxidant, it also protects folate against pressure and heat [68].

The HPP does not affect the vitamin content. For example in strawberry nectar, ascorbic acid was decreased only from 1129 to 1100 ppm after HPP treatment [7]. Strawberry coulis puree has high content of vitamin C and HP Treatment (400 MPa/30 min/20°C) caused 88.7% retention of total content. On other hand, vitamin C content in coulis only had 67% retention after thermal treatment (120°C/20 min/0.1 MPa).

An HP treatment does not cause any significant loss of vitamins B6 and B1. Butz and Tauscher [12] found that vitamin B1 concentration was 1.475 and 1.468 µg/mL, respectively in untreated and in treated samples at 600 MPa/30 min/20°C. HPP caused increase in vitamin concentration from 3.725 to 3.794 µg/mL in a model system after pressurization at 600 MPa/30 min/20°C.

In egg yolk, HP treatment caused insignificant effect on vitamin C concentrations even with treatments ranging from 400–1000 MPa, but vitamin C concentration tended to decrease with increasing boiling time. It was found that there was small increase in vitamin C levels in egg yolk and egg yolk ascorbate after pressurization. HP treatment caused extractive effects in foods when they were pressurized at 200, 300 and 400 MPa/20°C/30 min, it caused ascorbate retention of 92.6, 101.34 and 102.6%, respectively. Similar observations were made for thiamine retention in egg yolk as increase of 6.2 and 2.8% in thiamine retention after HP treatment of 600 and 800 MPa, respectively for 30 min at 20°C was found [37].

7.6.2 EFFECT OF HIGH PRESSURE ON MICROORGANISMS

High-pressure treatments are effective in inactivating most vegetative pathogenic and spoilage microorganisms at pressures above 200 MPa at chilled or process temperatures less than 45°C, but the rate of inactivation is strongly influenced by the peak pressure. Commercially, higher pressures are preferred as a means of accelerating the inactivation process, and current practice is to operate at 600 MPa, except for those products where protein denaturation needs to be avoided. A pressure treatment of 600–700 MPa readily kills vegetative cells of bacteria, yeasts and mold; while bacterial spores are more resistant [36, 55]. However, the best results of microbial inactivation can be achieved by the result of a combination of factors.

The cell membrane represents probably one of the major targets for pressure-induced inactivation of microorganisms. The membrane destruction causes the cells to collapse [58]. In addition, HP causes changes in cell morphology and biochemical reactions, protein denaturation and inhibition of genetic mechanisms.

7.6.2.1 Bacteria

The main bacteria that cause food poisoning are *Campylobacter* spp., *Salmonella* spp., *Listeria monocytogenes, Staphylococcus aureus, Escherichia coli* and *Vibrio* spp. Among these, *Listeria monocytogenes* and *Staphylococcus aureus* are probably most intensively studied species in terms of use of HP processing. Generally, Gram-positive bacterias are more resistant to heat and pressure than gram-negative bacteria, and cocci are more resistant than rod-shaped bacteria [27, 88]. Furthermore, it has been suggested that the complexity of the gram-negative cell membrane could be attributable to its HPP susceptibility [64].

The effect of HP treatment on the Gram-positive *Listeria monocytogenes* strain and the Gram-negative *Salmonella typhimurium* strain was determined in stationary phase cell suspensions. Pressure treatments were done at room temperature for 10 min. Increasing pressure resulted in an

TABLE 7.2   Pressure Required to Achieve a 5-log Cycle Inactivation Ratio for Certain Micro-Organisms, for 15 min Treatment [74]

Microorganism	Pressure, MPa
Escherichia coli O157:H7	680
Listeria monocytogenes	375
Salmonella enteritidis	450
Salmonella typhimurium	350
Staphylococcus aureus	700
Yersinia enterocolitica	275

exponential decrease of cell counts. *Salmonella typhimurium* suspended at low pH was more sensitive to pressure treatments [91].

Treating food samples using HP can destroy both pathogenic and spoilage microorganisms. However, there is a large variation in the pressure resistance of different bacterial strains and the nature of the medium can even affect the response of microorganisms to pressure (Table 7.2).

The growth phase of bacteria also plays a role in determining their pressure sensitivity or resistance. Cells in the stationary phase of growth are generally more pressure resistant than those in exponential phase [58], due to the synthesis of proteins that protect against a range of adverse conditions, such as high salt concentrations, elevated temperatures and oxidative stress [38]. It has also been shown that the highest resistance of stationary phase cells to HPP is partly due to the presence of the RpoS protein in *E .coli* and sigB in *Listeria monocytogenes* [81, 101], resulting in an increased bacterial stress response.

7.6.2.2   Bacterial Spores

The mechanism of inactivation of bacterial spores through high pressure has been suggested to have two steps: (a) high pressure will first induce spore germination, and (b) then inactivate the germinated spores. The step of germination is a crucial step, where the spore is monitoring its environment, when conditions are favorable for its growth; it germinates and goes through outgrowth, thus being converted back into a growing cell. During this process, there is a loss of resistance, and is therefore of great interest for a variety of sterilization techniques.

The elimination of bacterial endospores from food probably represents the greatest food processing and food safety challenges to the industry. It is well established that spores are most pressure-resistant life forms. In general, only very high pressures (800 MPa) can kill bacterial spores at ambient temperatures. Alternatively, other processing methods, applied in combination with HP, can be effective for elimination of bacterial spores, by achieving a synergistic or hurdle effect. In particular, HP treatment at elevated temperatures (e.g., HP treatment at up to 90°C) is very effective in the elimination of bacterial spores in foods. Pressure induced inactivation of bacterial spores is also markedly enhanced at temperatures of 50–70°C and perhaps also at or below 0°C.

The most heat-resistant pathogenic bacterium are *Clostridium botulinum* and spores of *C. botulinum*. Also *Bacillus cereus* has been widely studied because of its anaerobic nature and very low rate of lethality. It is recognized as a leading cause of bacterial food poisoning, with a variety of proteinaceous and starchy foods being implicated [97]. Spores of Clostridium spp. tend to be more pressure resistant than those of Bacillus spp. [73].

An alternative to using treatments combining heat and pressure for enhanced killing of bacterial spores is first to cause bacterial spore germination and then use HP to kill the much more pressure-sensitive vegetative cells. The effect of HHP on inactivation of *A. acidoterrestris* (spore forming spoilage bacterium) in orange, apple, and tomato juices showed that the effectiveness of treatment was increased with increase in pressure and temperature. At room temperature, there was no significant reduction of spores in all juice samples, which indicates that in order to use high pressure for spore inactivation; alternate treatment with mild heat is required [50].

7.6.2.3 Yeast and Molds

Many of the yeasts and molds of concern in the food industry are the spoilage agents of fruits beverages, jams, jellies, and vegetable preserves. In comparison, yeasts and molds are relatively HPP sensitive. However, ascospores of heat-resistant molds such as *Byssochlamys, Neosartorya* and *Talaromyces* are generally considered to be extremely HPP resistant [15, 88].

7.6.2.4 Viruses

The first virus to be pressure treated was the plant pathogen, tobacco mosaic virus. Later studies were conducted on HPP inactivation of animal and human viruses. The food industry is mainly concerned with food born viruses such as hepatitis A virus (HAV). Studies on viral inactivation by HPP revealed that some viruses are extremely pressure-resistant. For example, poliovirus is only inactivated by less than one log when subjected to 600 MPa for 1 hour [102]. Other viruses have been found to be very sensitive, such as *feline calicivirus* (FCV), which is completely inactivated by pressures as low as 275 MPa for 5 minutes [46].

The mechanisms of viral inactivation by HPP involve the dissociation and denaturation of the proteins of the virus's coat [87]; or in the case of enveloped viruses, damage to the envelope rather than the damage to viral nucleic acids. The pressure induced changes to the coat can be subtle alterations to capsid proteins or receptor recognition proteins, which can to lead loss of infectivity [26].

7.7   STERILIZATION BY COMBINED HIGH PRESSURE AND THERMAL ENERGY (HPHT)

High pressure high temperature (HPHT) processing or pressure-assisted thermal processing (PATP) involves the use of moderate initial chamber temperatures between 60°C and 90°C and then application of pressure, which causes internal compression heating at pressures of 600 MPa or greater, where in-process temperatures can reach upto 90°C to 130°C. The process has been proposed as a high-temperature short-time process, where both pressure and compression heat contribute to the process's lethality [51]. In this case, compression heat developed through pressurization allows instantaneous and volumetric temperature increase, which in combination with high pressure accelerates spore inactivation in low-acid media. For instance, pressurization temperatures of 90°C–116°C combined with pressures of 500–700 MPa have been used to inactivate a number of strains of Clostridium botulinum spores [28, 56].

Researchers have shown that certain bacterial endospores (*C. sporogenes, Bacillus stearothermophilus, B. licheniformis, B. cereus,* and *B. subtilis*) in selected matrices like phosphate buffer, beef, vegetable cream, and tomato puree [5, 32, 48, 61, 79, 82] can be eliminated after short-time exposure to temperatures and pressures above 100°C and 700 MPa, respectively. Some of these microbial spore inactivation approaches proposed combining [51, 72] of:

- Two low pressure pulses at 200–400 MPa (the first one for spore germination and the second for germinated cell inactivation);
- A low pressure pulse at 200 to 400 MPa for spore germination followed by a thermal treatment at 70°C for 30 min for vegetative cell inactivation;
- Package preheating above 75°C and pressurization at 620 to 900 MPa for 1 to 20 min; and

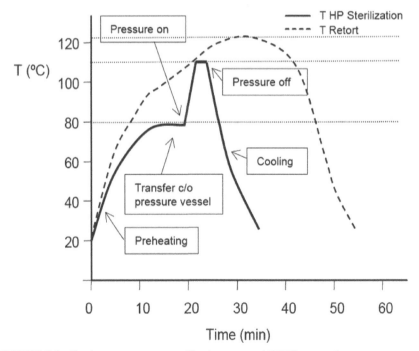

FIGURE 7.6   Product temperature profiles in retort and HPHT processing.

- Package preheating above 70°C and applying two or more pulses at 400 to 900 MPa for 1 to 20 min.

Three of the above-mentioned approaches have proven inconvenient from either a microbiological or an economical perspective. When applying low pressures between 200 and 400 MPa, combined with moderate temperature [cases (a) and (b)], residual dormant spores were detected after treatment [51, 97] making this option unlikely for a commercial process. Moreover, a high pressure multiple pulse approach [case (d)] is not recommended, as additional cycles decrease equipment lifetime and increase maintenance costs [22]. Hence, application of a single pulse above 600 MPa for 5 min or less [case (c)], combined with initial temperatures above 60°C, would be more cost-effective and a safer approach for industrial purposes [21].

7.7.1   ADVANTAGES OF HPHT PROCESSING

The main advantage of HPHT treatment is its shorter processing time (Figure 7.6) compared to conventional thermal processing in eliminating spore-forming microorganisms [57].

This shorter process time and ultimate pressurization temperatures lower than 121°C have resulted in higher quality and nutrient retention in selected products. For example, better retention of flavor components in fresh basil, firmness in green beans, and color in carrots, spinach and tomato puree have been found after HPHT processing [48, 49].

Nutrients such as vitamins C and A have also shown higher retention after HPHT processing in comparison to retort methods [57]. One more benefit of HPHT processing is its use to process non-pumpable foodstuffs like soups containing solid ingredients such as: noodles, barley, and/or cut-up vegetables and meat [21]. The high pressure low temperature processing provides direct product scale-up and higher efficiency for larger volumes of food, compared to thermal processing, due to "instant" hydrostatic pressure transmission.

Similarly, HPHT processing is suitable for larger package sizes, as compression heating to high temperatures is instantly achieved throughout the entire package volume. Nevertheless, the preheating step, or the period of time necessary to reach initial product temperature before pressurization, needs to be considered when evaluating overall processing time. A long preheating time, especially in a large container, may lower product quality retention at the end of the HPHT process.

Although the HPHT process can be seen as advantageous due to its shorter time, yet lower processing temperatures cannot yet be assured for *C. botulinum* in activation until optimal temperature/pressure/time combinations are identified.

7.7.2 HPHT EQUIPMENT

The HPHT process consists of: (a) preheating food packages in a carrier outside the vessel, (b) transferring the preheated carrier into the vessel and

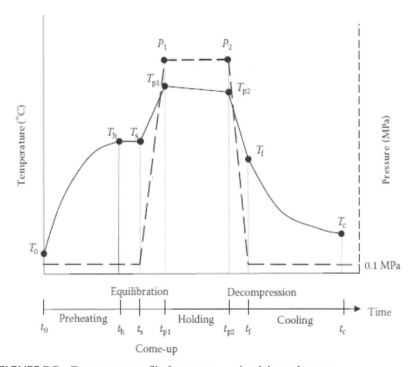

FIGURE 7.7 Temperature profile for pressure-assisted thermal process.

equilibrating up to an initial temperature, (c) pressurizing and holding at a target pressure, (d) releasing pressure, (e) removing carrier from vessel, and (f) cooling down products in the carrier and removing the products. Therefore, the temperature history inside an HPHT-processed food is determined by six main process time intervals [8] (Figure 7.7): (a) product preheating to a target temperature T_h, (b) product equilibration to initial temperature T_s, (c) product temperature increase to T_{p1} due to compression heating, (d) product cooling down to T_{p2} due to heat removal through the chamber, (e) product temperature decrease to T_f during decompression, and (f) product cooling to T_c.

Figure 7.7 shows a typical temperature profile during HPHT treatment, indicating the cooling down experienced in the holding process. Loss of heat is reflected in the difference between initial and final temperature during holding time ($T_{p2} < T_{p1}$), and temperature at the beginning and end of the pressurization–depressurization period ($T_f < T_s$). To achieve sterilization during pressurization, all parts of the treated food must at least reach T_{p1} at pressure P_1 (Figure 7.7), which is the maximum temperature targeted for the maximum pressure holding time. To achieve this goal, a number of variables must be controlled from the start [4]. Thus, understanding the effects of combined temperature and pressure on microbial inactivation distribution will require knowing the temperature within the pressure vessel at specific locations, and at all times during a high-pressure process [14].

In HPHT processing (or pressures over 400 MPa), pressure vessels can be built with two or more concentric cylinders of high tensile strength steel. The outer cylinders compress the inner cylinders so that the wall of the pressure chamber is always under some residual compression at the design operating pressure. In some designs, cylinders and frame are prestressed by winding layer upon layer of wire under tension. The tension in the wire compresses the vessel cylinder so that the diameter is reduced [40]. This special arrangement allows an equipment lifetime of over 100,000 cycles at pressures of at least 680 MPa. The preferred practice is to design high-pressure chambers with stainless steel food-contacting parts so that filtered (potable) water can be used as the isostatic compression fluid [28].

During pressurization at high temperature conditions, a temperature increase is produced in both the compression fluid and food [92].

However, since compression heating in the system steel vessel is almost zero [22, 92], there is heat loss toward the chamber wall. In theory, heat generated by compression is dissipated by a combination of conduction and convection within the pressurizing fluid in the chamber and transfer of heat across the chamber wall into the surroundings [14]. Heat dissipation may cause cooling down of the sample during both come up and holding time, which may there by decrease spore inactivation effectiveness [2, 23]. Thus, it is important to avoid heat loss through the chamber system. Modern systems are required to use several features for heat loss prevention by mainly: (a) adapting a dense polymeric insulating liner with a free moving piston at the bottom or valve to allow adequate pressure transmission; (b) preheating the inflowing pressurization fluid and pipes; and (c) preheating the vessel at a temperature higher than the initial fluid/sample temperature. Successful installation of these features can make the system close to adiabatic. In this way, preservation efficacy at chosen HPHT conditions is maximized.

7.8   APPLICATION OF HPP IN MILK AND MILK PRODUCTS

7.8.1   MILK

The various studies on the inactivation of pathogenic and spoilage microorganisms (naturally present or inoculated) by HPP have been performed in milk and have generally demonstrated that it is possible to obtain 'raw' milk pressurized at 400–600 MPa with a microbiological quality comparable to that of pasteurized milk (72°C, 15 s) [94], depending on the microbiological quality of milk [47, 65] but not sterilized milk due to HP resistant spores. For example, to achieve a shelf life of 10 days at a storage temperature of 10°C, a pressure treatment of 400 MPa for 15 min or 600 MPa for 3 min at 20°C is necessary [76].

In addition to the inhibition and destruction of microorganisms, HP influences the physico-chemical and technological properties of milk. When milk is subjected to HP, the casein micelles are disintegrated into smaller particles [86]. This disintegration is accompanied by an increase of caseins and calcium phosphate levels in the diffusible or serum phase of milk and by a decrease in the both non-casein nitrogen and serum nitrogen fractions, suggesting that the whey proteins become ready to sediment by centrifugation and perceptible at pH 4.6 [42].

The β-blactoglobulin is the most easily denatured serum protein by pressure treatment up to 500 MPa at 25°C, whereas denaturation of the immunoglobulins and ±-lactalbumin only occurs at highest pressures at 50°C. An application derived from this observation is the preservation of colostrum immunoglobulins as alternative to heat treatment, which induces immunoglobulins damage [29].

Studies carried out on free fatty acids (FFA) content (lipolysis of milk fat) in ewe's milk have shown that HP treatments between 100–500 MPa at 4, 25 and 50°C did not increase FFA content, even some treatments at 50°C showed lower FFA content than fresh raw milk [31]. This phenomenon is of great interest to avoid off flavors derived from lipolytic rancidity in milk.

7.8.2   CHEESE MANUFACTURING

The milk pasteurization destroys pathogenic but completely the spoilage microorganisms. It is the most important heat treatment applied to cheese milk to provide acceptable safety and quality. However, milk pasteurization adversely affects the development of many sensory characteristics of cheese, leading to alterations in texture and often delayed maturation [34]. HP technology can be used to increase the microbiological safety and quality of milk to produce high quality cheeses. The HP processing of milk at room temperature causes several protein modifications, such as whey protein denaturation and micelle fragmentation, and alters mineral equilibrium. These changes modify the technological aptitude of milk to make cheese, improving the rennet coagulation and yield properties of cheese milk [95].

7.8.3   ACCELERATION OF CHEESE RIPENING

Techniques accelerating the ripening process without affecting the quality of product provide a significant cost savings to cheese manufactures

[18]. The first attempt of the use of HP to accelerate cheese ripening was first shown in a patent [103]. Experimental cheddar cheese samples were exposed to pressure from 0.1 to 300 MPa for 3 days at 25°C after cheese making. Best results were obtained at 50 MPa, at which pressure a cheese with free amino acid amount and taste comparable to that of a 6-month-old commercial cheese was obtained [103].

HP treatments are able to accelerate cheese ripening by altering the enzyme structure, conforming changes in the casein matrix making it more prone to the action of proteases and bacterial lysis promoting the release of microbial enzymes that promote biochemical reactions [99]. HP treatments also increase pH (0.1 to 0.7 units) and modify water distribution of certain cheese varieties, promoting conditions for enzymatic activity.

7.8.4   RENNET COAGULATION TIME

Rennet coagulation time (RCT) is the time at which the milk coagulum becomes firm enough for cutting after rennet addition. RCT has been reported to reduce markedly at pressure exposure at of 200 MPa. The decreased RCT is related to a reduction in casein micelle size, leading to increased specific surface area and increased probability of inter-particle collision [3, 66]. However at high pressure (400 MPa), RCT again increases as denatured whey proteins are incorporated into the gel and their presences interferes with secondary aggregation phase; thereby, reduce the overall rate of coagulation.

7.8.5   INCREASED CHEESE YIELD

The maximum total solid recovered from the milk during cheese manufacturing is known as the yield of the cheese. Higher the recovered percentage of solids, the greater is the amount of cheese obtained, which has a positive effect in economic terms. During manufacturing of cheddar cheese [24] and semi-hard goat milk cheese [95], higher yields were reported when made from HP-treated milk. The yield of cheddar cheese made from HP treated milk (3 cycles of 1 min at 586 MPa) was 7% higher than that from raw or pasteurized milk. The treatment of milk at 200 MPa had no effect on wet curd yield [3], although denaturation of â-lactoglobulin was observed at 200 MPa; whereas at 300–400 MPa wet curd yield was significantly increased by upto 20% and reduced loss of protein in whey and the volume of whey.

Increased treatment time, up to 60 min at 400 MPa, increased wet curd yield and reduced protein loss in whey. The changes were greatest during the first 20 min of treatment. The increased cheese yield is primarily due to greater moisture retention, secondly due to incorporation of some denatured α-lactoglobulin. Additionally, the casein micelles and fat globules in HP-treated milk may not aggregate as closely as in untreated milk, therefore, allowing more moisture to be entrapped in the cheese [53].

7.8.6   YOGHURT

Firmness of yogurt made from high pressure treated milk has shown to increase with increasing pressure, due to disruption of casein micelles, resulting in a greater effective area for surface interaction. Yoghurt prepared from low fat milk and exposed to 300 MPa/10 min prevents after-acidification (developed acidity after packing) and significantly improved the shelf life. Acidification of yogurt milk with glucono-δ-lactone (GDL) at 200 MPa for 20 min resulted in fine coagulum, homogeneous gel than that of heat treated milk [35, 89].

7.8.7   CREAM, BUTTER, AND ICE CREAM

There are not enough studies carried out on the effects of HPP on cream, butter and ice cream. Pasteurized dairy cream samples (35 and 43% fat) were subjected to 100 to 500 MPa at 23.8°C for 1–15 min [11]. Using the freeze fracture technique and transmission electron microscopy, it was found that pressurization induced fat crystallization within the small emulsion droplets, mainly at the globule periphery. Fat crystallization was increased with the length of pressure treatment and was maximal after processing at 300–500 MPa. Moreover, the crystallization proceeded during further storage at 23°C after pressure release.

The two potential applications of this phenomenon are fast aging of ice cream mix and physical ripening of dairy cream for butter making. Whipping properties were improved when cream was treated at 600 MPa for 2 min [25], probably due to better crystallization of milk fat. When the treatment conditions exceed the optimum, an excessive denaturation of whey protein occurs and results in longer whipping time and destabilization of whipped cream. At <400 MPa, no noticeable effects on whipping properties of cream were found.

The HP processing at 450 MPa and 25°C for 10 to 30 minutes has showed a significant reduction in microbial load of a dairy cream (35%). Inactivation followed apparent first order kinetics, with a decimal reduction time of 7.4 min under the pressure treatment conditions [77].

7.9   CONCLUSIONS

The HPP technology has promised to meet the consumer demand for minimally processed foods, with high nutritional and sensory qualities, which increase self life. From a nutritional perspective, HPP is an excellent food processing technology which has the potential to retain compounds with health properties. Macro nutrients and most of micro nutrients do not appear to be affected by HPP, as in case of thermal treatment. The commercial production of pressurized food has become a reality in Japan, USA, and Spain and many more countries. This is the result of extensive scientific research, technological and technical advances in HPP equipment production and decrease in processing cost.

The range of commercially available HP processed products is relatively small, and the HPP technology has not been commercialized in many developing countries. The main drawback being the novelty of product, high equipment cost and the solid food which cannot be processed in continuous process but rather require batch or semi continuous equipment for its processing. HPP application can inactivate microorganisms and enzymes and modify structure with little or no effect on the nutritional and sensory quality of foods. The combined high pressure and high temperature technology can be claimed advantageous for its shorter processing time. The HPP has a more versatile application in food industries, and provide a unique point of difference for producers. HPP is a paradigm-shifting technology for the food industry that is on-trend with consumer interests. Its use will likely grow as cost declines and food manufacturers identify new applications where HPP can deliver product quality improvements that consumers appreciate and will pay for it.

7.10 SUMMARY

In recent years, consumers have moved on and have changed the selection criteria for food products; they have stressed more on quality and safety of food products. Conventional food sterilization and preservation methods often result in number of undesirable changes in food both in terms of chemical and nutritional value. However, high hydrostatic pressure (non-thermal food processing method) is one such technology that has the potential to fulfill both consumer and scientific requirements. The use of HHP has diverse applications in nonfood industries, but the extensive investigations have revealed the potential benefits of HHP as an alternative to heat treatments. The benefits are diverse in various areas of food processing such as: inactivation of microorganism's and enzymes, denaturation and alteration of functionality of proteins and structural changes to food materials.

In recent years, consumers have stressed more on quality and safety of food products. Conventional food sterilization and preservation methods often result in number of undesirable changes in food both in terms of chemical and nutritional value. However, high hydrostatic pressure can fulfill both consumer and scientific requirements. The benefits are diverse in various areas of food processing such as: inactivation of microorganism's and enzymes, denaturation and alteration of functionality of proteins and structural changes to food materials.

KEYWORDS

- cell morphology
- compression
- decompression
- elevated temperature
- enzyme inactivation
- high pressure processing
- high pressure vessel
- holding time
- isostatic
- microbial inactivation
- microbiological shelf life
- minimally processed foods
- native protein
- non-thermal sterilization
- olgomeric proteins
- pressure transmitting medium
- pressurization
- protein
- shelf life
- spore forming bacteria
- sterilization
- transient temperature
- vegetative cells

REFERENCES

1. Alzamora, S. M., Lopez-Malo, A. Y., & Tapia de Daza, M. S. (2000). Overview. In: *Minimally Processed Fruits and Vegetables-Fundamental Aspects and Applications*, edited by Alzamora, S. M., Tapia, M. S., & López-Malo, A. Gaithersburg, MD, USA: Aspen Publishers, Inc.

2. Ardia, A., Knorr, D., & Heinz, V. (2004). Adiabatic heat modeling for pressure build-up during high-pressure treatment in liquid-food processing. *Food and Bioproducts Processing, 82*, 89–95.

3. Arias, M., Lopez-Fandino, R., & Olano, A. (2000). Influence of pH on effect of high pressure on Milk. *Milchwissenschaft, 55*(4), 191–194.

4. Balasubramaniam, V., Ting, E., Stewart, C., & Robbins, J. (2004). Recommended laboratory practices for conducting high-pressure microbial inactivation experiments. *Innovative Food Science and Emerging Technologies. 5*, 299–306.

5. Balasubramanian, S., & Balasubramaniam, V. M. (2003). Compression heating influence of pressure transmitting fluids on bacteria inactivation during high pressure processing. *Food Research International, 36*(7), 661–668.

6. Balny, C., & Masson, P. (1993). Effects of high pressure on proteins. *Food Reviews International, 9*(4), 611–628.

7. Balny, C., Hayashi, R., Heremans, K., & Masson, P. (1992). High pressure and biotechnology. John Libbey Eurotext Montrouge, France.

8. Barbosa-Canovas, G. V., & Juliano, P. (2008). Food sterilization by combining high pressure and heat. In: *Food Engineering: Integrated Approaches,* edited by Gutierrez-Lopez, G. F., Barbosa-Canovas, G., Welti-Chanesand, J., & Paradas-Arias, E., New York: Springer. pp. 9–46.

9. Bridgman, P. W. (1912). Water, in the liquid and five solid forms, under pressure. JSTOR, *47*, 441–558.

10. Bridgman, P. W. (1914). The coagulation of albumen by pressure. *Journal of Biological Chemistry, 19*(4), 511–512.

11. Buchheim, W., & Abou El Nour, A. M. (1992). Induction of milk fat crystallization in the emulsified state by high hydrostatic pressure. *Fett Wissenschaft Technologie, 94*(10), 369–373.

12. Butz, P., & Tauscher, B. (1997). High pressure treatment of fruits and vegetables: problems and limitations. Coronet Books Inc., p. 435.

13. Butz, P., Endenharder, R., Fernandez Garcia, A., Fister, H., Merkel, C., & Tauscher, B. (2002). Changes in functional properties of vegetables induced by high pressure treatment. *Food Research International, 35*, 295–300.

14. Carroll, T., Chen, P., & Fletcher, A. 2003. A method to characterize heat transfer during high-pressure processing. *Journal of Food Engineering, 60*, 131–135.

15. Chapman, B., Winley, E., Fong, A. S. W., Hocking, A. D., Stewart, C. M.and Buckle, K. A. (2007). Ascospore inactivation and germination by high pressure processing is affected by ascospore age. *Innovative Food Science and Emerging Technologies, 8*, 531–534.

16. Cheftel, J. C., & Culioli, J. (1997). Effects of high pressure on meat: A review. *Meat Science, 46*(3), 211–236.

17. Cheftel, J. C. (1995). Review: High-pressure, microbial inactivation and food preservation. *Food Science and Technology International, 1*(2–3), 75–90.

18. Chopde, S. S., Deshmukh, A. M., Kalyankar, D. S., & Changade, P. S. (2014). High pressure technology for cheese processing: a review. *Asian Journal of Dairy and Food Research, 33*(4), 239–245.

19. Crossland, B. (1995). High pressure processing of vegetables. In: *High Pressure Processing of Foods,* edited by Ledward, D. A., Johnston, D. E., Earnshaw, R. G., & Hasting, A. P. M., Nottingham University Press, Nottingham, p. 7.

20. Cruss, W. V. (1924). *Commercial Fruit and Vegetable Products*. New York: McGraw-Hill Book Co. Inc.
21. De Heij, W. B. C., van den Berg, R. W., van Schepdael, L., Hoogland, H., & Bijmolt, H., (2005). Sterilization – only better. *New Food, 8*(2), 56–61.
22. De Heij, W. B. C., van Schepdael, L. J. M. M., Moezelaar, R., Hoogland, H., Matser, A. M., & van den Berg, R. W. (2003). High pressure sterilization: maximizing the benefits of adiabatic heating. *Food Technology, 57*(3), 37–42.
23. De Heij, W. B. C., Van Schepdael, L. J. M. M., van den Berg, R. W., & Bartels, P. V. (2002). Increasing preservation efficiency and product quality through control of temperature distributions in high pressure applications. *High Pressure Research, 22*, 653–657.
24. Drake, M. A., Harrison, S. L., Asplund, M., Barbosa Canovas, G., & Swanson, B. G. (1997). High pressure treatment of milk and effects on microbiological and sensory quality of Cheddar cheese. *Journal of Food Science, 62*(4), 843–845.
25. Eberhard, P., Strahm, W., & Eyer, H. (1999). High pressure treatment of whipped cream. *Agrarforschung, 6*(9), 352–354.
26. Elaine, P. B., Cynthia, M. S., & Dallas, G. H. (2011). Non-thermal processing technology for food. Chapter 5, In: *Microbiological Aspects of High Pressure Food processing*. Willey-Blackwell and Institute of Food Technologies, pp. 51–70.
27. Erkmen, O., & Karatas, S (1997). Effect of high hydrostatic pressure on *Staphylococcus aureus* in milk. *Journal of Food Engineering, 33*, 257–262.
28. Farkas, D. F., & Hoover, D. G. (2000). High pressure processing: Kinetics of microbial inactivation for alternative food processing technologies. *Journal of Food Science Supplement, 65*, 47–64.
29. Felipe, X., Capellas, M., & Law, A. R. (1997). Comparison of the effects of high pressure treatments and heat pasteurization on the whey proteins in goat's milk. *Journal of Agricultural and Food Chemistry, 45*(3), 627–631.
30. Fernandaz Garacia, A., Butz, P., & Tauscher, B. (2001). Effects of high pressure processing on carotenoid extractability, antioxidant activity, glucose diffusion, and water binding of tomato puree (*Lycopersiconesculentum Mill*). *Journal of Food Science, 66*, 1033–1038.
31. Gervilla, R., Ferragut, V., & Guamis, B. (2001). High hydrostatic pressure effects on color and milk-fat globule of ewe's milk. *Journal of Food Science, 66*(6), 880–885.
32. Gola, S., Foman, C., Carpi, G., Maggi, A., Cassara, A., & Rovere, P. (1996). Inactivation of bacterial spores in phosphate buffer and in vegetable cream treated with high pressures. In: *High Pressure Bioscience and Biotechnology, Vol. 13 (Progress in Biotechnology)*, edited by Hayashi, R., & Balny, C. Amsterdam: Elsevier, pp. 253–259.
33. Gomes, M., & Ledward, D. (1996). Effect of high-pressure treatment on the activity of some polyphenoloxidases. *Food Chemistry, 56*(1), 1–5.
34. Grappin, R., & Beuvier, E. (1997). Possible implications of milk pasteurization on the manufacture and sensory quality of ripened cheese: a review. *International Dairy Journal, 7*(12), 751–761.
35. Harate, F., Luedecke, L., Swanson, B., & Barbosa-Canvas, G. V. (2003). Low fat set yogurt made from milk subjected to combinations of high pressure and thermal processing. *Journal of Dairy Science, 86*, 1074–1082.

36. Hayakawa, I., Kanno, T., Yoshiyama, K., & Fujio, Y. (1994). Oscillatory compared with continuous high pressure sterilization on *Bacillus stearothermophilus* spores. *Journal of Food Science, 59*, 164–167.

37. Hayashi, R., Kawamura, Y., Nakasa, T., & Okinaka, O. (1989). Application of high pressure to food processing: pressurization of egg white and yolk, and properties of gels formed. *Agricultural and Biological Chemistry, 53*(11), 2935–2939.

38. Hill, C., Cotter, P. D., Sleator, R. D., & Gahan, C. G. M. (2002). Bacterial stress response in Listeria monocytogenes: jumping the hurdles imposed by minimal processing. *International Dairy Journal, 12*, 273–283.

39. Hite, B. (1899). The effect of pressure in the preservation of milk. *West Virginia Agricultural Experimental Station Bulletin, 58*, 15–35.

40. Hjelmqwist, J. (2005). Commercial high-pressure equipment. In: *Novel Food Processing Technologies,* edited by Barbosa-Canovas, G. V., Tapia, M. S., & Cano, M. P., Boca Raton, FL: CRC Press, pp. 361–373.

41. Indrawati, V., & Hendrickx, M. (2002). High pressure processing. In: *The Nutrition Handbook for Food Processors,* edited by Henry, C. J. K., & Chapman, C. Cambridge, England: Woodhead Publishing, pp. 433–461.

42. Johnston, D. E., Austin, B. A., & Murphy, R. J. (1992). Effects of high hydrostatic pressure on milk. *Milchwissenschaft, 47*(12), 760–763.

43. Jolibert, F., Tonello, C., Sagegh, P., & Raymond, J. (1994). The effects of high pressure on the polyphenol oxidase in fruits. *Bios Boissons., 25*(251), 27–37.

44. Jong, D. P. (2008). *Advanced Dairy Science and Technology.* Blackwell publishing Ltd., Oxford OX4–2DQ, UK.

45. Kauzmann, W., Bodanszky, A., & Rasper, J. (1962). Volume changes in protein reactions, II: Comparison of ionization reactions in proteins and small molecules. *Journal of the American Chemical Society, 84*(10), 1777–1788.

46. Kingsley, D. H., Chen, H., & Hoover, D. G. (2004). Inactivation of selected picornaviruses by high hydrostatic pressure. *Virus research, 102*, 51–56.

47. Kolakowski, P., Reps, A., Dajnowiec, F., Szczepek, J., & Porowski, S. (1997). Effect of high pressures on the microflora of raw cow's milk. In: *Proceedings of Process Optimization and Minimal Processing of Foods,* edited by Oliveira, J. C. & Oliveira, F. A. R. Leuven, Belgium: Leuven University Press, pp. 46–50.

48. Krebbers, B., Matser, A., Hoogerwerf, S., Moezelaar, R., Tomassen, M., & Berg, R. (2003). Combined high-pressure and thermal treatments for processing of tomato puree: evaluation of microbial inactivation and quality parameters. *Innovative Food Science and Emerging Technologies, 4*, 377–385.

49. Krebbers, B., Matser, A. M., Koets, M. and van den Berg, R. W. (2002). Quality and storage-stability of high-pressure preserved green beans. *Journal of Food Engineering, 54*(1), 27–33.

50. Kumar, R., Bawa, A. S., Kathiravan, T., & Nadanasabapathi, S. (2013). Inactivation of *Alicyclobacillusacidoterrestris* by non-thermal processing technologies: a review. *International Journal of Advanced Research, 1*(8), 386–395.

51. Leadley, C. (2005). High pressure sterilization: a review. Edited by Campden and Chorleywood. *Food Research Association, 47*, 1–42.

52. Ledward, D. A., Johnston, D. E., Earnshaw, R. G., & Hasting, A. P. M. (1995). *High Pressure Processing of Foods.* Nottingham University Press, Nottingham.

53. Lopez Fandino, R., Carrascosa, A. V., & Olano, A. (1996). The effects of high pressure on whey protein denaturation and cheese-making properties of raw milk. *Journal of Dairy Science, 79*(6), 900–926.

54. Ludikhuyze, L. R., Van den Broeck, I., Weemaes, C. A., Herremans, C. H., Van Impe, J. F., Hendrickx, M. E., & Tobback, P. P. (1997). Kinetics for Isobaric-Isothermal Inactivation of Bacillus subtilis α-Amylase. *Biotechnology Progress, 13*(5), 532–538.

55. Mackey, B. M., Forestie're, K., Isaacs, N. S., Stenning, R., & Brooker, B. (1994). The effect of high hydrostatic pressure on *Salmonella Thompson* and *Listeria monocytogenes* examined by electron microscopy. *Letters in Applied Microbiology, 19,* 429–432.

56. Margosch, D., Ehrmann, M. A., Ganzle, M. G., & Vogel, R. F. (2004). Comparison of pressure and heat resistance of *Clostridium botulinum* and other endospores in mashed carrots. *Journal of Food Protection, 67,* 2530–2537.

57. Matser, A. M., Krebbers, B., van den Berg, R. W., & Bartels, P. V., 2004, Advantages of High Pressure Sterilization on Quality of Food Products, *Trends in Food Science and Technology, 15*(2), 79–85.

58. McClements, J. M. J., Patterson, M. F., & Linton, M. (2001). The effect of growth stage and growth temperature on high hydrostatic pressure inactivation of some psychrotrophic bacteria in milk. *Journal of Food Protection, 64*(4), 514–522.

59. Mertens, B., & Deplace, G. (1993). Engineering aspects of high pressure technology in the food industry. *Food Technology, 47.*

60. Messens, W., Van Camp, J., & Huyghebaert, A. (1997). The use of high pressure to modify the functionality of food proteins. *Trends in Food Science and Technology, 8*(4), 107–112.

61. Meyer, R. S., Cooper, K. L., Knorr, D., & Lelieveld, H. L. M. (2000). High pressure sterilization of foods. *Food Technology, 54,* 67–72.

62. Miyagawa, K., & Suzuki, K. (1964). Studies on Taka-amylase A under high pressure: Some kinetic aspects of pressure inactivation of Taka-amylase A. *Biochemistry and Biophysics, 105*(2), 297–302.

63. Mozhaev, V. V., Lange, R., Kudryashova, E. V., & Balny, C. (1996). Application of high hydrostatic pressure for increasing activity and stability of enzymes. *Biotechnology and Bioengineering, 52*(2), 320–331.

64. Murchie, L. W., Cruz-Romero, M., Kerry, J. P., Linton, M., Patterson, M. F., Smiddy, M., & Kelly, A. L. (2005). High pressure processing of shellfish: a review of microbiological and other quality aspects. *Innovation in Food Science and Emerging Technologies, 6,* 257–270.

65. Mussa, D. M., & Ramaswamy, H. (1997). Ultra high pressure pasteurization of milk: kinetics of microbial destruction and changes in physico-chemical characteristics. *Lebensmittel Wissenschaft and Technologie, 30*(6), 551–557.

66. Needs, E. C., Capells, M., Bland, P., Manoj, P., MacDougal, D. B., & Gopal, P. (2000). Comparison of heat and pressure treatment of skimmed milk, fortified with whey protein concentrate for set yogurt preparation: Effects on milk proteins and gel structure. *Journal of Dairy Research, 67,* 329–348.

67. Noble, I., & Gomez, L. (1962). Vitamin retention in meat cooked electronically. Thiamine and riboflavin in lamb and bacon. *Journal of the American Dietetic Association, 41,* 217.

68. Oey, I., Van der plancken, I., Van Loey, A., & Hendrickx, M. (2008). Does high pressure processing influence nutritional aspects of plant based food system? *Trends in Food Science and Technology, 19*, 300–308.
69. Olsson, S. (1995). Production equipment for commercial use. In: *High Pressure Processing of Foods,* edited by Ledward, D. A., Johnston, D. E., Earnshaw, R. G., & Hasting, A. P. M., Nottingham University Press, Nottingham, p. 167.
70. Otero, L., Molina-Garcia, A. D., & Sanz, P. D. (2000). Thermal effect in foods during quasi-adiabatic pressure treatments. *Innovative Food Science and Emerging Technologies, 1*, 119–126.
71. Palou, E., Lopez-Malo, A., & Welti-Chanes, J. (2002). Innovative fruit preservation using high pressure. Chapter 43, In: *Engineering and Food for the 21st Century,* edited by Welti-Chanes, J., Barbosa-Canovas, G. V., & Aguilera, J. M., Food Preservation Technology Series, Boca Raton: CRC Press, pp. 715–726.
72. Patazca, E., Koutchma, T., & Balasubramaniam, V. M. (2007). Quasi-adiabatic temperature increase during high pressure processing of selected foods. *Journal of Food Engineering, 80*(1), 199–205.
73. Patterson, M. (1999). High-pressure treatment of foods. In: *The Encyclopedia of Food Microbiology,* edited by Robinson, R. K., Batt, C. A., & Patel, P. D., Academic Press, New York, pp. 1059–1065.
74. Patterson, M. F., Quinn, M., Simpson, R., & Gilmour, A. (1995). Sensitivity of vegetative pathogens to high hydrostatic pressure treatment in phosphate-buffered saline and foods. *Journal of Food Protection, 58*(5), 524–529.
75. Qiu, W., & Hanhu, J. (2001). Review on ultra-high pressure sterilization of foods and recent research progress. *Journal of Food Science, 22*(5), 81–84.
76. Rademacher, B., & Kessler, H. G. (1997). High pressure inactivation of microorganisms and enzymes in milk and milk products. In: *High Pressure Bio-Science and Biotechnology,* edited by Heremans, K. Leuven, Belgium: Leuven University Press, pp. 291–293.
77. Raffalli, J., Rosec, J. P., Carlez, A., Dumay, E., Richard, N., & Cheftel, J. C. (1994). High pressure stress and inactivation of *Listeriainnocua*in inoculated dairy cream. *Sciences des Aliments, 14*(3), 349–358.
78. Rasanayagam, V., Balasubramaniam, V. M., Ting, E., Sizer, C. E., Bush, C., & Anderson, C. (2003). Compression heating of selected fatty food materials during high pressure processing. *Journal of Food Science, 68*(1), 254–259.
79. Raso, J., Barbosa-Canovas, G., & Swanson, B. G. (1998). Sporulation temperature affects initiation of germination and inactivation by high hydrostatic pressure of *Bacilluscereus. Journal of Applied Microbiology, 85*, 17–24.
80. Rastogi, N. K., Raghavarao, K. S. M. S., Balsubramanim, V. M., Niranjan, K., & Knorr, D. (2007). Opportunities and challenges in high pressure processing of foods. *Critical Reviews in Food Science and Nutrition, 47*, 69–112.
81. Robey, M., Benito, A., Hutson, R. H., Pascaul, C., Park, S. F., & Mackey, B. M. (2001). Variation in resistance to high hydrostatic pressure and rpoS heterogeneity in natural isolates of Escherichia coli O157:H7. *Applied Environmental Microbiology, 67*, 4901–4907.
82. Rovere, P., Gola, S., Maggi, A., Scaramuzza, N., & Miglioli, L. (1998). Studies on bacterial spores by combined pressure-heat treatments: possibility to sterilize low

acid foods. In: *High Pressure Food Science, Bioscience and Chemistry*, edited by Isaacs, N. S. Cambridge: The Royal Society of Chemistry, pp. 354–363.

83. Royer, H. (1895). Action of high pressure on some bacteria. *Archive of Normal and Pathological Physiology, 7*, 7–12.

84. Sachez-Moreno, C., De-Ancos, B., Plaza, L., Elez-Martinez, P., & Cano, M. P. (2009). Nutritional approaches and health-related properties of plant foods processed by high pressure and pulsed electric fields. *Critical Reviews in Food Science and Nutrition, 49*, 552–576.

85. Sancho, F., Cambert, Y., Demazeau, G., Largeteau, A., Bouvier, J. M., & Narbonne, J. F. (1999). Effect of ultra-high hydrostatic pressure on hydrosoluble vitamins. *Journal of Food Engineering, 39*, 247–253.

86. Schmidt, D. G., & Koops, J. (1977). Properties of artificial casein micelles, Part 2. *Netherlands Milk and Dairy Journal, 31*, 342–357.

87. Silva, J. L., Luan, P., Glaser, M., Boss, E. W., & Weber, G. (1992). Effects of hydrostatic pressure on a membrane-enveloped virus: Immunogenicity of the pressure-inactivated virus. *Journal of Virology, 66*, 2111–2117.

88. Smelt, J. P. P. M. (1998). Recent advances in the microbiology of high pressure processing. *Trends in Food Science and Technology, 9*, 152–158.

89. Tanaka, T., & Hatanaka, K. (1992). Application of hydrostatic pressure to yogurt to prevent its after acidification. *Journal of Japanese Society of Food Science and Technology, 39*(2), 173–177.

90. Thakur, B., & Nelson, P. (1998). High pressure processing and preservation of food. *Food Reviews International, 14*(4), 427–447.

91. Tholozan, L. J., Ritz, M., Jugiau, F., Federighi, M., & Tissier, P. J. (2000). Physiological effects of high hydrostatic pressure treatments on *Listeria monocytogenes* and *Salmonella typhimurium. Journal of Applied Microbiology, 88*, 202–212.

92. Ting, E., Balasubramaniam, V. M., & Raghubeer, E. (2002). Determining thermal effects in high pressure processing. *Journal of Food Technology, 56*(2), 31–35.

93. Ting. E. Y., & Marshall, R. G. (2002). Production issues related to UHP food. Vhspter 44, In: *Engineering and Food for the 21st Century*, edited by Welti-Chanes, J., Barbosa-Canovas, G. V., & Aguilera, J. M., *Food Preservation Technology Series*, Boca Raton: CRC Press, pp. 727–738.

94. Trujillo, A. J., Capellas, M., Saldo, J., Gervilla, R., & Guamis, B. (2002). Applications of high-hydrostatic pressure on milk and dairy products: a review. *Innovative Food Science and Emerging Technologies, 3*, 295–307.

95. Trujillo, A. J., Royo, C., Guamis, B., & Ferragut, V. (1999). Influence of pressurization on goat milk and cheese composition and yield. *Milchwissenschaft, 54*(4), 197–199.

96. Urrutia-Benet, G. (2005). High-pressure low-temperature processing of foods: impact of metastable phases on process and quality parameters. PhD Thesis, Berlin University of Technology.

97. Van-Opstal, I., Bagamboula, C. F., Vanmuysen, S. C. M., Wuytack, E. Y., & Michiels, C. W. (2004). Inactivation of *Bacillus cereus* spores in milk by mild pressure and heat treatments. *International Journal of Food Microbiology, 92*(2), 227–234.

98. Verhoeven, D., Assen, N., Goldbohm, R., Dorant, E., Van'T Veer, P., Sturmans, F., Hermus, R., & Van den Brandt, P. (1997). Vitamins C and E, retinol, beta-carotene and dietary fiber in relation to breast cancer risk: a prospective cohort study. *British Journal of Cancer, 75*(1), 149.

99. Voigt, D. D., Chevalier, F., Qian, M. C., & Kelly, A. L. (2010). Effect of high-pressure treatment on microbiology, proteolysis, lipolysis and levels of flavor compounds in mature blue-veined cheese. *Innovative Food Science and Emerging Technologies, 11*, 68–77.

100. Weenberg, M., & Nyman, M. (2004). On possibility of using high pressure treatment to modify the physico-chemical properties of dietary fibers in white cabbage (*Brassica oleraceavarcapitita*). *Innovative Food Science and Emerging Technologies, 5*, 171–177.

101. Wemekamp-Kamphuis, H. H., Wouters, J. A., de Leeuw, P. P. L. A., Hain, T., Chakraborty, T., & Abee, T. (2004). Identification of sigma factors B-controlled genes and their impact on acid stress, high hydrostatic pressure, and freeze survival in Listeria monocytogenes EGD-e. *Applied Environmental Microbiology, 70*, 3457–3466.

102. Wilkinson, N., Kurdziel, A. S., Langton, S., Needs, E., & Cook, N. (2001). Residence of poliovirus to inactivation by hydrostatic pressure. *Innovative Food Science and Emerging Technologies, 2*, 95–98.

103. Yokoyama, H., Sawamura, N., & Motobayashi, N. (1992). Method for accelerating cheese ripening. *European Patent Application,* EP469–857(0), A1.

104. Zhang, T., (1992). Sterilization of mandarin orange by ultra-high pressure. *Science and Technology of Food Industry, 2*, 47–51.

105. Zimmerman, F., & Bergman, C. (1993). Isostatic high pressure equipment for food preservation. *Food Technology, 47.*

CHAPTER 8

MICROWAVE PROCESSING OF MILK: A REVIEW

BHUSHAN D. MESHRAM, A. N. VYAHAWARE, P. G. WASNIK, A. K. AGRAWAL, and K. K. SANDEY

CONTENTS

8.1 INTRODUCTION

Microwaves are largely a twentieth-century phenomenon for their use in foods. The first truly useful sources were devised in the 1940s during the wartime development of radars [139]. Microwaves are defined as a part of electromagnetic waves, which have frequency range between 300 MHz and 300 GHz corresponding to wavelength from 1 mm to 1 m. Microwave frequencies of 915 MHz and 2.45 GHz can be utilized for industrial, scientific, and medical applications [54, 80]. Microwaves have been applied in a broad range of food processing such as drying, tempering, blanching, cooking, pasteurization, sterilization, and baking. Microwave heating has considerable advantages over conventional heating methods, especially with regard to energy efficiency. Since heat is transferred from the surface of food to the interior by convection and conduction in conventional cooking method, it may result in a temperature gradient between outside and inside the food [49, 140]. In addition, it requires higher energy consumption and relatively long processing time [135]. In microwave heating, on the other hand, heat is generated (volumetric heating) inside the food in a short time when microwave penetrates through it [92, 96]. Microwaves have greater penetration depth, and this property coupled with volumetric heating can lead to rapid heating rate with short processing time; and also contribute to the minimization of temperature difference between the surface and interior of food [34, 140].

Microwave ovens are now becoming common house-hold appliances because the modern life requires simplified routines and standardization of foods with lesser preparation time and convenience in usage. Microwave energy has been widely used in the food processing industries due to its advantages. It has been widely used in drying, sterilization, pasteurization, tempering, baking, blanching, heating, solvent extraction, digestion, puffing and foaming [32]. The industrial and domestic use of microwaves has increased dramatically over the past few decades. While the use of large-scale microwave processes is increasing, recent improvements in the design of high-powered microwave ovens, reduced equipment manufacturing costs and trends in electrical energy costs offer a significant potential for developing new and improved industrial microwave processes [85].

Microwavable foods satisfy need for speed and palatability. In addition, to meet consumer demand for on-the-go eating, a growing number of microwave products are entering the market in single-serve portable packages [6]. Microwave technology offers a new range of opportunities as a substitute to thermal processing. It offers a technique of heating that requires neither the presence of a hot outer surface nor the need for long conduction lags [115]. An alternative to aseptic packaging, there is an ability of microwaves to heat-treat the products after packaging, which are considered as "electrical- isolators" like plastic, glass, ceramics, porcelain. This technique can be effectively used for traditional Indian dairy products (TIDPs) for in-pack sterilization. Also microwave processing results in excellent retention of nutritional and sensory value, besides having several advantages like savings in energy, cost, time, etc. [124, 136].

The use of microwave oven provides a convenient way to thaw, cook or reheat foods nowadays. Many studies have been conducted to assess the safety as well as possible nutrient loss associated with microwave cooking. The best available evidence supports that the use of microwave cooking resulted in foods with safety and nutrient quality similar to those cooked by conventional methods, provided that the consumers followed the given instructions [56].

In addition, microwave (MV) heating is effective to heat up the prepared (ready to eat, RTE) foods. Therefore, the microwave oven operated in the simple manner became an indispensable home appliance to cook RTE food. Western foods are often considered to be more suitable for microwave cooking and can be ascribed due to fact that those include a majority of baked food products and precooked meat patties that require preheating process, for example, oven-roasting or baking before consumption. In recent years, however, even in the countries like India, where the traditional cooking methods are still popularly used, the application of microwave heating for cooking has been significantly increased. This transforming trend strongly suggests that microwave heating have been widely adopted for cooking various food types.

This chapter discusses the effectiveness and potential of microwave heating technology for different food processing methods and briefly presents a literature review of experimental approaches of microwave heating.

8.2 PRINCIPLE OF OPERATION OF MICROWAVE TECHNOLOGY

8.2.1 GENERAL

According to https://en.wikipedia.org/wiki/Microwave, "*Microwaves are a form of electromagnetic radiation with wavelengths ranging from one meter to one millimeter; with frequencies between 300 MHz (100 cm) and 300 GHz (0.1 cm). This broad definition includes both UHF and EHF (millimeter waves), and various sources use different boundaries. In all cases, microwave includes the entire SHF band (3 to 30 GHz, or 10 to 1 cm) at minimum, with RF engineering often restricting the range between 1 and 100 GHz (300 and 3 mm). The prefix micro- in microwave is not meant to suggest a wavelength in the micrometer range. It indicates that microwaves are "small", compared to waves used in typical radio broadcasting, in that they have shorter wavelengths. The boundaries between far infrared, terahertz radiation, microwaves, and ultra-high-frequency radio waves are fairly arbitrary and are used variously between different fields of study."* Frequencies of 915 or 2,450 MHz are mainly used in commercial microwave ovens for food [116].

All microwave ovens have similar design that includes a magnetron device as a power source and a waveguide to bring radiation to a heating chamber. Microwave with 915 MHz frequencies, is used for industrial heating, and 2450 MHz, in domestic microwave oven worldwide. The mechanism through which microwave heating occurs is based on the oscillation and polarization of the charges and the molecules of a dielectric material under the influence of alternating electric field, resulting in the generation of heat. The main cause of heat production is due to dipolar rotation. Factors influencing the microwave heating are frequency, microwave power, speed of heating, mass of the material, moisture content, density and temperature of the material. The two major mechanisms, namely dipolar and ionic interactions, explain how heat is generated inside food [31, 32].

8.2.1.1   Dipolar Interaction

Once microwave energy is absorbed, polar molecules such as water molecules inside the food will rotate according to the alternating electromagnetic field. The water molecule is a "dipole" with one positively charged end and one negatively charged end (Figure 8.1). Similar to the action of magnet, these "dipoles" will orient themselves when they are subjected to electromagnetic field. The rotation of water molecules would generate heat for cooking.

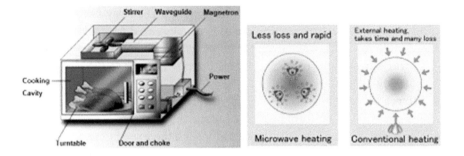

LEFT   **RIGHT**

FIGURE 8.1   Basic components of microwave oven (Left); microwave heats object internally (Right)..

8.2.1.2   Ionic Interaction

In addition to the dipole water molecules, ionic compounds (i.e., dissolved salts) in food can also be accelerated by the electromagnetic field and collides with other molecules to produce heat.

Hence the composition of a food will affect how it will be heated up inside the microwave oven. Food with higher moisture content will be heated up faster because of the dipolar interaction. As the concentration of ions (e.g., dissolved salts) increase, the rate of heating also increases because of the ionic interaction with microwaves. Although oil molecules are much less polar than water molecules and are non-ionic, yet food products with high oil content has a fast heating rate because the specific heat of oil is about less than half that of water.

8.2.1.3   Basic Components (Figure 8.1) [17]

- **Cooking cavity** is a space inside which the food is heated when exposed to microwaves.
- **Door and choke** allows the access of food to the cooking cavity. The door and choke are specially engineered that they prevent microwaves from leaking through the gap between the door and the cooking cavity.

- **Magnetron** is a vacuum tube in which electrical energy is converted to an oscillating electromagnetic field. Frequency of 2,450 MHz has been set aside for microwave oven for home use.
- **Power supply and control** controls the power to be fed to the magnetron as well as the cooking time.
- **Stirrer** is commonly used to distribute microwaves from the waveguide and allow more uniform heating of food.
- **Turntable** rotates the food products through the fixed hot and cold spots inside the cooking cavity and allows the food products to be evenly exposed to microwaves.
- **Waveguide** is a rectangular metal tube which directs the microwaves generated from the magnetron to the cooking cavity. It helps prevent direct exposure of the magnetron to any spattered food which would interfere with function of the magnetron.

8.2.1.4   Characteristics of Microwave Heating

a. **The Internal heating:** Microwave energy will reach the object to be heated at the same speed of light. Then it enters into the object as a wave, and by getting absorbed, the object generates heat (Figure 8.1, right) [17].

b. **Rapid heating:** In conventional heating (Figure 8.1, right), the object's temperature rises by spreading heat energy from the surface to inside (external heating). On the other hand, by microwave heating, the object will generate heat on its own by the penetration of the microwave. Not necessary to consider about the heat conduction. That is why rapid heating is possible by microwave. Although the object has to be large enough for the microwave to penetrate, yet the smaller objects will also be heated from inside as the depth of microwave penetration.

c. **High heating efficiency:** Microwave penetrates into the object at the speed of light. You get high heating efficiency because no need to consider the heating losses of air inside the heating furnace.

d. **Rapid response and temperature control**: Microwave penetrates into the object at the speed of light. So it allows rapid response. For example, you can start and stop the heating instantly. In addition, by the adjustment of microwave output, you can control the amount of heat energy generated inside the heated object.

e. **Heating uniformity**: Each part of the heated object generates heat, so even for those objects with complicated shape, it can be heated relatively uniform. To keep the heating uniformity, stirrer, turntable, and belt conveyor is used for heating blur related to wave length.

f. **Clean energy**: Microwave does not require a medium, because it propagates only by changes of electric fields and magnetic fields. It can propagate in a vacuum. It reaches the object and penetrates without heating the air. The heated object generates heat by absorbing microwave energy to convert it to heat energy. Therefore, it can be said as clean energy because it doesn't heat the air during process.

g. **Good operation and work environment**: Conventional heating requires a heat source, and the temperature rises not only of heated object, but also of heat source and the heating furnace. So the temperature of room equipped heating furnace goes high because of radiant heat. This is an operation and work environment issue. On the other hand, microwave heating only uses electricity to generate heat of the object. The temperature of the object only raises, not the furnace and there is no radiant heat, so it's possible to keep operational and good working environment.

8.3   MICROWAVE TECHNOLOGY: ADVANTAGES AND LIMITATIONS

8.3.1   ADVANTAGES

- Microwave penetrates inside the food materials and, therefore, cooking takes place throughout the whole volume of food internally, uniformly, and rapidly, which significantly reduces the processing time and energy.
- Since the heat transfer is fast, nutrients and vitamins contents, as well as flavor, sensory characteristics, and color of food are well preserved.

- Minimum fouling depositions, because of the elimination of the hot heat transfer surfaces, since the piping used is microwave transparent and remains relatively cooler than the product.
- High heating efficiency (80% or higher efficiency can be achieved).
- Suitable for heat-sensitive, high-viscous, and multiphase fluids.
- Low cost in system maintenance.

8.3.2   DISADVANTAGES

Today's uses range from these well-known applications over pasteurization and sterilization to combined process like microwave vacuum drying. The rather slow adoption of food industrial microwave applications may be due to following limitations:

- There is the conservatism of the food industry [35], and its relatively low research budget. Linked to this, there are difficulties in moderating the problems of microwave heating applications.
- In order to get good results, industry needs a high input of engineering intelligence.
- Different from conventional heating systems, where satisfactory results can be achieved easily by perception, good microwave application results often need a lot of knowledge or experience to understand and moderate effects like uneven heating or the thermal runway.
- Microwave heating as opposed to conventional heating needs electrical energy, which is its most expensive form.
- High initial capital investment, more complex technology devices, microwave radiation leakage problem.

8.4   MICROWAVABLE PACKAGING MATERIALS

Plastic containers are commonly used for microwave cooking. High density polyethylene (HDPE) can be used for foods with high water content, for foods with high fat or high sugar content, and for foods with high water content, as these foods may reach temperature above 100°C during microwave cooking. Paper and board can also absorb some microwave energy. However, it is not ideal for microwaved food because the strength of the paper would be affected when wet; and not all types of paper are suitable for microwave cooking.

When food is microwaved, heat is also retained in the glass. The degree of energy absorption depends on types of glass. Moreover, microwave energy can be superimposed at the center after passing through the glass containers, particularly the ones with small radius. Microwave oven has now been accepted as a modern domestic appliance for cooking and heating so the market needs innovative food packages which can be micro waved. Microwavable packages are made from a special aluminum laminates that can withstand high temperature, heat-processing treatments such as sterilization and microwave heating. Retort pouches and semi-flexible retortable thermoformed containers are the most popular microwavable packaging currently used. Cast polypropylene, polyester, polyamides, oriented polyamide, aluminum foils form the basic structure of a microwavable packing [56].

8.5 APPLICATIONS OF MICROWAVE TECHNOLOGY IN FOOD THE INDUSTRY

8.5.1 THAWING-TEMPERING

Tempering is the thermal treatment of frozen foods to raise the temperature from below −18°C to temperature just below the melting point of ice. At these temperatures, the mechanical product properties are better suited for further machining operation (e.g., cutting or milling). By using microwaves mostly at 915 MHz due to their larger penetration depth, the tempering time can be reduced to minutes or hours and the required space is diminished to one sixth of the conventional system [81]. Another advantage is the possibility to use the microwaves at low air temperatures, thus reducing or even stopping microbial growth.

Frozen meat, fish, vegetables, fruit, butter and juice concentrate are common raw materials for many food manufacturing operations. Frozen meat, as supplied to the industry, ranges in size and shape from complete hindquarters of beef to small breasts of lamb and poultry portions, although the majority of the material is 'boned-out' and packed in boxes approximately 15 cm thick weighing between 20 and 40 kg. Fish is normally in plate frozen slabs; fruit and vegetables in boxes, bags or tubs; and juice in large barrels. Few processes can handle the frozen material and it is usually either thawed or tempered before further processing [55].

Thawing is usually regarded as complete when all the material has reached 0°C and no free ice is present. This is the minimum temperature at which the meat can be boned or other products cut or separated by hand. Lower temperatures (e.g., −5 to −2°C) are acceptable for product that is destined for mechanical chopping, but such material is 'tempered' rather than thawed. The two processes should not be confused because tempering only constitutes the initial phase of a complete thawing process. Thawing is often considered as simply the reversal of the freezing process. However, inherent in thawing is a major problem that does not occur in the freezing operation. The majority of the bacteria that cause spoilage or food poisoning are found on the surfaces of food. During the freezing operation, surface temperatures are reduced rapidly and bacterial multiplication is severely limited, with bacteria becoming completely dormant below −10°C. In the thawing operation, these same surface areas are the first to rise in temperature and bacterial multiplication can recommence. On large objects subjected to long uncontrolled thawing cycles, surface spoilage can occur before the center regions have fully thawed.

Without doubt, thawing and tempering are most industrially widespread applications of microwave (MW) heating. There are about 400 systems in use in the United States for vegetables and fruits; and at least four in the United Kingdom for tempering of butter. Most of the studies were carried out in the 1970s and 1980s [3, 7], and these have analyzed the behavior and final characteristics of diverse types of meat during MW tempering. Tempered meat shows good final characteristics, and process time is greatly shortened. An attempt made by Cadeddu [18] to apply a microwave tempering tunnel to stretch Mozzarella curd was not successful.

Recently, other possible applications of this technology to for rice balls [60, 142], mashed potatoes [53], or cereal pellets or pieces [110] have been studied, with a few encouraging results in terms of the good physical and sensory properties. However, some studies [21, 61, 62] are mainly focused on reaching a better understanding of the relationship between the equipment (applied powers and cycles of work) and the product (dielectric properties, loads, and geometry).

8.5.2 HEATING OF PRECOOKED PRODUCTS

The heating of precooked products is principal practical application of microwave ovens, both in domestic use and in the catering industry, since a rapid, safe, and hygienic heating of the product can be obtained [15]. The objective of pre-cooking operation is to reduce preparation time for the consumer. In case of cereals, these operations consist basically of treating starch to reduce its gelatinization time during the final preparation of the food product. By toasting cereal flours, Wang et al. [138] obtained pre-cooked rice flour (with 13.4% moisture) after 11 min in a microwave oven (2450 MHz) and Chavan et al. [23] obtained pre-cooked wheat flour (with 14% moisture) after 11 min in a microwave oven (2,450 MHz), with good sensory and nutritional evaluations, which were used in combination with defatted soy flour with additional microwave treatment for 8 min, for porridge and soup products. The precooking process can be accelerated with the help of microwaves as has been established for precooking of poultry, meat patty and bacon [11, 35].

Recently, however, diverse studies have been carried out in an attempt not only to improve the physical and sensory characteristics of the heated product [46, 70], but also to obtain a greater uniformity in product heating [59, 104]. Jeong et al. (2004) [58] carried out microwave cooking of ground pork patties at 700 Watt up to 75°C and found the reduction in cooking time for high fat patties. Bilgen et al. [10] used microwave oven to prepare a white layer cake to achieve desired crumb qualities.

8.5.3 COOKING

Cooking with microwaves has recently become the most adaptable method all over the world. Microwave ovens are now used in about 92% of homes in the US. In this section, various reports on the effects of microwave on cooking parameters such as quality, taste and color retention for various food materials are reviewed. There are numerous reports on the baking of bread and cooking of rice and meat using microwaves. Microwave heating takes the product to the desired temperature in such a short time that product cooking does not take place; the product is hot, but has the appearance and flavor of the raw product [69].

There are several products used in the continuous study of this technology, such as fish [107], beans [88], egg yolk [86], and shrimp [49]. The nutritious characteristics of the food are quite well retained, but it does not reach the typical flavor of the cooked dish; thus it is necessary to combine microwave treatment with conventional technologies [25]. Microwave cooking is used industrially for chicken (1500 kg of chicken/hour) and for bacon (with an advantage over oil frying or infrared heating). Microwave oven is well suited for cooking the food in small quantities, especially for households, though not convenient for mass cooking. Daomukda et al. [30] studied the effect of different cooking methods on physico-chemical properties of brown rice. They concluded that the protein, fat and ash contents in rice cooked by microwave are retained at higher levels (8.49%, 2.45% and 1.42%, respectively) than conventional boiling and steaming methods. Microwave irradiation normally does not induce the Mallard reaction because of the short cooking times and low temperatures [143].

Sharma and Lal [114] reported that there was no significant change on the loss of B-complex vitamins during microwave boiling of cow and buffalo milk in comparison to conventional heating. Microwave cooking reduces cooking times of common beans and chickpeas [77]. In addition, microwave treatment was able to reduce cooking losses, increase the soluble: Insoluble and soluble: total dietary fiber ratios, but did not modify in-vitro starch digestibility. A higher protein concentration in soya milk was obtained by microwave heating of soya slurry than by the conventional methods of heating such as the use of boiling water [2]. Microwave oven heating of soya slurry, which was effective for protein extraction, also made the prepared tofu more digestible.

Iilow et al. [57] reported that microwave cooking resulted in minimum loss of vitamin C in white cabbage, cauliflower, Brussel sprouts and French beans in comparison to the traditional method of cooking. It also had a positive effect on the organoleptic quality of these vegetables. Microwave roasting of soaked soybean produced full-fat soya flour with high vitamin E without burnt color and browning [145]. Variation in the organic acids, sugars and minerals of raw and microwave cooked beetroot, broccoli, artichokes, carrots, cauliflower, fennel, potatoes, chilies, celery, spinach and courgettes have been reported by Plessi [39, 95].

Substantial reduction in the energy consumption was observed with controlled cooking (using microwave oven) of un-soaked rice (14–24%) and pre-soaked rice (12–33%) compared with normal cooking [73]. Conventional cooking of broccoli for 30, 60, 90, 120 and 300 s has been found to reduce total phenolic content by 31.6%, 47.5%, 55.9%, 61.7% and 71.9% in florets and 13.3%, 22.2%, 26.7%, 28.9% and 42.2% in stems; and there was no significant difference in the total phenolic content between microwave and conventionally cooked samples [1].

Because microwave oven is able to heat up foods using the energy of oscillating electromagnetic wave, it is possible to do selective and quick cooking. But the penetration depth of microwave is under about a few inches or below the surfaces of foods. So, if food size is small and the shape of food is flat, the uniform heating through overall volume is possible. It will lead less loss of moisture contents and the greatest energy savings, and the nutrition of foods will be preserved very well. But using conventional method, cooking of multiple foods containing particles of any shape and size together can be achieved through moist-heat method, but at the expense of moisture which keeps some of their nutrition. Therefore, new combination techniques, making the best use of the merits of microwave heating, should be studied.

8.5.4 BAKING

Baking is a thermal process that significantly changes physicochemical properties of dough. Baking process includes three phases: expansion of dough and moisture loss initiates in the first phase; in the second phase, expansion and the rate of moisture loss becomes maximal; and the third phase includes rise in product height and decrease in rate of moisture loss because the structure of the air cells within the dough medium collapses as a result of increased vapor pressure. Baking using microwave energy has been limited due to poor product quality compared to products baked by using conventional energy sources, which can be a reflection of the differences in the mechanism of heat and mass transfer [108]. In products such as breads, cakes and cookies, microwave baking can affect texture, moisture content and color of the final product, which represents a great challenge for research. Some researchers suggest adjustments in formulation and alterations in the baking process, while other investigators study the interactions between microwave energy and the ingredients of the formulation [94, 147].

Microwave baking of pastry products is already carried out on an industrial scale, whereas that of bread dough is still in the experimental phase [9, 19, 91]. Its application to batters in the production of tarts or cakes is also being studied [8, 29, 146]. Apart from the better retention of vitamins and nutrients, the fundamental effect is the greater volume that MW-baked pieces acquire, in the order of 25–30% more than that reached using classical baking. No differences were detected in the flavor. An additional effect is a minor development of molds in microwave-baked products. In the case of bread, an additional source of energy is needed to obtain desired color and texture of the crust.

In studies with bread during microwave heating, there was a rapid loss of moisture and, after microwaving, the mechanical strength of bread was increased greatly [22, 23, 51]. During cake production, after the mixing process, the cake must be deposited into cake pans and rapidly conveyed to the oven. Baking time is inversely related to baking temperature and the optimum baking conditions for cake baking are determined by the sweetener level of the formula, amount of milk used in the batter, fluidity of the batter, pan size and others [5, 144]. The effect of modifications in cake ingredients was studied by Tsoubeci [129]. Sucrose substitution with a blend of whey protein isolate (WPI) and maltodextrin and fat substitution with a special fiber were evaluated in a model cake system during baking by conventional and microwave methods. Substitution of sucrose with WPI and maltodextrin affected the dielectric properties of the batters. Power absorption predictions indicated that the power absorbed by microwave-heated cakes increased when sucrose was substituted. Removal of fat influenced the dielectric behavior of batters whereas substitution with the special fiber compensated for fat in both microwave and conventionally heated cakes. Takashima [126] patented a process to obtain a sponge cake free from bake shrinkage and good-looking voluminous appearance, through a batter prepared by adding a thermo-coagulation protein to a sponge cake premix containing as main ingredient a cereal powder consisting of starch and a pre-gelatinized starch cooked under heat with a microwave oven.

8.5.5 BLANCHING

Blanching is a thermal pre-treatment process, which is an essential step in several food processing techniques such as freezing, canning or drying, generally applied to inactivate enzymes that substantially affect to texture, color, flavor, and nutritive values of fruits and vegetables. Blanching is generally used for color retention and enzyme inactivation, which is carried out by immersing food materials in hot water, steam or boiling solutions containing acids or salts. Blanching has additional benefits, such as: the cleansing of the product, the decreasing of the initial microbial load, exhausting gas from the plant tissue, and the preheating before processing. A moderate heating process such as blanching may also release carotenoids and make them more extractable and bio-available. Blanching with hot water after the microwave treatment compensates for any lack of heating uniformity that may have taken place, and also prevents desiccation or shriveling of delicate vegetables. And while microwave blanching alone provides a fresh vegetable flavor, the combination with initial water or steam blanching provides an economic advantage. This is because low-cost hot water or steam power is used to first partially to raise the temperature, while microwave power, which costs more, does the more difficult task of internally blanching the food product. Microwave blanching of herbs such as marjoram and rosemary was carried out by soaking the herbs in a minimum quantity of water and ex-posed to microwaves [120].

Microwave blanching was observed for maximum retention of color, ascorbic acid and chlorophyll contents than that of water and steam blanching. Microwave blanched samples were found to have better retention of quality parameter than that of microwave dried samples without blanching [119]. In comparison to the traditional method of blanching in hot water, microwave blanching of vegetables has the advantage of avoiding the loss of nutrients (vitamins, etc.) and pigments, which are lixiviated to some extent in the blanching water, ash has been observed for tomatoes [4], carrots [65], beans [64], soybeans [52], asparagus [66], mushrooms [134], and strawberries [141]. Additionally, there is no production of waste water.

Recent studies reveal that enzymatic browning can be limited to some extent in several products such as bananas [20] and potatoes [112]. Nevertheless, some disadvantages of MW application to the blanching

process are: the surface dehydration of the product, especially in leaf vegetables such as spinach and cabbages; caramelization in fruits [20, 27]; and
the heterogeneous heating of the products, depending on size and shape,
which can lead to severe alterations caused by undesirable thermal effects
Some authors [25] propose the combination of microwave blanching with
other techniques (steam or hot water) in order to minimize product alterations. Microwave blanching operations can be used to inactivate enzymes
in fresh vegetables and fruits that lead to premature food spoilage at freezing temperatures [115].

Ramaswamy and Fakhouri [101] studied the microwave blanching of carrot slices and french-fry style sweet potatoes. Premakumar
and Khurdiya [98] stated that banana puree prepared from microwave-
blanched fruits has higher nutritional and organoleptic qualities as compared to conventional method of blanching. Severini et al. [112] stated that
microwave blanching of cubed potatoes can be achieved at 600 watts for
5 min. Brewer and Begum [12] carried out blanching of broccoli, green
beans and asparagus and found the reduction in peroxidase activity.

8.5.6 FOOD STERILIZATION AND PASTEURIZATION

Pasteurization and sterilization are done with the purpose of destroying or
inactivating microorganisms to enhance the food safety and storage life
[13, 33, 89]. Solid products are usually sterilized after being packed, so no
metallic materials can be used in packaging when microwaves are used in
this process. This factor limits the use of this technique in food sterilization.
Possible non-thermal effect on destruction of microorganisms under microwave heating has been reported; the polar and /or charged moieties of proteins (i.e., COO-, and NH4$^+$) can be affected by the electrical component of
the microwaves [35]. Also, the disruption of non-covalent bonds by microwaves is a more likely cause of speedy microbial death [67]. Academic and
industrial approaches to microwave pasteurization or sterilization cover the
application for precooked food like yogurt or pouch packed meals as well as
the continuous pasteurization of fluids like milk [34–36].

Due to the rapid heating of the product, a better retention of the nutritional properties in comparison with the current technologies can be
obtained [90]. Nevertheless, this is still only being applied on a laboratory

scale in many cases, the main reasons for which are high costs and the uncertainty about microbial inactivation. The latter is associated with the heterogeneity in product heating, which does not ensure that all points of the food reach the required temperature to induce microbial death [14, 17]. In this sense, the possible combined application of hydrogen peroxide and a microwave treatment to inactivate some microorganisms has been studied [71]. The 915-MHz microwave-circulated water combination (MCWC) heating technology was validated for a macaroni and cheese product using inoculated pack studies. This study suggests that the MCWC heating technology has potential in sterilizing packaged foods [37, 47, 100].

Pasteurization can be achieved by novel thermal (RF and Ohmic heating) and non-thermal technologies (high hydrostatic pressure, UV treatment, pulsed electric field, high intensity ultrasound, ionizing radiation and oscillating magnetic field) without affecting the color, flavor or nutritive value of food materials [93, 103]. In Ohmic heating, the heating occurs due to the electrical resistance caused by the food materials when a current is passed through them. For a pulsed electric field process, a very high voltage is applied for a very short time through the fluid. This generates mild heat and cell disruption of microorganisms occurs due to electroporation. During a high hydrostatic pressure process, pressures of 100 to 1000 MP a are applied and as a result, large microorganisms or enzymes consisting of large molecules were affected. This technique is used for the aroma components for which the sensory and nutritional qualities need to be maintained. The advantages of a high hydrostatic pressure process are: the release of minimal heat, homogeneous nature of the process and its applicability to packaged materials. Most of the novel and non-thermal techniques provide energy savings up to 70% compared to the traditional cooking methods [93].

Advantage of using microwaves in microorganism deactivation are the possibly high and homogeneous heating rates, also in solid foods and the corresponding short process times, which can yield a very high quality. For both processes, it is extraordinarily important to know or even to control the lowest temperature within the product, where the microorganism destruction has the slowest rate. Since both calculation and measurement of temperature distribution are still very difficult, this is one reason that up to now microwave pasteurization and sterilization can be found very seldom in industrial use, and then only for batch sterilization operation.

8.5.7 FOOD DEHYDRATION

Dehydration process removes moisture from food materials without affecting the physical and chemical composition. It is also important to preserve the food products and enhance their storage stability which can be achieved by drying. Dehydration of food can be done by various drying methods such as: solar (open air) drying, smoking, convection drying, drum drying, spray drying, fluidized-bed drying, freeze drying, explosive puffing and osmotic drying [26]. The application of microwaves to food drying has also received widespread attention recently [35, 44, 45]. The heat generated by microwaves induces an internal pressure gradient that involves vaporization and expelling of the water toward the surface. This greatly accelerates the process, when compared to hot air or infrared dehydration [125], in which the drying rate is dependent on the diffusion of water inside the product toward the surface.

Generally, microwave drying can be subdivided into two cases: the drying at atmospheric pressure and under applied vacuum conditions. Until now, more common in the food industry are combined microwave-air-dryers that can again be classified into a serial or parallel combination of both methods. In the serial process, mostly the microwaves are used to finish partly dried food, where an intrinsic leveling effect is advantageous. Well-studied and still applied examples for a serial hot air and microwave dehydration are pasta drying and the production of dried onions [35].

In the studies on MW applications to drying, changes in drying kinetics of fruits and vegetables have been analyzed, as have the effects of MW power on final product properties such as texture, structure, rehydration capability, and color [42, 59, 68, 72, 78, 79, 99, 106, 128]. A large amount of work has been carried out on potatoes [40, 41], carrots [75, 97], apples and banana [38, 68, 74, 105]. Uprit and Mishra [131] studied microwave convective drying and storage of soy-fortified paneer (SFP) and found that hot air temperature of 53.5° C and microwave power of 111.5 W gave good quality dried SFP cubes of uniform texture and surface, unblemished and with clear color. The dried SFP cubes rehydrated well and had a shelf life of 118 days under accelerated storage conditions (38 ± 2°C, 90% relative humidity). Cui et al. [28] studied the microwave-vacuum drying kinetics of carrot slices.

The desire to eliminate the existing problems in drying and to achieve fast and effective thermal processing has resulted in the increasing interest in the use of microwaves for food drying [123]. CSUF [27] has pioneered the development of microwave vacuum (MIVAC) dehydration. MW-related (MW-assisted or MW-enhanced) combination drying is a rapid dehydration technique that can be applied to specific foods, particularly to fruits and vegetables. The advantages of MW-related combination drying include: shorter drying time, improved product quality, and flexibility in producing a wide variety of dried products [147]. Sangwan et al. [109] dried blanched and chopped onions into microwave oven at 800 W powers for 3 to 4 min and prepared onion powder showed superior quality. Kalse et al. [63] also studied the osmo-microwave drying of onion slices. Sharma et al. [113] developed a small scale microwave dryer for the purpose of drying agricultural produce and carried out drying of garlic cloves in it with superior quality. The energy consumption for drying of pumpkin slices using microwave, air and combined microwave-air-drying treatments has been studied [145]. It was concluded that high energy consumption was observed for air oven drying compared to combined microwave-air-drying treatment and, the lowest energy consumption among treatments was observed during microwave drying.

However, there is one key problem with the above-mentioned techniques. Because of non-uniform heating, the uneven distribution of microwave field can occur. In addition, the overheating and quality deterioration can take place [145]. To overcome these problems, the microwave drying technique has been combined with various other methods. The MW freeze drying (MFD) and MW vacuum drying (MVD) are good examples, wherein drying is assisted by microwaves to produce high quality foods. Especially, conventional fluidized bed dryer combined with microwave heating is good choice for drying products containing fine particles. In the future, various hybrid methods will emerge.

8.6 APPLICATIONS OF MICROWAVE TECHNOLOGY IN THE DAIRY INDUSTRY

Milk is traditionally pasteurized in a heat exchanger before distribution. The application of microwave heating to pasteurize milk has been well studied and has been a commercial practice for quite a long time.

The success of microwave heating of milk is based on established conditions that provide the desired degree of safety with minimum product quality degradation. Since the first reported study on the use of a microwave system for pasteurization of milk [50, 76], several studies on microwave heating of milk have been carried out. The majority of these microwave-based studies have been used to investigate the possibility of shelf-life enhancement of pasteurized milk, application of microwave energy to inactivate milk pathogens, assess the influence on the milk nutrients or the non-uniform temperature distribution during the microwave treatment [71].

Microwave application allows pasteurization of glass, plastic, and paper products, which offers a useful tool for package treatment. The food products that best respond to MW pasteurization treatment are: pastry, prepared dishes, and soft cheeses [16]. The technique has also been tested on milk [118, 130] and fruit juices [41, 117] in devices suitable for continuous treatment. Valero et al. [132] pasteurized raw milk by submitting it to continuous-flow microwave treatment at 80 or 92°C for 15 seconds and found no adverse effect on flavor and chemical composition. The microwave pasteurization of cow milk and its nutritional quality has also been studied by Valsechi et al. [133].

Mishra and Pandey[82] standardized the process using domestic microwave oven for pasteurization of raw milk samples and concluded that milk can be pasteurized in 3.3 min in microwave without any appreciable changes in its physico-chemical properties and it can be stored for 14 hours at room temperature (20–25°C) and for 21 days at refrigeration temperature (5°C). Albert et al. [2] studied the effect of microwave pasteurization on the composition of milk and found no changes in amino acid composition. Geczi et al. [43] concluded that microwave and the convection heat treatments are equivalent in pasteurization and the microwave heat treatment is suitable for primary processing of freshly milked milk [54]. Tochampa et al. [127] also concluded that microwave heating is a good alternative to HTST pasteurization of milk. Wang et al. [138] studied the assessment of microwave sterilization of foods using intrinsic chemical markers. Guan et al. [48] used a pilot-scale 915 MHz microwave-circulation water combination (MCWC) sterilization system to treat macaroni and cheese entrees and observed that the MCWC system provided desired sterility (with a F_0 value of 7 min) within one fourth of the time required by conventional retort methods to produce shelf-stable products. Vyawahare and Meshram [137] studied effect of microwave treatment on physic-chemical, sensory and shelf life of date-*burfi* and observed that microwave treatment is effective in enhancing the keeping quality of date-*burfi* by four and seven days more at 30±1°C and 10±1°C, respectively as compared to non-microwave treated date-*burfi*, which had shelf-life of 3 days at 30±1°C while 14 days at 10±1°C. The microwave treated date-*burfi* was found to be acceptable up to seven days when stored at 30±1°C and twenty-one days in case of storage at 10±1°C.

8.6.1 EFFECT ON MILK NUTRIENTS

Milk is a rich source of vitamins and heat treatment affects some of these nutrients. The effects of microwave heating on several vitamins in cows' milk have been studied by many researchers. Sierra et al. and Unnikrishnan et al. [118, 130] research in milk B1, B2 and B6 vitamins included continuously operating microwave and conventional (tube heat exchanger) heating methods. They found that 3.4% and 0.5% fat milk at 90°C with a heat treatment method did not cause vitamin loss. Most studies report an insignificant loss in vitamin A, carotene, vitamin B1 or B2 in microwave-pasteurized milk, while losses of approximately 17% for vitamin E and 36% for vitamin C have been found. They compared the heat stability of vitamins B1 and B2 in milk between continuous microwave heating and conventional heating having the same heating, holding, and cooling steps. No significant losses in the vitamins were reported during microwave heating at 90°C without holding period, while vitamin B2 was found to decrease by 3%–5% during 30–60 s of holding.

Microwave pasteurization of milk was reported to result in lower levels of denaturation of whey proteins compared to conventional thermal processes and that the denaturation of β-lacto-globulin was almost similar in both processes. Moreover, the process yielded lower microbial counts and lower lactose isomerization [75].

The inactivation of *Streptococcus fecalis, Yersinia enterocolitica, Campylobacter jejuni, and Listeriamono-cytogenes* in milk by microwave energy has been reported by Choi et al. [24]. The complete inactivation of *Y. enterocolitica, C. jejuni,* and *L. monocytogenes* occurred at 8, 3, and 10 min when the cells were heated at a constant temperature of 71.1°C using microwaves.

8.6.2 MICROWAVE BASED DAIRY PRODUCTS

The effects of microwave heating on several TIDP have been studied by many researchers. Dairy products have limited shelf life and deterioration of concentrated dairy product, for example, *Burfi* occurs mainly due to improper traditional packaging techniques, lack of modified technologies and unawareness of hygienic practices especially in *Halwais*. Hence there is a need of an instant and affordable technology such as microwave technology which will be beneficial for dairy industry for extension of shelf life of products. Solanki et al. [122] carried out storage study of microwave treated *Burfi* by subjecting *Burfi* samples to microwaves at different power-time combinations. The shelf life study indicated that there was faster deterioration in non-microwave treated control sample at 30±1°C as compared to microwave-treated *Burfi* samples and the treated samples had 6 days more shelf life as compared to control, for example, non-microwave treated sample. Naresh et al. [87] also carried out the research to know the suitability of microwave heating as a post-heat treatment process to extend the shelf life of *Peda* and inferred that the shelf life of microwave treated (25 and 30 sec) *Peda* samples increased to 15 days and 18 days at room temperature; and 93 days and 126 days at refrigeration temperature as against 12 days and 71 days, respectively for non-microwave treated control sample.

Schlipalius et al. [111] developed the method for the preparation of microwave puffed cheese snack. Monsalve et al. [84] prepared shelf-stable butter containing microwave popcorn. Singh et al. [121]reported the technology of *Ghee* production by microwave oven. Mishra and Pandey [83] also prepared *Ghee* from cream (60% fat) and butter (80% fat) by microwave process and concluded that ghee can be successfully prepared in microwave oven within 25±0 and 19±0.5 minutes for cream and butter, respectively without appreciable losses in vitamin A content and such ghee can be stored up to 13–14 months and 9–10 months for cream and butter, respectively. Rao and Pagote [102] reported the preparation of rennet coagulated milk cake using microwave heating.

8.7 FUTURE ASPECTS OF MICROWAVE HEATING

In the past few years, there has been a surge of interest in the application of microwave heating for industrial purposes. This is primarily due to the worldwide energy crisis and the growing acceptance of and familiarity with microwave ovens. It is well known that conventional means of heat processing irreversibility alter the flavor, color, and texture of many foods. From the point of flavor alone, it is desirable to improve our present methods of processing foods. Some improvements have been made in the direction with a number of food products by use of high-temperature short time processing. Research has been conducted in recent years to ascertain whether it is possible to improve the color, flavor and retention of nutrients in processed foods by means other than heat for processing. Among the alternative means that have been considered to obtain this objective in the food processing is the utilization of several of the radiations of the electromagnetic spectrum.

8.8 CONCLUSIONS

The successful applications of microwave heating technology for processing of various foods have been discussed in this chapter. The microwave heating technology for pasteurization and sterilization contributed to effectively destroy pathogenic microorganisms and significantly reduce processing time without serious damage in overall quality of liquid food as compared to traditional methods. The use of microwave heating for food processing applications such as blanching, cooking, and baking has a great effect on the preservation of nutritional quality of food. Furthermore, microwave heating requires significantly less energy consumption for dehydrating food than conventional method.

In these days, the potential of continuous flow microwave heating at commercial scale and the combination heating methods supplemented with conventional thermal treatment for uniform heating of particulate foods has been widely investigated due to inherent advantages of microwave heating. Although microwave heating technology for a variety of food processing applications provide significant advantages with respect to lethal effect on pathogens, processing time, and energy consumption; yet several other quality aspects of food products processed using conventional methods are still better than microwave in terms of color, texture, and other organoleptic properties of food products. Although microwave energy has wide applications in various food processes, yet it needs significant research aimed at improvements in certain areas.

Specifically, methods to obtain final food products with better senso-rial and nutritional qualities need to be explored. Improving the energy efficiency in rice cooking and obtaining good quality product in bread baking are examples of other potentially challenging areas. Microwave processing of food materials needs to be carried out to a great extent at a pilot scale level than at laboratory conditions so that the results might be useful for industrial applications. In spite of the complex nature of microwave-food interactions, more research needs to be carried out for a better understanding of the process. Therefore, the investigation of parameters which can influence the workability of microwave heating such as dielectric, physical, and chemical properties of food products should be carried out [143].

8.9 SUMMARY

In recent years, microwave heating has been increasingly popular all over the world, in particular for modern household food-processing applica-tions, due to increased economic merits in many developing countries such as steady economic growth, high disposable income, etc. This trend also seems to be associated with increased awareness about the ben-efits of nutritious and healthy foods as well as functionalities of certain phytochemicals in diets, which may act as nutraceuticals. Microwave heating is known for its operational safety and nutrient retention capac-ity with minimal loss of heat-labile nutrients such as B and C vitamins, dietary antioxidant phenols and carotenoids. This review was aimed to provide a brief yet comprehensive update on prospects of microwave heating for food processing applications, its use is limited such as only for cooking, baking, drying, tempering, pasteurization, sterilization, etc. Nevertheless, many investigations carried out worldwide proved that microwave treatment could be used in the dairy industry for pasteuriza-tion and sterilization purposes with special emphasis on the benefits at household level and its impact on quality in terms of microbial and nutri-tional value changes. Food products undergo deterioration mostly due to improper perseveration techniques. Therefore, there is a need of an instant, advanced and affordable technology such as microwave technol-ogy which will be beneficial for dairy as well as food industry for shelf life extension of products.

KEYWORDS

- baking
- blanching
- cooking
- cooking cavity
- dairy products
- dipolar interaction
- door and choke
- drying
- foaming
- food dehydration
- food processing
- ionic interaction
- magnetron
- microwave heating
- milk
- pasteurization
- power supply and control
- puffing
- shelf life
- sterilization
- tempering
- thawing
- waveguide

REFERENCES

1. Alajaji, S. A., & El-Adawy, T. A. (2006). Nutritional composition of chickpea (*Cicerarietinum L.*) as affected my microwave cooking and other traditional cooking methods. *J Food Compos Anal.*, *19*, 806–812.
2. Albert, Cs., Zs. Mandoki, Zs. Csapo-Kiss, & J. Csapo (2009). The effect of microwave pasteurization on the composition of milk. *Acta Univ. Sapientiae, Alimentaria.* *2*(2), 153–165.

3. Aronowicz, J. (1975). In-line microwave tempering upgrades quality of sliced meats. *Food Process.*, *36*(12), 54–55.

4. Begum, S., & Brewer, M. S. (2001). Chemical, nutritive, and sensory characteristics of tomatoes before and after conventional and microwave blanching and during frozen storage. *J. Food Qual.*, *24*(1), 1–15.

5. Bennion, E. B., & Bamford, G. S. T. (1997). *The Technology of Cake Making.* 6th edition, Chapman and Hall, London.

6. Bertrand, K. (2005). Microwave foods. *Food Technol.*, *59*(1), 30–40.

7. Bezanson, A. (1975). Thawing and tempering frozen meat. *Proceedings of the Meat Industry Research Conference, Raytheon Co., Waltham, Massachusetts,* American Meat Science Association: Illinois, USA, pp. 51–62.

8. Bhattacharjee, S., Ray, P. R., Bandyopadhyay, A. K., & Ghatak, P. K. (2003). Application of microwave heat treatment in enhancing the keeping quality of milk. *J. Dairying, Foods, & H. S.*, *22*(1), 55–58.

9. Bilbao, C., Albors, A., Gras, M., Andres, A., & Fito, P. (2000). Shrinkage during apple tissue air-drying: Macro and microstructural changes. In: *Proceedings of the 12th International Drying Symposium (IDS2000).* Eindhoven, The Netherlands, Paper No. 330.

10. Bilgen, S., Y. Coskuner, & E. Karababa (2004). Effects of baking parameters on the white layer cake quality by combined use of conventional and microwave ovens. *J. Food Processing Preservation*, *28*, 89–102.

11. Bookwalter, G. N., Shukla, T. P., & Kwolek, W. F. (1982). Microwave processing to destroy *Salmonellae* in corn-soy-milk blends and effect on product quality. *J. Food Sci.*, *47*(5), 1683–1686.

12. Brewer, M. S., & Begum, S. (2003). Effect of microwave power level and time on ascorbic acid content, peroxidase activity and color of selected vegetables. *J. Food Processing Preservation*, *27*, 411–426.

13. Brody, A. L. (2012). The coming wave of microwave sterilization and pasteurization. *Food Technol.*, March, 78–80.

14. Buffler, C. R. (1992). Microwave cooking and processing. In: *Engineering Fundamentals for the Food Scientist.* Chapman and Hall: New York.

15. Buffler, C. R. (1993). *Microwave Cooking and Processing: Engineering Fundamentals for the Food Scientist.* New York: Van Nostrand Reinhold.

16. Burfoot, D., & Foster, A. M. *Microwave Reheating of Ready Meals.* MAFF Microwave Science Series.

17. Burfoot, D., & James, S. J. (1992). Developments in microwave pasteurization systems for ready meals. *Process Technol.*, 6–9.

18. Cadeddu, S. (1981). Using microwave techniques in the production of Mozzarella cheese. In: *Proceedings Second Biennial Marchall International Cheese Conference,* Madison, Wisconsin, Marshall Italian and specialty cheese by California Dairy Research Foundation (CDRF) and Rhodia, Inc.: California, USA, pp. 176–179.

19. Campana, L. E., Sempe, M. E., & Filgueira, R. R. (1993). Physical, chemical, and baking properties of wheat dried with microwave energy. *Cereal Chem.*, *70*(6), 760–762.

20. Cano, M. P., Marın, A., & Fuster, C. (1990). Effects of some thermal treatments on polyphenol oxidase activities of banana (Musa cavendishii var. Enana). *J. Sci. Food Agric.*, *51*, 223–231.

21. Chamchong, M., & Datta, A. K. (1990a). Thawing of foods in a microwave oven. I. Effect of power levels and power cycling. *J. Microwave Power Electromagn. Energy*, *34*(1), 9–21.

22. Chavan, K. D., & Kulkarni, M. B. (2006). Effect of solar radiation and microwave heating on microbiological quality of *khoa*. *Indian J. Dairy Sci.*, *59*(5), 291–295.

23. Chavan, R. S., & Chavan, S. R. (2010). Microwave baking in food industry: A review. *International Journal of Dairy Science*, *5*(3), 113–127.

24. Choi, H. K., Marth, E. H., & Vasavada, P. C. (1993). Use of microwave energy to inactivate *Yersiniaenterocolitica and Campylobacter jejuni* in milk. *Milchwissenschaft*, *48*, 134–136.

25. Chung, J. C., Shu, H. H., & Der, S. C. (2000). The physical properties of steamed bread cooked by using microwave-steam combined heating. *Taiwanese J. Agric. Chem. Food Sci.*, *38*(2), 141–150.

26. Cohen, J. S., & Yang, T. C. S. (1995). Progress in food dehydration. *Trends in Food Science and Technology*, *6*, 20–25.

27. CSUF (1996). Microwave vacuum dehydration technology. *Indian Food Packer.*, Sept., 45–46.

28. Cui, Z. W., Xu, S. Y., & Sun, D. W. (2004). Microwave: Vacuum drying kinetics of carrot slices. *J. Food Engi.*, *65*, 157–164.

29. Daglioglu, O., Tasau, M., & Tuncel, B. (2000). Effects of microwave and conventional baking on the oxidative stability and fatty acid composition of puff pastry. *J. Am. Oil Chem. Soc.*, *77*(5), 543–545.

30. Daomukda, N., Moongngarm, A., Payakapol, L., & Noisuwan, A. (2011). Effect of Cooking Methods on Physicochemical Properties of Brown Rice. *6*, V1-1–V1-4, Singapore: IACSIT Press.

31. Dar, B. N., Ahsan, H., Wani, S. M., Kaur, D., & Kaur, S. (2010). Microwave heating in food processing: A review. *Beverage and Food World*, March, 36–40.

32. Dayananda, K. R., & Kumar, T. K. (2004). Application of microwaves in food industries. *Beverage and Food World*, Feb, 47–48.

33. De La Fuente, M. A., Carazo, B., & Juarez, M. (1997). Determination of major minerals in dairy products digested in closed vessels using microwave heating. *J. Dairy Sci.*, *80*, 806–811.

34. Dealler, S., Rotowa, N., & Lacey, R. (1990). Microwave reheating of convenience meals. *Br. Food J.*, *92*(3), 19–21.

35. Decareau, R. V. (1985). *Microwaves in the Food Processing Industry*. Academic Press: New York.

36. Dincov, D. D., Parrott, K. A., & Pericleous, K. A. (2004). Heat and mass transfer in two-phase porous materials under intensive microwave heating. *J. Food Engi.*, *65*, 403–412.

37. Dumuţa-Codre, A., Rotaru, O., Giurgiulescu, L., Boltea, F., Crişan, L., & Neghelea, B. (2010). Preliminary research regarding the microwaves influence on the milk microflora. *Analele Universităţii din Oradea–Fascicula Biologie Tom.*, *17*(1), 103–107.

38. Erle, U., & Schubert, H. (2001). Combined osmotic and microwave-vacuum dehydration of apples and strawberries. *J. Food Eng.* 2001, *49*(2/3), 193–199.

39. Everini, C., Baiano, A., & Pilli, T. D. (2003). Microwave blanching of cubed potatoes. *J. Food Processing Preservation*, *27*, 475–419.

40. Fathima, A., Begum, K., & Rajalakshmi, D. (2001). Microwave drying of selected greens and their sensory characteristics. *Plant Foods Hum. Nutr.*, *56*(4), 303–311.
41. Fox, K. (1994). Innovations in citrus processing. *Fluss. Obst.*, *61*(11), 338–340 and *Fruit Process.*, *4*(11).
42. Funebo, T., & Ohlsson, T. (1998). Microwave-assisted air dehydration of apple and mushroom. *J. Food Eng.*, *38*, 353–367.
43. Geczi, G., Nagy, P. I., & Sembery, P. Primary Processing of the animal food products with microwave heat treatment. http://www.agir.ro/buletine/1311.pdf
44. George, M. (1997). Industrial microwave food processing. *Food Rev.*, *24*(7), 11–13.
45. Gould, G. W. (1995). *New Methods of Food Preservation*. Blackie Academic and Professional: London, UK.
46. Grooper, M., Ramon, O., Kopelman, I. J., & Mizrahi, S. (1997). Effects of microwave reheating on surimi gel texture. *Food Res. Int.*, *30*(10), 761–768.
47. Guan, D., Gray, P., Kang, D. H., Tang, J., Shafer, B., Ito, K., Younce, F., & Yang, T. C. S. (2003). Microbiological validation of microwave circulated water combination heating technology by inoculated pack studies. *J. Food Sci.*, *68*(4), 1428–1432.
48. Guan, D., Plotka, V. C. F., Clark, S., & Tang, J. (2002). Sensory evaluation of microwave treated macaroni and cheese. *J. Food Processing and Preservation, 26*, 307–302.
49. Gundavarapu, S., Hung, Y. C., & Reynolds, A. E. (1998). Consumer acceptance and quality of microwave-cooked shrimp. *J. Food Qual.*, *21*(1), 71–84.
50. Hamid, M. A. K., Boulanger, R. J., Tong, SC., Gallop, R. A., & Pereira, R. R. (1969). Microwave pasteurization of raw milk. *J. Microwave Power*, *4*, 272–275.
51. Haynes, L. C., & Locke, J. P. (1995). Microwave permittivity of cracker dough, starch and gluten. *J. Microwave Power and Electromagnetic Energy, 30*(2), 124–131.
52. Ho, C. D., Jong, G. Y., & Yong, H. C. (1999). Effect of microwave blanching on the improvement of the qualities of immature soybean. *J. Kor. Soc. Food Sci. Nutr.*, *28*(6), 1298–1303.
53. Hoke, K., Klima, L., Gree, R., & Houska, M. (2000). Controlled thawing of foods. *Czech J. Food Sci.*, *18*(5), 194–200.
54. Hoogenboom, R., Wilms, T. F. A., Erdmenger, T., & Schubert, U. S. (2009). Microwave-assisted chemistry: a closer look at heating efficiency. *Aust J Chem.*, *62*, 236–243.
55. http://unctad.org/en/docs/ditccom20032_en.pdf (2003). Organic fruit and vegetables from the tropics: market, certification and production information for producers and international trading companies.
56. http://www.cfs.gov.hk/english/program/program_rafs/files/microwave_ra_e.pdf (2005). Microwave cooking and food safety. Risk Assessment Studies Report 19, Hong Kong.
57. Iilow, R., Regulska, I. B., & Szymozak, J. (1995). Evaluation of losses of vitamin C in some vegetables cooked by use of conventional procedures and in a microwave oven. *Bromatologiai Chemia Toksykologiczna*, *28*(4), 317–321.
58. Jeong, J. Y., Lee, E. S., Paik, H. D., Chol, J. H., & Kim, C. J. (2004). Microwave cooking properties of ground pork patties as affected by various fat levels. *J. Food Sci.*, *69*(9), C708–C712.
59. John, C. S., & Otten, L. (1989). Thin layer microwave drying of peanuts. *Can. Agric. Eng.*, *31*(2), 265–270.

60. Juliano, B. O. (1985). Production and utilization of rice. In: *Rice Chemistry and Technology*, (2nd edn.), edited by Juliano, B. O. St. Paul: American Association of Cereal Chemists, pp. 1–16.

61. Jun, S. K., Chang, H. L., & Ouk, H. (1998). Effects of height for microwave defrosting on frozen food. *J. Kor. Soc. Food Sci. Nutr.*, *27*(1), 109–114.

62. Jun, S. K., Kwang, J. P., Chang, H. L., & Ji, S. I. (1998). Physicochemical properties of dried anchovy (Engraulis japonica) subjected to microwave drying. Kor. *J. Food Sci. Technol.*, *30*(1), 103–109.

63. Kalse, S. B., Patil, M. M., & Jain, S. K. (2012). Osmo-microwave drying of onion slices. *Beverage and Food World*, May, 14–17.

64. Kaur, C., & Kapoor, H. C. (2001). Effect of different blanching methods on the physicochemical qualities of frozen French beans and carrots. *Indian J. Food Sci. Technol.*, *38*(1), 65–67.

65. Kidmose, U., & Kaack, K. (1999). Changes in texture and nutritional quality of green asparagus spears (*Asparagus officinalis,* L.) during microwave blanching and cryogenic freezing. *B. Soil Plant Sci.*, *49*(2), 110–116.

66. Kidmose, U., & Martens, H. J. (1999). Changes in texture, microstructure and nutritional quality of carrot slices during blanching and freezing. *J. Sci. Food Agric.*, *79*(12), 1747–1753.

67. Koutchma, T., Le Bail, A., & Ramaswamy, H. S. (2001). Comparative experimental evaluation of microbial destruction in continuous-flow microwave and conventional heating systems. *Can Biosyst Eng.*, *43*, 3.1–3.8.

68. Krokida, M. K., & Maroulis, Z. B. (1999). Effect of microwave drying on some quality properties of dehydrated products. *Drying Technol.*, *17*(3), 449–466.

69. Ku, H. S., Ball, J. A. R., & Siores, E. (2010). Review- Microwave processing of materials: Part III. http://citeseerx.ist.psu.edu/viewdoc/download?doi=10.1.1.514.96 68&rep=rep1&type=pdf.

70. Kuo, W. L., Li, C. L., & Suey, P. C. (1998). Sensory and physicochemical characteristics of precooked, microwave-reheated, low-fat sausage. *J. Chin. Soc. Anim. Sci.*, *27*(1), 129–142.

71. Kutchma, T. (1998). Synergistic effect of microwave heating and hydrogen peroxide on inactivation of microorganisms. *J. Microw. Power Electromagn. Energy*, *33*(2), 77–87.

72. Laguerre, J. C., Tauzin, V., & Grenier, E. (1999). Hot air and microwave drying of onions: A comparative study. *Drying Technol.*, *17*(7–8), 1471–1480.

73. Lakshmi, S., Chakkaravarthi, A., Subramanian, R., & Singh, V. (2007). Energy consumption in microwave cooking of rice and its comparison with other domestic appliances. *J Food Eng.*, *78*, 715–722.

74. Lewicki, P. P., Witrowa, D., & Sawczuk, A. (2001). Convective drying of apples and carrot assisted with microwaves. *Zywnosc*, *8*(2), 28–42.

75. Lopez-Fandino, R., Villamiel, M., Corzo, N., & Olano, A., (1996). Assessment of the thermal treatment of milk during continuous microwave and conventional heating. *J Food Prot.*, *59*, 889–892.

76. Manjunatha, H., Prabha, R., Ramachandra, B., Krishna, R., & Shankar, P. A. (2012). Bactericidal effect of microwave on isolated bacterial cells in milk, *paneer* and *khoa*. *J. Dairying, Foods, & H. S.*, *31*(2), 85–90.

77. Marconi, E., Ruggeri, S., Paoletti, F., Leonardi, D., & Carnovale, E. (1998). Physicochemical and Structural modifications in chickpea and common bean seeds after traditional and microwave cooking processes. *3rd European Conference on* grain legumes: *Opportunities for High Quality, Healthy and Added Value Crops to European Demands. Valladolid, Spain,* 14–19 November, pp. 358–359.

78. Martın, M. E., Fito, P., Martınez-Navarrete, N., & Chiralt, A. (1999). Combined air-microwave drying of fruit as affected by vacuum impregnation treatments. *Proceedings of the 6th Conference on Food Engineering, Dallas, TX, American Institute of Chemical Engineers (AIChe).*

79. Maskan, M. (2001). Drying, shrinkage and rehydration characteristics of kiwifruits during hot air and microwave drying. *J. Food Eng., 48*(2), 177–182.

80. Meredith, R. J. (1998). *Engineers' Handbook of Industrial Microwave Heating.* Institution of Electrical Engineering, London.

81. Metaxas, A. C. (1996). *Foundation of Electro-Heat.* John Wiley and Sons.

82. Mishra, B. K., & Pandey, R. K. (2008). Quality of pasteurized milk by microwave process. *Indian J. Dairy Sci., 61*(1), 13–18.

83. Mishra, B. K., & Pandey, R. K. (2011). Application of microwave process in the manufacture of Kheer. *Indian J. Dairy Sci., 64*(1), 16–20.

84. Monsalve, A. E., Peterson, G. V., & Palkert, P. E. (1999). *Shelf Stable Butter Containing Microwave Popcorn Article and Method of Preparation.* United States Patent.

85. Mudgett, R. E. (1982). Electrical properties of foods in microwave processing. *Food Technol.,* Feb, 109–115.

86. Murcia, M. A., Martınez Tome, M., del Cerro, I., Sotillo, F., & Ramırez, A. (1999). Proximate composition and vitamin E levels in egg yolk: losses by cooking in a microwave oven. *J. Sci. Food Agric., 79*(12),1550–1556.

87. Naresh, L., Venkateshaiah, B. V., Arun Kumar, H., & Venkatesh, M. (2009). Effect of microwave heat processing on physico-chemical, sensory and shelf life of *peda. Indian J. Dairy Sci., 62*(4), 262–266.

88. Negi, A., Boora, P., & Khetarpaul, N. (2001). Effect of microwave cooking on the starch and protein digestibility of some newly released moth bean (Phase *olusaconitifolius Jacq.)* cultivars. *J. Food Compos. Anal., 14*(5), 541–546.

89. Nott, K. P., & Hall, L. D. (1999). Advances in temperature validation of foods. *Trends in Food Science and Technology, 10,* 366–374.

90. Ohlsson, T. (1991). Microwave processing in the food industry. *Eur. Food Drink Rev., 7,* 9–11.

91. Ozmutlu, O., Sumnu, G., & Sahin, S. (2001). Assessment of proofing of bread dough in the microwave oven. *Eur. Food Res. Technol., 212*(4), 487–490.

92. Patil, P. B., Sajjanar, G. M., Biradar, B. D., Patil, H. B., & Devarnavadagi, S. B. (2010). Technology of *Hurda* production by microwave oven. *J. Dairying, Foods, & H. S., 29*(3/4), 232–236.

93. Pereira, R. N., & Vincente, A. A. (2010). Environmental impact of novel thermal and non-thermal technologies in food processing. *Food Research International, 43,* 1936–1943.

94. Picouet, R. A., Fernandez, A., Serra, X., Sunol, J. J., & Arnau, J. (2007). Microwave heating of cooked pork patties as a function of fat content. *J. Food Sci., 72*(2), E57–E63.

95. Plessi, M. (1995). Estimation of organic acids, sugars and minerals in widely consumed vegetables: Effect of microwave cooking. *Rivistadella Societa Italiana dell Alimentazione, 24*(1), 23–33.

96. Pozar, D. M. (2009). *Microwave engineering.* John Wiley and Sons.

97. Prabhanjan, D. G., Ramaswamy, H. S., & Raghavan, G. S. V. (1995). Microwave assisted convective air drying of thin layer carrots. *J. Food Eng., 25,* 283–293.

98. Premakumar, K., & Khurdiya, D. S. (2002). Effect of microwave blanching on the nutritional qualities of banana puree. *J. Food Sci. Technol., 39*(3), 258–260.

99. Raghavan, G. S. V., & Silveira, A. M. (2001). Shrinkage characteristics of strawberries osmotically Dehydrated in combination with microwave drying. *Drying Technol., 19*(2), 405–414.

100. Rakcejeva, T., Zagorska, J., Dukalska, L., Galoburda, R., & Eglitis, E. (2009). Physical-chemical and sensory characteristics of Cheddar cheese snack produced in vacuum microwave dryer. *Chemical Technol., 3*(52), 16–20.

101. Ramaswamy, H. S., & Fakhouri, M. O. (1998). Microwave blanching: Effect on peroxidase activity, texture and quality of frozen vegetables. *J. Food Sci. Technol., 35*(3), 216–222.

102. Rao, K. J., & Pagote, C. N. (2012). Traditional Indian milk products – Value addition and functionality. *Indian Food Industry, 31*(3), 31–37.

103. Report No. 2. London: MAFF Publications, PB0524, (1991). Característic as nutrición aisesensoriais de sopacremosasemi-instantânea à base de farinhas de tri-goesojadesengordurada. *Pesquisa Agropecuária Brasileira, 29*(7), 1137–1143, ISSN-0100–204X.

104. Richardson, P. S. (1992). Product design for microwave reheating. *Indian Food Ind., 11*(2), 24–32.

105. Roberts, J. S., & Gerard, K. A. (2004). Development and evaluation of microwave heating on apple mash as a pretreatment to pressing. *J. Food Process Engi., 27,* 29–46.

106. Ruiz, G., Martınez-Monzo, J., Barat, J. M., Chiralt, A., Fito, P. (2000). Applying microwaves in drying of orange slices. *Proceedings of the 12th International Drying Symposium (IDS2000),* Eind-hoven, The Netherlands, Paper No. 239.

107. Sahin, S., & Sumnu, G. (2002). Effects of microwave cooking on fish quality. *Int. J. Food Prop., 4*(3), 501–512.

108. Sakonidoua, E. P., Karapantsiosa, T. D., & Raphaelides, S. N. (2003). Mass transfer limitations during starch gelatinization. *Carbohydrate Polymers, 53*(1), 53–61.

109. Sangwan, A., Kawatra, A., & Sehgal, S. (2010). Nutritional evaluation of onion powder dried using different drying methods. *J. Dairying, Foods, & H. S., 29*(2), 151–153.

110. Schwab, E. C., & Brown, G. E. (1993). *Microwave Tempering of Cooked Cereal Pellets or Pieces.* Minneapolis, MN: General Mills, US Patent 05182127.

111. Scilipalius, L. E., Myers, J. J., & Frey, J. P. (1989). *Process for producing a microwave puffed cheese snack.* United States Patent.

112. Severini, C., de Pilli, T., Baiano, A., Mastrocola, D., & Massini, R.(2001). Preventing enzyme at ice browning of potato by microwave blanching. *Sci. Aliments, 21*(2), 149–160.

113. Sharma, G. P., Jain, S. K., & Verma, R. C. (2012). Drying of garlic cloves using microwaves. *Beverage and Food World,* Dec., 25–26.

114. Sharma, R., & Lal, D. (1998). Influence of various heat processing treatments on some B vitamins in buffalo and cow's milk. *J Food Sci. and Technol.*, *35*(6), 524–526.

115. Sharma, S. (2007). Microwave technology – Application in food processing. *Indian Food Packer.*, Jan-Feb, 74–79.

116. Sieber, R., Eberhard, P., & Ruegg, M. (1989). Microwave treatment of food especially milk and milk products-A literature overview. *Food Technol.*, *22*(9), 198–203.

117. Sieber, R., Eberhard, P., & Gallmann, P. U. (1996). Heat treatment of milk in domestic microwave ovens. *Int. Dairy J.*, *6*, 231–246.

118. Sierra, I., Vidal-Valverde, C., & Olano, A. (1999). Effects of continuous flow microwave treatment and conventional heating on the nutritional value of milk as shown by influence on vitamin B1 retention. *Eur. Food Res. Technol.*, *209*, 352–354.

119. Singh, J., G. Maurya, & P. R. Patel (2009). Technology of *gheer* production by microwave oven. *J. Dairying, Foods, & H. S.*, *28*(2), 119–121.

120. Singh, M., Raghavan, B., & Abraham, K. O. (1996). Processing of marjoram (*Marjonahortensis Moench.*) and rosemary (*Rosmarinusofficinalis* L.): Effect of blanching methods on quality. *Nahrung*, *40*, 264–266.

121. Singh, S., & Rai, T. (2004). Process optimization for diffusion process and microwave drying of *Paneer J. Food Sci. Technol.*, *41*(5), 487–491.

122. Solanki, P., Dabur, R. S., & Masoodi, F. A. (2002). Storage study of microwave treaded *burfi. Indian Food Packer.*, Nov–Dec, 153–156.

123. Soni, M., Nigam, S., Patidar, K., Yadav, C. P. S., Jain, D., & Jain, S. (2010). Microwave drying: A promising drying technique. *J. Pharmacy Res.*, *3*(11), 2790–2793.

124. Steed, L. E., Truong, V. D., Simunovic, J., Sandeep, K. P., & Kumar, P., Cartwright, G. D., & Swartzel, K. R. (2008). Continuous flow microwave–assisted processing and aseptic packaging of purple-fleshed sweet-potato purees. *J. Food Sci.*, *73*(9), E455–E462.

125. Steele, R. J. (1987). Microwave in the food industry. *CSIRO Food Res. Q.*, *47*(4), 73.

126. Takashima, H. (2005). *Sponge Cake Premix and Method of Manufacturing Sponge Cake By Using Said Premix.* United States Patent 6,884,448.

127. Tochampa, W., Jittrepotch, N., Kongbangkerd, T., Kraboun, K., & Rojsuntornkitti, K. (2011). Study of microwave heating time on chemical and microbiological properties and sensory evaluation in sweet fermented glutinous rice. *Int. Food Res. J.*, *18*, 239–248.

128. Torringa, H. M., Nijhuis, H. H., & Bartels, P. V. (1993). Preservation of vegetables by microwave drying: Comparison of conventional with microwave drying. *Voedingsmiddelentechnologie*, *26*(8), 12–15.

129. Tsoubeci, M. N. (1994). Molecular interactions and ingredient substitution in cereal based model systems during conventional and microwave heating. PhD Dissertation, University of Minnesota, United States, Minnesota.

130. Unnikrishnan, V., & Vedavathi, M. K. (1993). A note on microwave heating for determination of total solids in milk. *Indian J. Dairy Sci.*, *46*(6), 283–285.

131. Uprit, S., & Mishra, H. N. (2003). Microwave convective drying and storage of soy-fortified *Paneer. Trans I. Chem.*, *81*, 89–96.

132. Valero, E., Villamiel, M., Sanz, J., & Martınez Castro, I. (2000). Chemical and sensorial changes in milk pasteurized by microwave and conventional systems during cold storage. *Food Chem.*, *70*(1), 77–81.

133. Valsechi, O. A., Horii, J., & De Angelis, D. (2004). The effect of microwaves on microorganisms. Arquivos do. *Instituto Biologico,* São Paulo, *71*(3), 399–404.

134. Van Mourik, L. E., & Bartels, P. V. (1999). Microwave blanching of mushrooms may improve yield. *Voedings middle entechnologie, 32*(10), 11–14.

135. Varith, J., Dijkanarukkul, P., Achariyaviriya, A., & Achariyaviriya, S. (2007). Combined microwave-hot air drying. *J Food Eng., 81,* 459–468.

136. Venkateshaiah, B. V., & Naresh, L. (2004). Application of microwave processing in dairy and food industry. *Indian Dairyman, 56*(5), 29–34.

137. Vyawahare, A. N. (2014). *Effect of Microwave Treatment on Physico-Chemical, Sensory and Shelf Life of Date Burfi.* MTech thesis submitted to Maharashtra Animal and Fishery Science University (MAFSU), Nagpur (MS), India.

138. Wang, S. H., Clerici, M. T. P. S., & Sgarbieri, V. C. (1993). Característic assensoriaise nutriciona is de mingau de preparorápido à base de farinhas de arroz e sojadesengordurada e leiteempó. *Alimentos e Nutrição, 5*(1), 77–86.

139. Whittaker, G. (2004). Microwave chemistry. *School Sci. Review, 85*(312), 87–94.

140. Witkiewicz, K., & Nastaj, J. F. (2010), Simulation Strategies in Mathematical Modeling of Microwave Heating in Freeze-Drying Process. *Drying Technol., 28,* 1001–1012.

141. Wrolstad, R. E., Lee, D. D., & Poei, M. S. (1980). Effect of microwave blanching on the color and composition of strawberry concentrate. *J. Food Sci., 45,* 1573–1577.

142. Yanai, K., Miura, M., Nakamura, R., Nishinomiya, T., Harada, T., & Kobayashi, S. (2001). Effect of freezing, storage and thawing conditions on quality of frozen cooked rice. *J. Japan Soc. Food Sci. Technol., 48*(10), 777–786.

143. Yanai, H., Negishi, H., & Asagiri, M. (2005). IRF-7 is the master regulator of type-I interferon-dependent immune responses. *Nature, 7,* 772–777.

144. Yeo, H. C. H., & Shibamoto, T. (1991). Chemical comparison of flavors in microwaved and conventionally heated foods. *Trends Food Sci Tech., 2,* 329–332.

145. Young, L., & Cauvain, S. P. (2007). *Technology of Bread making.* 2nd Edition. Springer, Germany.

146. Zhang, M., Tang, J., Mujumdar, A. S., & Wang, S. (2006). Trends in microwave related drying of fruits and vegetables. *Trends Food Sci Technol., 17,* 524–534.

147. Zhang, H., & Huang, J. (2001). Study on the microwave technology to inhibit the angel cake mold. *Food Mach., 2,* 18–19.

CHAPTER 9

MILK SILOS AND OTHER MILK STORAGE SYSTEMS

VANDANA CHOUBEY

CONTENTS

9.1   INTRODUCTION

Dairy industry employ various tasks in processing and handling various type of milk and milk products and therefore different types of storage system is required for various processes like transport, storage, culturing, intermediate process, maturing process, etc. The use of storage tanks starts from initial till final processing of the milk as milk that is collected at procurement center is chilled and kept in bulk milk cooler, then the milk is transferred to processing unit in insulated tankers. In processing unit, milk is received at RMRD and stored in raw milk storage tanks, which is then clarified and heat treated and stored in holding tanks. Then this milk is forwarded to various sections and stored in intermediate tanks for manufacturing different products.

Design of process and storage tanks for liquid foods must take into consideration: the ease of cleaning and method of cleaning, whether manual or automatic [6]. Storage tanks are made of different shapes: round, oval, cylindrical, different sizes and different design based on the process requirements like some tanks are insulated to maintain temperature, some are double jacketed to attain and maintain temperature, and some time hermetically sealed sterile tanks are also used. Tanks generally employ agitator, temperature indicator, manhole, watch glass and openings for product inlet and outlet pipelines and CIP pipelines.

Most commonly used tanks in dairy are cylindrical in shape that may be either vertical (Figure 9.1) or horizontal depending upon the requirement with adjustable leg with leveling screw to balance any irregularities of floor. All stainless steel welding joints should be continuous with welding material of similar composition [4]. Inner and outer shell of tanks, fittings and mountings and those entire surfaces that come in direct contact with product are generally made of stainless steel excluding the light, sight glass and gasket. Care should be taken so that all the surfaces which come in contact with product must be smooth and seamless so that during processing and after CIP no milk traces gets deposited at joints or welds. Sharp edges or any parts which are harder to clean must be avoided. The inner shell bottom should have slope towards the outlet of the tank to help complete removal of milk. The outer shell is also made of stainless steel and all welds and joint on outer shell must be smooth. Generally stiffeners are used between inner and outer shell and also as supporting unit for the base of the tank. For this purpose, stiffeners can be made of mild steel but should be painted with epoxy primer to avoid rusting. In addition, the tank-wall and tank-ceiling separations should be large enough to allow access to cleaning operations [3].

Construction materials for food processing and auxiliary system equipment that are in contact with foods or cleaning agents should have certain characteristics [5]:

1. *Resistance to corrosive action* of foods or chemicals (cleaning and sanitation agents).

FIGURE 9.1   Milk tanks: cylindrical in shape and vertical.

2. *Suitable surface finish* discouraging buildup of dirt that can accumulate with excessive surface rugosity.

3. *Good mechanical behavior* according to performance of mechanical functions, such as structural strength, resistance to abrasion and physical or thermal shocks, and pressure charges.

4. Easy assemblage and fastening operations using common methods (screw threads, welding, etc.) should not require special techniques. Possible forming of materials into desired shapes, into undulated surface sheets (e.g., plate heat exchangers), sheets and plates, rods, pipes, elbows, etc.

Materials not in contact with food or cleaning agents should meet many of the specifications required for machine construction, such as adequate rigidity and mechanical strength [1].

This chapter deals with types of tanks, components of tanks, design and construction of tanks and uses of different tanks in different sections of plant.

9.2 TYPES OF MILK TANKS

9.2.1 SILO/STORAGE TANKS

These tanks are basically meant for collection, reception and storage of large quantity of milk within the limited space. The capacity ranges from 25,000 to 2,50,000 Liters depending on the requirements. These tanks are placed outdoors to reduce the cost of building.

The silo tanks are usually of double wall, where the inner shell is made of stainless steel with conical head and flat bottom that slopes down with an inclination of 6% towards the outlet of tank to facilitate complete drainage. Outer shell is either made of stainless steel or mild steel with anticorrosion paint for cost cutting purpose. Outer shell also has a conical head and flat bottom. Insulating material is placed between inner and outer shell. Mineral wool is most commonly used insulating material. Accurate calculations for the exact insulating material thickness for silo stored outside the building depend upon the specific climatic condition.

Agitators are used for uniform mixing action in the entire volume to prevent cream separation by gravity without aeration of milk and fat breakage. Generally propeller agitators are used. Electrodes are provided in the tank wall at the top of the tank to indicate complete filling and prevent overfilling by closing inlet valve, at the bottom to indicate low level and in drainage line to indicate complete milk removal from the tank. Temperature of the milk in tank is indicated in control panel. An electric transmitter is used to transmit signal to central monitoring station.

FIGURE 9.2 Milk processing tank.

9.2.2 MILK PROCESSING TANKS

Processing tanks (Figure 9.2) are employed for production of various dairy products by changing the properties of milk, which may require maintenance of certain temperature management and stirring of the components. These are used for treatment, inoculation, maturation, homogenization, assertion, crystallization tanks for whipping cream, fermentation of yogurt and other dairy products. There are many types of process tanks that are designed for specific application. For example, yogurt and cream tanks are commonly manufactured with conical heads and bottoms and cooled using dimpled jackets with vessel shell and bottom cone to maintain optimum processing and operating temperature.

Process tanks are made up of stainless steel and with or without insulation employing various types of agitators depending on the viscosity of the product, process, and monitoring and control equipment. Capacity varies from 160 to 20,000 liters. Heating and cooling system on the shell and the bottom of the tank may be provided with maximum allowable pressure of 3 bars for optional heating/cooling arrangements.

9.2.3 BUFFER TANKS

Buffer tanks are used as intermediate storage tanks between two operations to store the product for short time before it continues in the process line. For example: after heating and cooling, milk is stored in buffer tank from where it goes to filling machine and in case of any break-down milk is stored in buffer tank till the production starts again. These tanks are usually insulated with the mineral wool between the two shells to maintain the temperature of stored product and the inner and outer shell are made of stainless steel. Buffer tanks generally have level indicator, temperature indicator, agitator and cleaning system.

9.2.4 BALANCE TANKS

Balance tanks are usually placed at the inlet of the pump so that the constant level of the product is maintained at the pump inlet. Product is free from air; and pressure in the suction side is uniform to maintain constant

FIGURE 9.3   Bulk milk cooler.

flow. Balance tank is equipped with float lever connected to an eccentrically pivoted roller that operates inlet valve of tank that operates opening and closing of valve. When float moves, downward valve is opened and upward valve is closed. The inlet of the tank is placed at bottom to avoid foam formation and aeration by splashing. However, some deaeration takes place as any air present in the product at entry will rise in tank thus helping the pump to operate more gently.

9.3 BULK MILK COOLER

Bulk milk cooler (Figure 9.3) is used for rapid cooling if milk volume exceeds 2,000 to 40,000 liters. These tanks are generally closed elliptical with top man hole. Tank inner shell, outer shell, piping, fittings, dipsticks, outlet inlet valve, blank flanges, filter body, lockable cover, agitator all are made of stainless steel. The gasket of good grade, for example, nitrile or neoprene rubber material should be used.

The tank should be equipped with agitator for uniform mixing and proper distribution of fat, non-foam inlet and outlet valve, inspection window, level indicator, manhole with locking arrangement, top cover lifting handle and ladder. Temperature sensors must be provided to sense the temperature and transmit signal to digital indicator. Digital indicator is installed in the control panel. Milk cooler are manufactured using CFC

FIGURE 9.4 Milk tanker.

free polyurethane insulation. The tank is also fitted with condensing unit, automatic washing system and control panel to ensure optimal cooling and storage of milk.

9.4 MILK TRANSPORT TANKS/TANKERS

In order to transport larger quantities of milk, tankers are used (Figure 9.4). Tanker with pump, hose, flow meter and other necessary installations is known as "Bulk Milk Pick Up Tank." The capacity varies from 500 to 12,000 Liters or more. Tankers are made of stainless steel in the inner shell with a thick layer of insulation. The outer layer is made up of either carbon steel, aluminum or stainless steel. Presently outer layer of SS is mostly used.

All inside shell welding should be smooth and polished as well as the corners and the edges must be smooth and round so that no residue rests on the inner surface. A manhole with fill connection, vent and cover assembly is required at the top of tanker for inspection, loading and cleaning. A manhole of suitable size must be provided for proper cleaning or an automatic cleaning device must be installed. A sanitary cover assembly is equipped with sanitary rubber gasket and locking device to form a tight seal.

Unloading is usually performed by sanitary pump. A stainless type sanitary valve is placed at the lower side than bottom of tank for complete removal of milk. The pumping of milk is carried through hose pipe from tankers to silo. Hose pipes are either made of rubber or plastics. The strength of the hose pipe should be such that it neither punctures nor collapses as pump starts. A dummy or plug must be placed at both the opening of hose pipe to close it when not in use to keep it free from dirt and dust. The rise in temperature of the milk in tanker is very less than that of the milk in cans. Hence, the milk stored in tankers can be stored at same temperature for longer period of time. The only problem is that if only a small quantity of bad milk is eventually added in the tanker, then it will spoil the entire milk in that tanker. Therefore, care must be taken before loading milk in the tankers from different places.

FIGURE 9.5   Aseptic tank.

9.4.1 ASEPTIC TANKS

Aseptic tanks (Figure 9.5) are intermediate storage tanks between UHT and packing machine. These are hermetically sealed tanks to ensure a certain level of sterility through thermal treatment followed by storage and filling in sterile condition and sterile packaging. A sterile product has a long shelf life at normal room temperature and requires neither cooling nor freezing.

The tank is sterilized by steam at a minimum temperature of 125°C for a period of time. It is then cooled by water circulating through the cooling jacket. During cooling, sterile air is fed into the tank to prevent vacuum formation. During production, sterile air fills the tank space above the product level. The pressure is automatically controlled to maintain the feed pressure required by the filling machine in operation. As an option, it can be equipped with an agitator. This is recommended for products that can separate in the tank during storage (e.g., chocolate milk) and to even out the product temperature. A valve cluster module with control panel directs product flow, sterile air, cleaning liquids, and steam. During production, a steam barrier (110°C) is applied to protect the product from contamination. After the filling machines, the end valve cluster prevents reinfection. The tank is cleaned in place by a central CIP system. Since tank operation includes high-temperature sterilization followed by cooling, the tank is designed to be completely implosion-proof. One of the three stainless legs is equipped with a load cell which measures the content of the tank and shows the reading on the panel. Tank operation is fully automated and production interlocks are included for safety reasons. The operator only has to initiate the process steps: tank sterilization, production and CIP. The tank is operated from its own programmable control in the control panel.

Aseptic tanks are used for supplying milk to several filling machines in case of any breakdown, total shutdown is not required. Aseptic tanks must have special safety armature and high hygiene sensors for control process.

9.5   COMPONENTS OF TANKS

9.5.1   MANHOLE

An oval shaped or round shaped manhole should be equipped at the front end of the tank for cleaning and inspection. Manhole (Figure 9.6) should be provided with leak proof inside and outside swing insulated stainless steel door with locking and tightening device. The door should open inside but with arrangement to take out when required. The gasket should be of good quality air tight and made up of either neoprene or nitrile rubber.

FIGURE 9.6   Manhole.

9.5.2   SPACE FOR CLEANING EQUIPMENT

Removable type stainless cleaning equipment such as spray head or rotary head must be provided at the top of the tank for spraying of cleaning solution over complete inner surface for effective cleaning during CIP. The equipment shall have SS at outer connections.

9.5.3   LIGHT AND VIEW PORTS

A temperature sensor (Figure 9.7) will record the temperature of milk. Light and view port (Figure 9.8) assemblies should be provided at approximately 150 mm of head space when the tank is full. Light assembly must be of stainless steel with lamp holder and view port assembly must also be of stainless steel placed at the top in such a manner that the whole view inside the tank is clear and level mark is easily readable. These fittings should be with few fittings part and simple so as to dissemble it easily for daily cleaning.

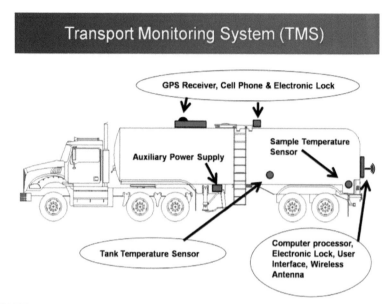

FIGURE 9.7 Temperature sensor. (Source: http://www.rs.uky.edu/regulatory/milk/milktransport/project_slides/index.php)

FIGURE 9.8 View port.

9.5.4 SAMPLING COCK

In order to take samples of the tank material a sampling valve must be provided. Sampling cock should be of stainless steel with sanitary design of size more than 5 mm and can be provided anywhere on the outer nozzle pipe or generally located on the manhole cover to avoid an additional fitting in insulation.

9.5.5 AIR VENT

An air vent of stainless steel must be provided on the top of the tank of not less than 150 mm size to prevent vacuum during CIP and emptying and excess pressure during filling of the tank. The air vent must be screened with removable wire mesh cover to prevent entrance of insects. A cover/ lid should also be bolted down in an air vent to prevent entry of dirt, dust or other particle falling from the top.

9.5.6 THERMOMETER

Thermometers are provided in the tank to indicate the temperature of the product in the tank. Stainless steel along the inclined thermo-well is welded in the inner and outer shell for thermometer fitting or special stainless steel jackets shall be provided for mounting thermometer on the front of the tank.

9.5.7 LIQUID LEVEL INDICATOR GAUGE

Level indicators are used to give a nearly approximate value of the product in tanks and should be calibrated. The gauge may be of tubular type in which the liquid level rises inside the glass tube carrying a graduated scale. The product in the tank can easily be read against the scale. Or the inner shell of the tank should be calibrated in such a way that there is a clear marking visible at opposite side through sight glass.

9.5.8 AGITATOR

Agitators (Figure 9.9) are necessary to obtain uniform composition, assure uniform distribution of fat and get better cooling efficiency. The speed of the agitator should be slow to ensure uniform mixing and non-separation of fat. Agitator speed depends on size of blade and tank, nature of product and way of agitation. Agitators may be vertical or horizontal:

Vertical agitators have long SS shaft and impellers driven by induction motor and an oil drip proof reduction gear box. The shafts are made of single piece solid rod of SS and need a step bearing in the bottom of

FIGURE 9.9   Agitator.

the tank as the shaft is long and removable. The shaft bearing shall be so located that it does not interfere the drainage of the milk from the tank. Vertical agitator has blades close to bottom thus allowing agitation of even small volume of product.

Horizontal agitators are located close to manhole for easy cleaning and replacement without entering the tank. They have oil less bearing and rotary seals that requires neither lubrication nor packing not fit for low volumes but greater accessibility and cleaning.

9.5.9   LADDER AND TOP PLATFORM

A SS ladder (Figure 9.10) should be welded for easy access to the top of the tank for inspection and maintenance and a platform made of dimpled SS shall be provided on the top of the tank for providing easy and enough space to sight glass and other assembly on top of the tank.

9.6   CLEANING OF TANKS: CLEAN-IN-PLACE (CIP) SYSTEM

Internal cleaning of food equipment can be manual or automatic. In hand cleaning, equipment should be designed to facilitate disassembling for cleaning and subsequent reassembling [2]. Manual cleaning, however, requires a great deal of time and labor. On the other hand, automatic cleaning is carried out without disassembling the equipment, resulting in great

FIGURE 9.10 Ladder.

savings in cleanup labor cost and time. This procedure is referred to as a clean-in-place (CIP) system. When applying a CIP system, following considerations should be taken into account [8–11]:

- The food processing plant, in which the CIP system is installed, must exhibit hygienic design. The design solution for equipment, including construction materials, should permit the installation of this system. In other words, if the CIP system is installed in a running process plant, it must be assumed that similar or better hygienic levels will be achieved.
- Careful selection of cleaning products in conjunction with type of soil removed and materials used to construct food equipment.
- Impact of the CIP system installation on total cost must be estimated, since supplementary capital investment and other operation cost will be needed. Installation of the CIP system must be profitable and economically feasible.

CIP systems are designed according to the product (nature, composition, and quantity), the most suitable cleaning frequency, and the equipment being cleaned (process or storage tanks), pipes, pumps or food processing equipment, such as heat exchangers and evaporators. Thus, the cleaning program should use the most adequate cleaning and sanitizing agents, and the frequency of application should be determined [12]. The selection of the best distribution system (spray-balls, rotating jets, etc.) depends on how the equipment will be cleaned. The main function of these devices is to distribute the cleaning agent uniformly over the entire surface being cleaned. Other designs for cleaning product distribution devices are spray rings and spray cane, used in evaporators, dryers, vacuum chambers, and other equipment of irregular design. All of these distribution devices, including spray-balls, allow the cleaning of more or less difficult points [11].

9.7 SUMMARY

Dairy industry employs various tasks in processing and handling of milk and milk products. Therefore different types of storage system are required for various processing operations like transport, storage, culturing, intermediate process, maturing process, etc. This chapter deals with types of tanks, components of tanks, design and construction of tanks and uses of different tanks in different sections of plant.

KEYWORDS

- agitator
- air vent
- aseptic tank
- balance tank
- buffer tank
- bulk milk cooler
- CIP
- ladder
- level indicator
- light port
- man hole
- process tank
- sampling cock
- silo
- spray head
- tanker
- temperature sensor
- UHT
- view port

REFERENCES

1. Baquero, J., & Llorente, V. (1985). *Equipments for Chemical and Food Industry (Equipos para la IndustriaQuímica y Alimentaria)*. Madrid, Ed. Alhambra.
2. Farrall, A. W. (1976). *Food Engineering Systems, Volumes I and II*. Westport, CT: AVI Publishing, Inc.

3. Hall, C. W., & Davis, D. C. (1979). *Processing Equipment for Agricultural Products.* 2nd edition. Westport, CT: AVI Publishing, Inc.
4. Hall, C. W., & Farrall, A. W. (1986). *Encyclopedia of Food Engineering.* Westport, CT: AVI Publishing Inc.
5. Jowitt, R. (1980). *Hygienic Design and Operation of Food Plant.* Chichester, UK: Ellis Horwood Ltd.
6. Kessler, H. G. (1987). *Food Engineering and Dairy Technology.* Verlog A. Kessler, Freising.
7. McKenna, B. B. (1984). *Engineering and Food, Volumes I and II.* London: Elsevier Applied Science Publishers.
8. Seiberling, D. A. (1979). Process/CIP engineering for shelf-life improvement. *American Dairy Review, 41*(10), 14–22, 64–67.
9. Seiberling, D. A. (1986). Clean-in-place and sterilize-in-place applications in the parenteral solutions process. *Pharmaceutical Engineering, 6*(6), 30–35.
10. Seiberling, D. A. (1992). Alternatives to conventional process/CIP design for improved cleanability. *Pharmaceutical Engineering, 12*(2), 16–26.
11. Seiberling, D. A. (1997). CIP Sanitary process design. In: *Handbook of Food Engineering Practice,* edited by Valentas, K. J., Rotstein, E., & Singh, R. P. CRC Press, pp. 581–631.
12. Troller, J. H. (1993). *Sanitation in Food Processing.* London: Academic Press.

CHAPTER 10

MILK COOLING METHODS: IMPORTANCE AND POTENTIAL USE

BHAVESH B. CHAVHAN, A. K. AGRAWAL, S. S. CHOPADE, and G. P. DESHMUKH

CONTENTS

10.1 INTRODUCTION

The first operation in a dairy plant is reception, chilling and storage of milk. Raw milk is pumped from the dump tank to the storage tank through a filter and chillers. The purpose of storage tank is to hold milk at low temperature so as to maintain continuity in milk processing operations and prevent any deterioration in the quality during holding and processing period. The milk may arrive at a chilling center or dairy plant in cans. After unloading the cans, milk is chilled and stored in storage tanks. Storage tanks are used to store raw or even pasteurized milk. Milk leaves the udder at body temperature of about 38°C. The bacterial load may grow rapidly and bring about curdling and other undesirable changes if milk is held at the ambient temperature. Freshly drawn raw milk should be promptly cooled and held at 4°C till processing to preserve it against bacterial deterioration. Milk may be held in chilled condition (<4°C) in the tank for up to 72 hours between reception and processing. Normally the milk storage capacity should be equivalent to one day's intake [4].

This chapter presents methods of milk cooling.

10.2   IMPORTANCE OF COOLING

Milk leaves the udder at body temperature containing only a few microorganisms. Normally milk contains bacteria coming from the animal's udder, milk vessels and operators. When the milk leaves the udder, bacteria grow well at the ambient temperature (20–40°C) and milk starts deteriorating. Bacterial growth factor goes down to 1.05 at 5°C and 1.00 at 0°C. Critical temperature for bacterial growth is 10°C. The growth factor at 10°C is 1.80 which rises to 10.0 at 15°C. Hence, freshly drawn raw milk should be promptly cooled to 5°C or below and held at that temperature till it is processed [6].

The growth factor increases rapidly, if growth is not checked immediately by chilling the milk. Chilling is necessary after receiving milk at collection/chilling center. Chilled milk can easily and safely be transported without having appreciable deteriorative changes due to microbial growth. Thus, raw milk is chilled to: (a) limit the growth of bacteria, (b) minimize micro-induced changes, and (c) maximize its shelf life. However, chilling of milk involves additional expense, which increases the cost of processing. Importantly, chilling process does not kill microorganisms nor it renders milk safe for human consumption. It is only a means of checking the growth of microorganisms for a certain period.

10.3   EFFECTS ON MICROBIAL GROWTH

Generally, milk is cooled immediately after milking to below 10°C within 4 hours to prevent/retard the multiplication of thermophilic and mesophilic bacteria including disease producing and food poisoning organisms until the milk reaches the dairy. The extent of control of growth of microorganisms is dependent on type of organisms. *Staphylococci* do not grow below 10°C. Growth stops for most types of *E. coli, B. proteus* and *Micrococci* between 0°C and 5°C. If milk is stored cold for too long time, there can be an undesirable increase in psychotropic organisms, which produce extremely heat resistant lipases and proteases.

The time factor is critical in arresting bacterial growth in fresh milk. As milk from the udder of healthy cows has a low bacterial count. There is a lag phase immediately after milking, for around 4 hours, before bacterial multiplication begins to grow. The quicker milk is cooled, the better is the quality. The milk is cooled immediately after milking to 4°C or below and held at that temperature till it is processed. The effect of storage temperature on microbial growth in raw milk is shown in Table 10.1.

10.3.1 EFFECTS ON STORAGE QUALITY OF MILK

Fresh raw milk is cooled to 4°C to extend its shelf-life (freshness). At this temperature, the activity of enzymes, the growth of microorganisms and metabolic processes are slowed down. As a result, prolonged holding of chilled milk is bound to cause significant deteriorative alterations in keeping quality of milk [2]. In addition, cooling causes a considerable dissociation of b-casein, calcium and phosphate ions and proteases from the casein micelles [3]. The milk loses its suitability for cheese making, coagulation times are increased and the curd tension of the coagulum is less.

TABLE 10.1 Effects of Storage Temperature on Microbial Growth in Raw Milk

Raw milk storage temperature (°C) for a period of 18 hours	Bacterial growth factor*
0	1.00
5	1.05
10	1.80
15	10.00
20	200.00
25	1,20,000.00

* Final count = (bacterial growth factor) x (Initial count of bacteria).

Chemical and biochemical processes are considerably slowed down by cooling. However, milk, which has been stored, sometime has a bitter off-flavor. Enzymes and microorganisms can cause chemical changes, which are accompanied by a low pH value and change in nitrogen-containing compounds. Psychrophilic microorganisms cause proteolysis of casein and, together with enzymes, also that of albumin. Protein breakdown products (polypeptides) are formed. Certain bacteria are responsible for the hydrolysis of fats causing rancid flavor development. Several enzymes such as oxidize catalyze and reductase are active for a long time, even at 0°C. Hence, if the time between milk reception and processing is 2 to 3 days, the storage temperature should be kept between 2 and 5°C for minimum effect on keeping quality of milk.

10.3.2   EFFECTS ON PHYSICO-CHEMICAL PROPERTIES OF MILK

10.3.2.1   Failure to Rennet/Acid Coagulation

The failure of casein to coagulate at 2°C either at pH 4.7 or after rennet treatment has been utilized in the development of continuous cheese making process, where the milk is either acidified or renneted at 2°C and the temperature is subsequently raised to about 15.6°C or 30°C to affect coagulation.

10.3.2.2   Failure to Coagulate at Iso-Electric Point

Milk fails to coagulate at 2°C after adjusting to the iso-electric point (pH) of casein. At 2–3°C, there is an increase in the diffusible inorganic salts and a change in the casein micelle structure. Some micellar casein is converted to a non-micellar or soluble form (e.g., b-casein). At 2°C, the pH of the milk must be reduced to 4.3 to affect complete casein coagulation, whereas at 30°C the recovery of the casein at pH 4.6 was nearly complete. Also the properties of casein obtained by acid precipitation at 2°C and pH 4.3, and at 30°C and pH 4.6 were slightly different.

10.3.2.3   Increase in Viscosity

Storage of milk at 2 to 5°C, both raw and pasteurized, caused an increase in the viscosity of the product which may be related to changes in the protein system, since viscosity is influenced largely by the colloidal components of milk. Probably, conversion of colloidal calcium partly to soluble form may uncoil the casein micelle. The change in viscosity with storage at low temperature (2 to 5°C) was greatest during the first 24 hours and reaches maximum after about 72 hours.

10.3.2.4   Decrease in Cheese Curd Firmness

The cold aging of milk increased the rennet coagulation time at 30°C. The increased coagulation time was inversely related to the ratio of colloidal calcium-phosphate, and could be reversed by heating to 40°C for 10 minutes or by addition of calcium chloride to the milk prior to cold aging [8].

10.3.2.5   Increased Hydrolytic Rancidity

Cold storage of milk below 7°C is associated with an increase in the rate of development of rancidity. Cooling tends to dissociate the casein micelle and increases the total available lipase in the milk system. Subsequent treatment to milk (warming, agitation, etc.) bring lipases into contact with fat globules and liberate free fatty acids to produce rancidity in milk.

10.3.2.6   Increased Foaming

Cold milk foams readily. Milk proteins concentrate in the lamellae of the foam where b-lacto globulin acts as a surface active agent. Foams are formed by the preferential adsorption of surface active materials at an air-liquid interface with orientation of the material to form an air bubble.

10.3.2.7   Physical Structure of Fat Globules

Crystal structure and size vary as a function of both cooling rate and cooling temperature and regulate the hardness of the milk fat. More fat passes into the solid state by direct cooling than in stepwise cooling. The sensitivity of the fat globule membrane to shear and subsequent release of free fat is greater in milk that has a higher proportion of solid to liquid fat. Thus, milk rapidly cooled, to 0–5°C, is more sensitive to shear damage than that is cooled more slowly and in a stepwise manner [5].

10.4   EQUIPMENT AND METHODS OF CHILLING (APPENDIX A)

Cooling is the predominant method of maintaining milk quality during collection. The most important factor next to hygienic production of milk is the time between completion of milking and reducing the temperature low enough to restrict bacterial growth. Whatever the method of cooling, the faster the temperature is reduced from 37°C at milking, the better will be the resultant milk quality. Selection of a suitable method and equipment for prompt cooling, for example, chilling milk is dependent upon the available facilities at the moment keeping in view the volume of milk handling and time for cooling and keeping it cold till reaches for processing. Various methods of cooling of milk are described in the following subsections.

10.4.1   CAN (CONTAINER) IMMERSION

The fresh milk immediately after milking is placed in a container (preferably metal can) through strainer. The cans are gently lowered into a tank containing cooled water. The water level in the tank should be lower than the level of milk in cans to prevent water entering into the milk. Cooling of milk will slowly take place and if the water is cold enough, the milk temperature will be reduced low enough to allow the milk to be marketed/processed. The milk inside the cans may be stirred with the help of plunger for uniform quick cooling. In this method, a much smaller refrigeration unit is needed. The cans are kept cooled at the desired temperature (5–7°C) and the capacity of the unit is 200–280 liters of milk.

10.4.2   SURFACE COOLER

An improvement of water cooling is a metal surface cooler, where water flows through the inner side and milk flows over the outer surface in a thin layer (Figure 10.1). The milk is distributed over the outer surfaces of the cooling tubes from the top by means of a distributor pipe and flows down in a continuous thin stream. The cooling medium (mostly chilled water) is circulated in the opposite direction through inside of the tubes. A well designed water cooler will reduce milk temperature almost instantaneously. The cooled milk is received below in a receiving trough, from which it is discharged by gravity or a pump. It can be either an individual unit or cabinet type. The latter consists of two or more individual units, compactly assembled and enclosed in a cabinet. It is usually larger than those used on the farm/chilling center.

10.4.2.1 Advantages

- Transfers heat rapidly and efficiently.
- Relatively in-expensive.
- Aerates the milk and thus improves the flavor.

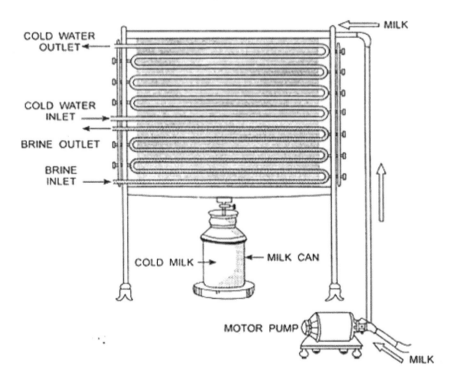

FIGURE 10.1 Surface cooler.

10.4.2.2 Disadvantages

- Requires constant attention of flow rate.
- Greater chances for air-borne contamination.

10.4.3 PLATE HEAT EXCHANGER (PHE)

The PHE is very effective equipment for chilling of milk in commercial dairy plants. It is widely used for large scale cooling of milk of 5000 to 60,000 liters/day at the chilling centers. This method of chilling is efficient, more hygienic, involves less manual labor and cost effective. Several stainless steel plates are mounted on a solid stainless steel frame in which the milk to be chilled and chilling water flow alternatively and counter-currently (Figure 10.2).

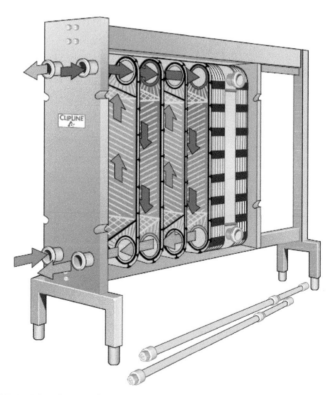

FIGURE 10.2   Plate heat exchanger.

The number and size of plates in the exchanger depend upon the capacity of the plant. It is widely used for large scale cooling of milk of 5000 to 60,000 liters/day at the chilling centers. They are efficient, compact and easily cleanable. In chillers, the gasket plates are tightly held between the plates. These plates are so arranged that milk flows on one side of plate and cooling medium (usually chilled water) on the other. There is a counter current flow between the milk and the chilled water through alternate plates. It helps in efficient transfer of heat to the cooling medium resulting in quick chilling of milk. The chilled milk flows from the plate cooler to the insulated storage tank at 4°C. A mechanical refrigeration system (IBT) is needed [7].

10.4.4   TUBULAR COOLER

This consists of two concentric tubes: inner tube usually carries the milk to be chilled while cold water is passing through the hollow space in between the pipes. The length and diameter of both the tubes are determined according to the capacity of the plant. The flow of the milk and chilled water is in opposite direction, for example, counter-current. The tubular cooler is efficient, where milk is not exposed to atmosphere.

10.4.5   MILK COOLER

This consists of a double jacketed vat fitted with a mechanical agitator. It has provision for circulation of chilled water, which comes from the chilled water tank. Normally, milk is chilled and subsequently stored at low temperature until transported to processing units for further processing. Bulk milk coolers are generally installed at chilling centers. Bulk tank coolers are run by mechanical refrigeration system, which cools the milk rapidly. These coolers maintain the temperature automatically during storage. Milk can be poured directly from milking pails into the tanks. This method is suitable for handling 500–2500 liters/day. It is widely used at village level milk collection centers. From the bulk milk cooler (BMC), the milk is pumped to the insulated tankers for transportation to dairy plants. The BMC uses horizontal or vertical cylindrical tanks with inner jacket and insulated body on the other side. There is provision of inner shell of the tank or direct expansion refrigerant coil for cooling. Milk is directly poured into the tank or pumped into the tank. Milk remains in contact with the inner shell of the tank for cooling it to 4°C. The agitator is provided for uniform cooling.

10.4.6 ROTOR FREEZE

In this system, evaporating unit cools the water, which in turn cools the milk in can. Several cans of milk can be cooled at a time. Rotor freeze provides spray of chilled water outside the cans obtained by mechanical refrigeration system and passing through the perforated tubes around the neck of the can. With this system, milk temperature is brought down to 10°C from 35°C within 15 minutes.

10.4.7 BRINE COOLING

The direct expansion coil is used to cool brine, which is then circulated by a pump around the product to be cooled. Brine system of cooling may be of: (a) brine circulating type, (b) brine storage type, and (c) congealing-tank type. This system has the advantage of being safe with ammonia and of causing less damage in case of a leakage and the temperature can be easily controlled. It allows heavy refrigerating loads of short duration to be carried with a system having a much smaller compressor than direct expansion system. The overall thermal efficiency of a brine system is usually less than the direct expansion system on account of the one extra heat transfer and the added radiation losses.

10.4.8 ICE COOLING

Ice, produced by commercial ice plants, is used in some countries to cool milk. The use of ice for cooling is generally fairly expensive and not particularly effective due to the problems in getting an optimum and rapid heat transfer from the liquid milk to the solid ice.

10.4.9   CAN COOLING

In this method, ice is placed in a metal container, known as ice gum or ice cone, which is inserted into the can of milk. This permits more effective heat exchange rate by giving off latent heat of ice and sensible heat of melted water, but reduces the volume of milk that can be carried in the milk can. When ice is completely melted in the ice cone and there is no more heat transfer, the water is thrown and fresh ice pieces are put in. The process of cooling milk by this method continues even during transportation from collection centers to processing unit.

10.4.10   DIRECT ADDITION OF ICE

Sometimes cooling of milk is done by direct putting ice into the milk. While this achieves an effective transfer of energy, and reasonably rapid cooling, it has a major disadvantage of diluting the milk with water, which will require removal at subsequent processing or the sale of adulterated milk.

10.4.11   MECHANICAL COOLING

Mechanical refrigeration system is the most effective means of arresting bacterial growth by lowering milk temperature to around 40°C. A typical flow diagram in the cooling process is shown in Figure 10.3. This system of cooling can be utilized in the following manners:

1. Household refrigerator

This is a practical method for small volume of milk (say from 1 or 2 animals, approx. 5 liters), where the farmers have a refrigerator. The milk in metal container immediately after milking is placed in a domestic household refrigerator, where the milk will slowly cool to the temperature of the refrigerator.

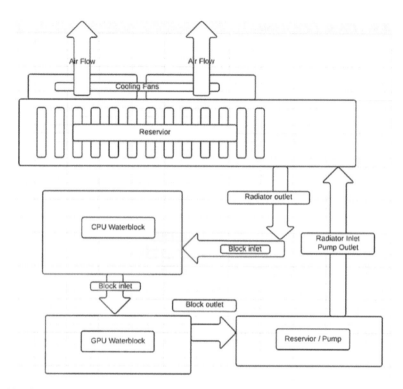

FIGURE 10.3   A typical flow diagram for a cooling system.

2. Surface/immersion cooler

Under direct expansion system, a mechanical refrigerator compressor and condenser (usually air cooled) produces a liquid refrigerant (Freon or ammonia). The liquid refrigerant while passing through an expansion system causes a rapid reduction in temperature [1]. Evaporating unit of a refrigeration unit is submerged directly into cans. Evaporator coil is fitted with an agitator. Milk is agitated for quick and proper transfer of heat from milk to refrigerant.

3. Cabinet cooler

It has a series of surface coolers installed closely together in a vertical position. Capacity of cabinet cooler to cool the milk depends upon the number of sections in surface coolers. This type of cooler requires very small floor space for installation.

10.4.12 EXPANSION BULK TANK

Direct expansion bulk tank, ranging in size from 500 L to 20,000 L, is an energy efficient system of cooling the milk to 40° C within the acceptable period of 4 hours. It is used directly on farms where, medium to large sized herds are milked or at collection/chilling center.

10.4.13 ICE AROUND METAL CANS OF MILK

It is the simplest form of cooling milk, in which ice slabs are stacked around the metal cans of milk on the delivery vehicle and the system relies on heat transfer by contact.

10.4.14 ICE BANK

The ice bank is a widely used for fast cooling of milk. This method of cooling reduces the size of the refrigeration compressor (hence, power requirement) by building up a reserve of ice over a long period. In ice bank, cooling is done through a plate heat exchanger or a surface type cooler with chilled water being the cooling medium. The chilled water is pumped from the ice bank through the heat exchanger and back to the ice bank. Ice banks have considerable flexibility in size and range from a small, self-contained portable unit to a large, using a multiple ammonia compressors, water condensers and associated cooling towers.

10.4.15 INTERNAL TUBULAR COOLER

It is a continuous cooling system consisting of a stainless steel tube of about 2.5–5.0 cm in diameter surrounded by a similar tube, forming a concentric cylinder. Several such tubes may then be connected in series to obtain sufficient cooling. The cooling medium flows in a counter current to the milk flow.

10.4.16 VAT/TANK COOLING

It is suitable for batch cooling, especially of small quantity. It consists of a tank within the tank, with the space between the two being used for circulation of the cooling medium, by either pump or main pressure. An agitator is provided to agitate the milk for rapid cooling.

10.5   CONCLUSIONS

Production of milk is vary widely scattered and at vast distances from the places of high consumptions or processing plant. The hygienic conditions and environment of milk production are still not up to the desired slandered. High ambient temperature throughout the year is an additional disadvantage since the bacterial growth is very rapid in the temperature of milk, as produced, is not brought down immediately after production. Absence of cooling or delayed cooling of milk after production increases the bacterial load considerably. Besides, bacterial multiplication is quite rapid as temperature of storage increases from 4° to 35°C. Therefore, it is essential to cool immediately after milking to maintain quality of milk, as cooling and transporting in bulk to processing plant may take 8 hours or more from time of milking. This stage of cooling/chilling the milk at the production center is most important factor.

10.6   SUMMARY

The dairy industry as it stands today, needs to improve the overall quality of the milk. The bacterial load is a reflection of the hygienic quality of milk. This aspect has certainly been ignored due to lack of facility for proper on-farm cooling of milk immediately after milking. During summer, the high ambient temperature result in high bacterial growth. As the battle for quality is won or lost at the village level itself, it is crucial to control the bacterial growth at the initial stages by cooling immediately after milking. Generally during summer season, cooling of raw and process milk is very important. Now-a-days, several methods and equipment are used such as: can immersion, surface cooler, ice cooling, plate heat exchanger, and rotor freeze.

KEYWORDS

- bacterial load
- brine cooling
- bulk milk coolers
- cabinet cooler
- chilling
- cooling equipment
- cooling of milk
- holding period
- hygienic quality of the milk
- ice bank
- immersion cooler
- internal tubular cooler
- on-farm cooling
- plate heat exchanger
- processing period
- raw milk
- rotor freeze
- storage of milk

REFERENCES

1. Ahmad, T. (2008). *Dairy Plant Engineering and Management*. 8th Edition, Kitab Mahal, 22-A, Sarojini Naidu Marg, Allahabad, India. pp. 189–190.
2. Bidii, K. P. (2015). Factors influencing sustainability of small scale dairy farming: a case of small scale dairy farmers in Cherangany, sub-county, Transnzoia county, Kenya. PhD dissertation, University of Nairobi, Kenya.
3. Holt, C. (2004). An equilibrium thermodynamic model of the sequestration of calcium phosphate by casein micelles and its application to the calculation of the partition of salts in milk. *European Biophysics Journal*, *33*(5), 421–434.
4. http://www.agrimoon.com/wp-content/uploads/Market-Milk-1.0.pdf.
5. http://dairy-technology.blogspot.in/2014/01/chilling-of-milk.html.
6. http://ecoursesonline.iasri.res.in/mod/page/view.php?id=147886.
7. *Tetra Pack: Hand Book of Dairy Processing*. S-221, 86 Lund, Swedan.
8. Walstra, P. (2005). *Dairy Science and Technology*. 2nd Edition, CRC Press, Taylor and Francis Group. pp. 601–603.

APPENDIX A   COOLING METHODS

Milkcooling tank

Travel plastic ice cooler

PART III

ENERGY USE IN DAIRY ENGINEERING: SOURCES, CONSERVATION, AND REQUIREMENTS

CHAPTER 11

USE OF RENEWABLE ENERGY IN THE DAIRY INDUSTRY

JANAKKUMAR B. UPADHYAY and RUCHI PATEL

CONTENTS

11.1   INTRODUCTION

Energy is one of the most important resources to sustain our lives and plays an important role in the growth and development of any economy. The requirement is increasing with increase in the population of the world. There is direct correlation between the development and amount of energy used. The demand for energy is increasing every day due to change in life-style of the people. The demand of energy continuously increasing but supply is limited. This situation is called energy crisis [10].

Energy is an essential input for industrial activities. Energy consumed in dairy's operations is in two major forms: Thermal energy and Electrical. Thermal energy consumption in the form of steam is far greater than the consumption of electrical energy in the form of power. Moreover, the quantum and manner in which steam is consumed, in most cases has direct relationship with the quantity of power consumed. To overcome problem the use of renewable energy has great scope for its commercial use in the dairy processing operations and it is estimated that renewable energy could contribute to at least half of all electric power in each of the large economies by 2050. At current rate of consumption and production, coal reserves in India would last for about 130 years and at current rate of consumption and production, oil in India would last only for about 20 to 25 years [4]. The world's average energy consumption per person is equivalent to 2.2 tons of coal. In industrialized countries, it is four times more than the world average energy consumption [7].

This chapter presents an overview of use of renewable energy in the dairy industry

11.2 ENERGY SCENARIO IN INDIA

The energy consumption in India is fourth biggest after China, USA and Russia [22]. The total primary energy consumption from coal (55%), Natural gas (10%), Diesel (1%), Nuclear (3%), Hydro (20%) and renewable (11%) [11]. About 70% of India's electricity generation capacity is from fossil fuels, with coal accounting for 40% followed by crude oil and natural gas at 28% and 6% respectively [24]. For oil and gas, India will become ever more dependent on imports from a few distant part of the world. Today, major electricity generation takes place at central power stations which utilize coal, oil, water, gas or fossil nuclear materials as primary fuel sources. They are not renewable, less efficient (65–75%) and expensive. Renewable energy is that energy which comes from the natural energy flows on earth. Unlike conventional forms of energy, renewable energy will not get exhausted. Renewable energy is also termed as "green energy," "clean energy," "sustainable energy" and "alternative energy" [6].

11.3 CLASSIFICATION OF ENERGY

11.3.1 PRIMARY AND SECONDARY ENERGY

Primary energy sources are those that are either found or stored in nature. Common primary energy sources are coal, oil, natural gas, and biomass (such as wood). Other primary energy sources available include nuclear energy from radioactive substances, thermal energy stored in earth's interior, and potential energy due to earth's gravity.

Primary energy sources are mostly converted in industrial utilities into secondary energy sources; for example coal, oil or gas converted into steam and electricity. Primary energy can also be used directly. Some energy sources have non-energy uses, for example coal or natural gas can be used as a feedstock in fertilizer plants.

11.3.2 COMMERCIAL ENERGY AND NON-COMMERCIAL ENERGY

11.3.2.1 Commercial Energy

The energy sources that are available in the market for a definite price are known as commercial energy. By far the most important forms of commercial energy are electricity, coal and refined petroleum products. Commercial energy forms the basis of industrial, agricultural, transport and commercial development in the modern world. In the industrialized countries, commercialized fuels are predominant source not only for economic production, but also for many household tasks of general population. Examples: Electricity, lignite, coal, oil, natural gas, etc.

11.3.2.2 Non-Commercial Energy

The energy sources that are not available in the commercial market for a price are classified as non-commercial energy. Non-commercial energy sources include fuels such as firewood, cattle dung and agricultural wastes, which are traditionally gathered, and not bought at a price used especially in rural households.

11.4   RENEWABLE AND NON-RENEWABLE ENERGY

Renewable energy is energy obtained from sources that are essentially inexhaustible. Examples of renewable resources include wind power, solar power, geothermal energy, tidal power and hydroelectric power. The most important feature of renewable energy is that it can be harnessed without the release of harmful pollutants.

Non-renewable energy is the conventional fossil fuels such as coal, oil and gas, which are likely to deplete with time.

11.4.1   WHAT IS RENEWABLE ENERGY?

Renewable energy is one of the cleanest sources of energy options with almost no pollution or carbon emissions and has the potential to significantly reduce reliance on coal and other fossil fuels. By expanding renewable energy, world can improve air quality, reduce global warming emissions, create new industries and jobs, and move world towards a cleaner, safer, and affordable energy future.

Renewable energy is an energy obtained from natural and persistent flows of energy occurring in the immediate environment [30].

Examples of renewable resources include wind power, solar power, geothermal energy, tidal power and hydroelectric power.

11.5   RENEWABLE ENERGY IN INDIA

India has a vast supply of renewable energy resources, and it has one of the largest programs in the world for deploying renewable energy products and systems. According to the Ministry of New and Renewable Energy (MNRE), a renewable energy system converts the energy found in sunlight, wind, falling water, sea waves, geothermal heat or biomass into a form, we can use such as heat or electricity. Among these, the largest share is 70% from wind power, 14% small hydropower, 2% from solar, and 14% from biomass and waste. [11].

11.5.1 WIND ENERGY

Wind energy is basically harnessing of wind power to produce electricity. The kinetic energy of the wind is converted to electrical. When solar radiation enters the earth's atmosphere, different regions of the atmosphere are heated to different degrees because of earth curvature. This heating is higher at the equator and lowest at the poles. Since air tends to flow from warmer to cooler regions, this causes winds, and it is harnessed in windmills and wind turbines to produce power.

Wind power is not a new development as this power, in the form of traditional windmills for grinding corn, pumping water, sailing ships have been used for centuries. Now, wind power is harnessed to generate electricity in a larger scale with better technology.

11.5.1.1 Wind Energy Technology

The basic wind energy conversion device is the wind turbine. Although various designs and configurations exist, these turbines are generally grouped into two types:

1. **Vertical-axis wind turbines**, in which the axis of rotation is vertical with respect to the ground (and roughly perpendicular to the wind stream),
2. **Horizontal-axis turbines**, in which the axis of rotation is horizontal with respect to the ground (and roughly parallel to the wind stream.)

The subsystems include a blade or rotor, which converts the energy in the wind to rotational shaft energy; a drive train, usually including a gearbox and a generator, a tower that supports the rotor and drive train, and other equipment, including controls, electrical cables, ground support equipment, and interconnection equipment.

- A wind energy system usually requires an average annual wind speed of at least 15 km/h.

- Small wind turbines can range in size from 20 watts to 100 kilowatts (kW).
- A 60 to 120-foot tower (5 to 10 stories) is common for small wind energy systems.
- The Amount of energy which the wind transfer to rotor depends on the air, rotor area and wind speed.

$$\text{Wind Power} = P = 0.5 \times A \times \rho \times V^3 \tag{1}$$

where, P = power, A = rotor area (m^2), ρ = air density (kg/m^3), and V = wind velocity (m/s).

11.5.1.2   Wind Energy in India

India has 19051 MW of installed capacity and has a potential of utilization up to 102772 MW. Some of the major wind energy plants are located in Tamil Nadu (7160 MW), Gujarat (3093 MW) and Maharashtra (2976 MW) [9].

11.5.2   SOLAR ENERGY

India being situated between the tropic of cancer and the equator, has an average temperature of 25–27.5°C and receives 260–300 clear sunny days per year making it the best solar resource in the world [18]. Earth receives on an average of 5–7 kWh (kilowatt-hour) solar radiation per square meter per day. The sun provides a virtually unlimited supply of energy. The energy from the sun is virtually free once the initial cost of the system has been recovered. The use of solar energy can, not only bridge the gap between the demand and supply of electricity but it also displaces conventional energy, which usually results in a proportional decrease in GHG emissions.

11.5.2.1   Solar Energy in India

The highest annual solar radiation is received by Rajasthan whereas the north-eastern parts of the country receive the least [17]. India has an installed power capacity of 1686 MW, making it sixth largest consumer in the world. Major plants are located in Gujarat, Rajasthan, Jodhpur, Tamil Nadu and Orissa [9].

11.5.2.2   Solar Water Heaters

SWH Systems for industrial and commercial applications are better known by the type of solar collector used. Based on the type of collectors, SWHS are divided into following three types:

- Flat Plate Collectors (FPC);
- Evacuated Tube Collectors (ETC); and
- Solar Concentrator.

11.5.2.3   Flat Plate Collector

A black absorbing surface (absorber) inside the flat plate collectors absorb solar radiation and transfer the energy to water flowing through it. The solar radiation is absorbed by flat plate collectors, which consist of an insulated outer metallic box covered on the top with glass sheet. Inside there are blackened metallic absorber (selectively coated) sheets with built in channels or riser tubes to carry water. The absorber absorbs the solar radiation and transfers the heat to the flowing water. It heats the fluid up to a 40–60°C.

11.5.2.4   Evacuated Tube Collector

The collector is made of double layer borosilicate glass tubes evacuated for providing insulation. The outer wall of the inner tube is coated with selective absorbing material. This helps absorption of solar radiation and transfers the heat to the water, which flows through the inner tube. ETC is highly efficient with excellent absorption (>93%) and minimum emittance (<6%) as the tubes are round and sunrays are striking the tubes at right angles thus minimizing reflection. The entire system is controlled and monitored by an automatic control panel. There is no scaling in the glass tubes thus, suitable for areas with hard water.

11.5.2.5   Solar Concentrator

Solar Concentrator is a device, which concentrates the solar energy incident over a large surface onto a smaller surface. The concentration is achieved by the use of suitable reflecting or refracting elements, which results in an increased flux density on the absorber surface as compared to that existing on the concentrator aperture. In order to get a maximum concentration, an arrangement for tracking the sun's virtual motion and accurate focusing device is required. Thus, a solar concentrator consists of a focusing device, a receiver system and a tracking arrangement. High temperature can be achieved using solar concentrators, and hence they have potential applications in both thermal and photovoltaic utilization of solar energy at high delivery temperatures.

11.5.2.6   Solar Photovoltaic (PV)

Photovoltaic is the technical term for solar electric. Photo means "light" and voltaic means "electric". PV cells are usually made of silicon, an element that naturally releases electrons when exposed to light. Amount of electrons released from silicon cells depend upon intensity of light incident on it. The silicon cell is covered with a grid of metal that directs the electrons to flow in a path to create an electric current. This current is guided into a wire that is connected to a battery or DC appliance. Typically, one cell produces about 1.5 watts of power. Individual cells are connected together to form a solar panel or module, capable of producing 3 to 110 Watts power. Panels can be connected together in series and parallel to make a solar array, which can produce any amount of Wattage as space will allow. Modules are usually designed to supply electricity at 12 Volts. PV modules are rated by their peak Watt output at solar noon on a clear day.

11.5.3   BIO-ENERGY

One third contributor of energy to India is biomass which comprises of solid biomass, which is an organic, non-fossil material of biological origins. Biogas which is principally methane and carbon dioxide is produced by anaerobic digestion of biomass and combusted to produce heat. Biogas is a clean and efficient fuel, generated from cow-dung, human waste or any kind of biological materials derived through anaerobic fermentation process. The biogas consists of 60% methane with rest mainly carbon-dioxide. Biogas is a safe fuel for cooking and lighting. By-product is usable as high-grade manure.

11.5.3.1 Components of Typical Biogas Plant

A digester in which the slurry (dung mixed with water) is fermented, an inlet tank for mixing the feed and letting it into the digester, gas holder/dome in which the generated gas is collected, outlet tank to remove the spent slurry, distribution pipeline(s) to transport the gas into the kitchen, and a manure pit, where the spent slurry is stored. Biomass fuels account for about one-third of the total fuel used in the country. It is the most important fuel used in over 90% of the rural households and about 15% of the urban households. Using only local resources, namely cattle waste and other organic wastes, energy and manure are derived. Thus the biogas plants are the cheap sources of energy in rural areas.

Currently, India has 3697 MW installed capacity and it results in a saving of about Rs. 20,000 crores every year [11]. Following is a list of some states with most potential for biomass production: Andhra Pradesh (200 MW), Bihar (200 MW), Gujarat (200 MW), Karnataka (300 MW), Maharashtra (1,000 MW), Punjab (150 MW), Tamil Nadu (350 MW), Uttar Pradesh (1,000 MW) [9].

11.5.4 HYDRO ENERGY

Energy from small hydro is the oldest. It is most reliable of all renewable energy sources. The potential energy of falling water, captured and converted to mechanical energy by waterwheels, powered the start of the industrial revolution.

Wherever sufficient head, or change in elevation, could be found, rivers and streams were dammed and mills were built. Water under pressure flows through a turbine causes it to spin. The turbine is connected to a generator, which produces electricity.

Hydroelectric power for large-capacity plants has been estimated to be 148,700 MW. For small plants, a total capacity is 15,384 MW [29]. India utilizes twelve primary hydroelectric power plants: Bihar, Punjab, Uttaranchal, Karnataka, Uttar Pradesh, Sikkim, Jammu and Kashmir, Gujarat, and Andhra Pradesh.

11.5.5 WAVE ENERGY

Sea waves are the result of the concentration of energy from various natural sources like sun, wind, tides, ocean currents, moon, and earth rotation. Waves originate from wind and storms far out to sea and can travel long distances without significant energy loss, and hence power production is much steadier and more predictable. Unlike the wind and solar, power from sea waves continues to be produced round the clock.

Wave energy contains roughly 1000 times the kinetic energy of wind. Hence it allows smaller devices to produce power. Wave energy varies as the square of wave height whereas wind power varies with the cube of air speed. Water being 850 times as dense as air results in much higher power produced from wave averaged over time. Theoretically it is possible to extract 40 MW of power per km of coast where there are gentle waves (say 1 m height) and 1000 MW per km of coast where the wave height is 5 m.

Kinetic energy from waves can be used to power a turbine. As the wave rises into a chamber, the rising water forces the air out of the chamber. The moving air spins a turbine which can turn a generator. When the wave goes down, air flows through the turbine and back into the chamber through doors that are normally closed thus generating power even when wave is receding.

11.5.6 TIDAL ENERGY

Tidal energy is another form of ocean energy. Tides are generated by the combination of the moon and sun's gravitational forces. Greatest effects of tides are in spring when sun and moon combine forces. Bays and inlets amplify the tide. Cycles of low and high tides occur twice a day. When tides come into the shore, they can be trapped in reservoirs behind dams. Then when the tide drops, the water behind the dam can be let out just like in a regular hydroelectric power plant. In order for the tidal energy to be practicable for energy production, the height difference needs to be at least 5 m is needed.

11.5.7 GEOTHERMAL ENERGY

The top most part of the earth is the crust. Below the crust of the earth is a layer called mantle. The top layer of the mantle is a hot liquid rock called magma. The crust of the earth floats on this liquid magma mantle. When magma breaks through the surface of the earth in a volcano, it is called lava.

For every 100 meters you go below ground, the temperature of the rock increases about 3°C. So, at a depth of about 3000 meters below ground, the temperature of the rock would be hot enough to boil water. Deep under the surface, water sometimes makes its way close to the hot rock and turns into boiling hot water or into steam. The hot water can reach temperatures of more than 148°C. When this hot water comes up through a crack in the earth, it is known as hot spring.

Some of the areas have so much steam and hot water that it can be used to generate electricity. Holes are drilled into the ground and pipes lowered into the hot water. The hot steam or water comes up through these pipes from below ground. A geothermal power plant is like in a regular power plant except that no fuel is burned to heat water into steam.

11.6   USE OF RENEWABLE ENERGY IN THE DAIRY INDUSTRY

Renewable energy is one of the most promising and important opportunities for value added products in dairying [23]. The type of renewable energy technology used in dairying depends on the type of energy required, access to the renewable energy sources and the design of the dairy facilities and processes. There are number of renewable energy sources which can easily be integrated in the dairy industry such as solar energy for cooling and heating purpose, bio-energy for process heat for dairy operation, etc.

and this energy can generate power at competitive cost. Adoption of renewable energy sources in dairying can help in reducing hydrocarbon emission.

11.6.1 APPLICATION OF SOLAR ENERGY IN THE DAIRY INDUSTRY

Now, a day technology has been developed in such a way that the solar energy is commercially feasible to collect. The cost of solar energy is static or rather decreasing. Solar energy system as non-convectional sources is being developed for various industrial applications such as heating of water for cleaning, washing, and/or as boiler feed water, etc. Presently, the Indian dairy industry has to bear increased cost of energy/liter of processed milk due to increased cost of traditional energy inputs. So, the use of solar energy in dairy processing operation is the best option to overcome convectional energy sources.

11.6.1.1 Solar Water Heating Systems (SWH)

Solar water heating considered as the most cost-effective alternatives for industry and household application. A dairy unit requires many heating operation which are at present carried out using steam from boiler. Thus, solar energy for various heating application is already being used in some of the dairy plants for reducing fuel bill [32]. A solar water heater (SWH) is a combination of an array of collectors, an energy transfer system and a thermal storage system. In active solar water heating systems, a pump is used to circulate the heat-transferring fluid through the solar collectors. The amount of hot water produced from a SWH critically depends on design and climatic parameters such as solar radiation, ambient temperature and wind speed. SWH system heating water at 60–80°C or even higher temperature which can be conveniently used for crate washing, cleaning, CIP, pre-heating of boiler feed water, etc. [19].

Solar Energy in Pasteurization

There is tremendous scope of utilizing solar energy in dairy processing such as pasteurization of milk. Solar panels/concentrator based milk pasteurizer system is developed to meet the demand of pasteurization. It was observed that base temperature of solar heated water reached up to 100°C, which have easily attained pasteurization temperature ranging from 65–75°C in two-three hours [31].

Solar Steam Generation

Low temperature steam is extensively used in sterilization processes and desalination evaporator supplies. Parabolic trough collectors (PTCs) are high efficient collectors commonly used in high temperature applications to generate steam. PTCs use 3 concepts to generate steam [14] the steam flash, direct and the unfired-boiler. The Heat Transfer fluid that circulates inside the solar field (primary cycle) is heated and transferred to heat exchanger, including super heaters, evaporators and pre heaters, where steam is generated for the power cycle or other applications [1]. This steam can be used for sterilization and pre-heating of air and also in drying operation.

Solar Energy for Cooling Purpose

Photovoltaic Operated Refrigeration Cycle, Photovoltaic (PV) involves the direct conversion of solar radiation to direct current (DC) electricity using semiconducting materials. In concept, the operation of a PV-powered solar refrigeration cycle is simple. Solar photovoltaic panels produce DC electrical power that can be used to operate a dc motor, which is coupled to the compressor of a vapor compression refrigeration system [15]. This system is feasible for cooling of milk in chilling center.

Solar Absorption Refrigeration System (NH₃-Water)

Solar Absorption Refrigeration system was designed to operate with the ammonia water mixture for a maximum capacity of 8 kg of ice/day. It consists of a compound parabolic collector (CPC) with a cylindrical receiver acting as the generator/absorber, a condenser, a storage tank, an expansion valve, a capillary tube and an evaporator. The system operates exclusively with solar energy and no moving parts are required [20]. The Electric heating requires 61 MJ of energy to produce 9 kg of ice and solar energy produces 7–10 kg of ice after receiving 28–30 MJ of radiation energy.

Solar powered Li-Br-water vapor absorption refrigeration (SVAR) system having rated capacity of 5 TR (17.5 kW) is available. The SVAR system consists of array of heat pipe evacuated tube collectors (HP-ETC), generator, evaporator, absorber, condenser and heat exchanger. The HP-ETC produces hot water at a temperature of 65–95°C which is used for the supply of thermal energy at the generator of the system. The values of actual COP of the SVAR system ranged from 0.24 to 0.66. The thermal energy supplied at generator ranged from 31.01 to 60.69 kW. The refrigerating effect is produced by SVAR system is 11.19 to 22.31 kW. The use of solar energy for the operation of VAR has a scope for cold storage of fruits and vegetables where temperature requirement is not very low [13].

Solar Drying and Dehydration Systems

Currently, electricity is always used to heat the air and as an additional energy source. Conventional drying systems using fossil fuels as a source of combustion, while solar dryer use solar irradiation for drying in industries, such as brick, crops, fruits, coffee, wood, textiles, leather, green malt and sewage sludge [25]. There are two main groups of dryer, high and low temperature dryers. Almost all high-temperature dryers use fossil fuels or electricity for the heating process. While the low temperature dryers use fossil fuels or solar energy, a low temperature generated by solar energy is ideal for use in the preheating process [12].

Direct Solar Drying

Direct drying consists of use of incident radiation only, or incident radiation plus reflected radiation. Most solar drying techniques that use only direct solar energy also use some means to reflect additional radiation onto the product to further increase its temperature. An example of direct absorption dryer is the hot box dryer. The aim of this type of a dryer is mainly to improve product quality by reducing contamination by dust, insect infestation, and animal or human interference. It consists of a hot box with a transparent top and blackened interior surfaces. Ventilation holes in the base and upper parts of slide walls maintained a natural air circulation. The farmers can dehydrate vegetables when these are available in plenty and at low cost. Dehydrated vegetables can be sold in the off-season when prices of vegetables are high and farmers can generate more income [21].

Indirect Solar Drying

Indirect solar dryer is generally known as conventional dryer. In this case, a separate unit termed as solar air heater is used for solar energy collection for heating of entering air. The air heater is connected to a separate drying chamber where the product is kept. The heated air is allowed to flow through wet material. Here, the heat for moisture evaporation is provided by convective heat transfer between the hot air and the wet material. The drying is basically by the difference in moisture concentration between the drying air and the air in the vicinity of product surface. A better control over drying is achieved in indirect type of solar drying systems and the product obtained is of good quality [27].

Solar Energy for Pumping Dairy Fluids

An SPV pump is a DC or AC, surface-mounted or submersible or floating pump that runs on power from an SPV array. It may use to run a hot water pump, chill water pump, milk pump and CIP (cleaning in place) pump. The array is mounted on a suitable structure and placed in a shadow free open space with its modules facing south and inclined at local latitude. A typical SPV pumping system consists of an SPV array of 200–3000 W capacity, mounted on a tracking/non-tracking type of structure. The array is connected to a DC or AC pump of matching capacity. SPV pumps are used to draw water for irrigation as well as for drinking. The SPV array converts sunlight into electricity and delivers it to run the motor and pump. The water can be stored in tanks for use during non-sunny hours, if necessary [8].

Solar Energy to Lightning Dairy Offices and Premises

SPV lighting systems are becoming popular in both the rural and urban areas of the country. In rural areas, SPV lighting systems are being used in the form of portable lanterns, home-lighting systems with one or more fixed lamps, and street-lighting systems. A solar street-lighting system (SLS) is an outdoor lighting unit used to illuminate a street or an open area usually in dairy, garden, road approach to dairy and chilling center [8]. A compact fluorescent lamp (CFL) is fixed inside a luminary which is mounted on a pole. The PV module is placed at the top of the pole, and a battery is placed in a box at the base of the pole. The module is mounted facing south, so that it receives solar radiation throughout the day. A typical street-lighting system consists of a PV module of 74 W capacities, a flooded lead–acid battery of 12 V, and a CFL of 11 W rating. The CFL automatically lights up when the surroundings become dark and switches off around sunrise time [6].

Solar Energy for Electrifying (Electric Fences)

Solar Electric fences are widely used in dairy to prevent stock or predators from entering or leaving an enclosed field. These fences usually have one or two 'live' wires that are maintained at about 500 volts DC. These give a painful, but harmless shock to any animal that touches them. This is generally sufficient to prevent stock from pushing them over. They require a high voltage but very little current and they are often located in remote areas where the cost of electric power is high. These requirements can be met by a photovoltaic system involving solar cells, a power conditioner and a battery [5].

11.6.2 APPLICATION OF BIO-GAS IN THE DAIRY INDUSTRY

Currently, most dairy digester produced biogas is used on site for energy generation. Electrical Production is generally the primary use of the produced biogas although heat is frequently also produced for use in the anaerobic digester either as part of a combined heat and power system (CHP) or separate dedicated boiler systems.

11.6.2.1 Biogas As Electricity and Heat Generation

Anaerobic digester systems have been used for decades at municipal wastewater facilities, and more recently, have been used to process industrial and agricultural wastes. These systems are designed to optimize the growth of the methane-forming (methanogenic) bacteria that generate CH_4. Typically, using organic wastes as the major input, the systems produce biogas that contains 55–70% CH_4 and 30–45% CO_2. On dairy farms, the overall process includes the following:

Manure Collection and Handling

Key considerations in the system design include the amount of water and inorganic solids that mix with manure during collection and handling.

Pre-treatment: Collected manure may undergo pre-treatment prior to introduction in an anaerobic digester. Pre-treatment include screening, grit removal, mixing, and/or flow equalization is used to adjust the manure or slurry water content to meet process requirements. A concrete or metal collection/mix tank may be used to accumulate manure, process water and/or flush water. Proper design of a mix tank prior to the digester can limit the introduction of sand and rocks into the anaerobic digester itself. If the digestion processes requires thick manure slurry, a mix tank serves a control point where water can be added to dry manure or dry manure can be added to dilute manure. If the digester is designed to handle manures mixed with flush and process water, the contents of the collection/mix tank can be pumped directly to a solids separator. A variety of solids separators, including static and shaking screens are available and currently used on farms.

Anaerobic digestion: An anaerobic digester is an engineered containment vessel designed to exclude air and promote the growth of methane bacteria. The digester may be a tank, a covered lagoon, or a more complex design, such as a tank provided with internal baffles or with surfaces for attached bacterial growth. It may be designed to heat or mix the organic material. Manure characteristics and collection technique determine the type of anaerobic digestion technology used. Some technologies may include the removal of impurities such as hydrogen sulfide (H_2S), which is highly corrosive.

By-product recovery and effluent use: It is possible to recover digested fiber from the effluent of some dairy manure digesters. This material can then be used for cattle bedding or sold as a soil amendment. Most of the ruminant and hog manure solids that pass through a separator will digest in a covered lagoon, leaving no valuable recoverable by-product.

Biogas recovery: Biogas formed in the anaerobic digester bubbles to the surface and may accumulate beneath a fixed rigid top, a flexible inflatable top, or a floating cover, depending on the type of digester. The collection system, typically plastic piping, then directs the biogas to gas handling subsystems.

Biogas handling: Biogas is usually pumped or compressed to the operating pressure required by specific applications and then metered to the gas use equipment. Prior to this, biogas may be processed to remove moisture, H_2S, and CO_2. Depending on applications, biogas may be stored either before or after processing, at low or high pressures.

Biogas use; Recovered biogas can be used directly as fuel for heating or it can be combusted in an engine to generate electricity. Biogas, a mixture consisting primarily of methane and carbon dioxide, is produced from dairy wastes through anaerobic digestion, a natural process that breaks down organic material in an oxygen-free environment [16].

11.6.2.2   Biogas for Refrigeration

Refrigeration accounts for about 15% to 30% of the energy used on dairy farms; most of this is for compressors used for chilling milk. Since dairy cows are milked daily, a steady source of energy is required for refrigeration needs.

Many dairies use well water for pre-chilling, chilled water or glycol can be produced from biogas-fired absorption or adsorption chillers and used in milk pre-coolers. Milk cooling using absorption and adsorption chillers also presents a potential opportunity to use waste heat captured from a biogas-driven generator set. Use of this waste heat could significantly reduce the on-farm electrical refrigeration load [16].

11.7   SUMMARY

Today, India is number one in milk production, producing about 138 million tons per annum. Out of these approximately 20% of the total milk production is handled by the organized sectors. Most of the milk processing operations, room conditioning for milk product packaging and cold stores for milk and milk products are operating on grid electric supply. Energy is one of the critical inputs for economic development of any Country.

In order to overcome the present energy scenario problems, energy should be conserved and since we are consuming disproportionate amount of energy that day is not far when all our non-renewable resources will expire forcing us to rely just on renewable sources. To overcome this problem, the use of renewable energy mainly solar and bio energy in the dairy is generally found for hot water supply to boiler, and hot water generator for processing of milk or for CIP cleaning. Use of renewable energy has great scope for its commercial use in the dairy processing operations and it is estimated that renewable energy could contribute to about half of all the types of energy used in each of the large economies by 2050.

KEYWORDS

- anaerobic digestion
- bio energy
- biogas for electricity and heat generation
- biogas recovery
- biogas refrigeration
- dairy
- energy
- energy scenario
- evacuated tube collectors (ETC)
- flat plate collectors (FPC)
- geothermal energy
- hydro energy
- India
- non renewable energy
- renewable energy
- renewable energy in dairy
- solar absorption refrigeration system
- solar concentrator
- solar dryer

- solar electric fence
- solar energy
- solar light
- solar pump
- solar water heater
- tidal energy
- wave energy
- wind energy

REFERENCES

1. Alguacila, M., Prietoa, C., Rodriguez, A., & Lohr, A. (2014). Direct steam generation in parabolic trough collectors. *Energy Procedia, 49,* 21–29.
2. BEE, (2013). Application of non-conventional and renewable energy sources http://www.emea.org/pdf, 147–161.
3. BEE, (2015). Application of non-conventional and renewable energy sources http://www.emea.org/pdf, 263–290.
4. Bhattacharya, T. (2014). Sustained growth of green energy economics. *International Journal of Inventive Engineering and Sciences, 2,* 16–18.
5. Carr, A., Fletcher, S., O'Mara, K., & Rayner, M. (1999). *Solar cell principle and applications.* Australian CRC for renewable energy Ltd.
6. Date, V. (2010). Financial feasibility of solar power project with reference to rural electrification of 39 talukas in Karnataka. NICMR.
7. Desai, H., & Zala, A. (2010). An overview on present energy scenario and scope for energy conservation in dairy industry. *National Seminar on Energy Management and Carbon Trading in Dairy Industry,* pp. 1–7.
8. Desai, D., Rao, J., Patel, S., & Chauhan, I. (2013). Application of solar energy for sustainable dairy development. *European Journal of Sustainable Development, 4,* 131–140.
9. Dhingra, R., Jain, A., Pandey, A., & Mahajan, S. (2014). Assessment of renewable energy in India. *International Journal of Environmental Science and Development, 5,* 459–462.
10. Garg, P. (2012). Energy scenario and vision 2020 in India. *Journal of Sustainable Energy and Environment, 3,* 7–17.
11. Gera, R. K., Rai, H. M., Parvej, Y., & Soni, H. (2013). Renewable energy scenario in India: opportunities and challenges. *Indian Journal of Electrical and Biomedical Engineering, 1,* 10–16.
12. Handayani, N. A., & Ariyanti, D. (2012). Potential of solar energy applications in Indonesia. *Journal of Renewable Energy Development, 1,* 33–38.

13. Juda, A. (2014). *Performance Evaluation of Solar Powered Vapor Absorption Refrigeration System*. MTech Thesis. Anand Agricultural University, Anand.
14. Kalogirou, S., Lloyd, S., & Ward, J. (1997). Modeling optimization and performance evaluation of a parabolic trough collector steam generation system. *Solar Energy, 60*, 49–59.
15. Klein, S. A., & Reindl, D. T. (2005). Solar Refrigeration. *American Society of Heating, Refrigerating and Air-Conditioning Engineers, 47*, 26–30.
16. Krich, K., Augenstein, D., Batmale, J. P., Benemann, J., Rutledge, B., & Salour, D. (2005). *Biomethane from Dairy Waste*. A Sourcebook for the Production and Use of Renewable Natural Gas in California.
17. Md-Aquil, A., Shadab, K., Shadman, H. Q., & Tiwari, G. (2014). The contemporary scenario of Indian renewable energy sector. *International Research Journal of Environment Science, 3*, 82–89.
18. Meisen, P., & Quéneudec, E. (2006). Overview of sustainable renewable energy potential of India. http://www.geni.org/globalenergy/research/renewableenergypotential-ofindia/pdf.
19. Mistry, M., & Gurjar, M. (2010). Application of solar water heater and its feasibility as a CDM project in dairy industry. National Seminar on Energy Management and Carbon Trading in Dairy Industry, pp. 121–123.
20. Moreno-Quintanar, G., Rivera, W., & Best, R. (2011). Development of a solar intermittent refrigeration system for ice production. *World Renewable Energy Congress*.
21. Nahar, N. M. (2009). Processing of vegetables in a solar dryer in arid areas. International Solar Food Processing Conference.
22. Panwar, V., & Kaur, T. (2014). Overview of Renewable Energy Resources of India. *International Journal of Advanced Research in Electrical, Electronics and Instrumentation Engineering, 3*, 1–9.
23. Rathor, N. S. (2010). Scope of renewable energy sources in dairy industries for energy conservation. National Seminar on Energy Management and Carbon Trading in Dairy Industry, pp. 23–26.
24. Saradhi, I. V., Pandit, G. G., & Puranik, V. D. (2009). Energy supply, demand and environmental analysis – a case study of Indian energy scenario. *International Journal of Civil and Environmental Engineering, 1*, 115–120.
25. Schnitzer, H., Christoph, B., & Gwehenberger, G. (2007). Minimizing greenhouse gas emissions through the application of solar thermal energy in industrial processes, approaching zero emissions. *Journal of Cleaner Production, 15*, 71–86.
26. Sharma, A., Chen, C. R., & Vu Lan, N. (2009). Solar-energy drying systems: A review. *Renewable and Sustainable Energy Reviews, 13*, 1185–1210.
27. Sontakkel, M. S., & Salve, S. P. (2015). Solar drying technologies: A review. *International Refereed Journal of Engineering and Science, 4*, 29–35.
28. State-wise potential of various renewable energy technologies, (2013). http://data.gov.in/sites/default/files/Statewise_Potential_of_Various_Renewable_Energy_T echnologies_1.xls.
29. Sukhatme, S. P. (2012). Can India's future needs of electricity be met by renewable energy sources? A revised assessment. *Current Science, 103*, 79–86.
30. Twidell, J., & Weir, T. (2006). *Renewable Energy Resources*. Second edition, Taylor and Francis, 2 Park Square, Milton Park, Abingdon.

31. Zahira, R., Akif, H., Amin, N., Azam, M., & Haq, Z. (2009). Fabrication and performance study of a solar milk pasteurizer. *Pakistan Journal of Agricultural Science, 46*, 162–170.
32. Zala, A. M., Shah, D. R., & Birendrakumar (2007). Advancement in utilities in dairy plants. National seminar on revamping dairy engineering: education and industry in global context.
33. www.anaerobic-digestion.com.
34. www.mospi.gov.in (2015). *A Report on Energy Statistics*. Ministry of Statistics And Program Implementation Overnight of India, New Delhi.
33. www.nddb.org (2015). Dairy Plant Management Training Program Report. 16th Feb–20th March, 2015, NDDB, Anand.
34. www.suscon.org/cowpower/biomethaneSourcebook/Chapter_1.pdf.
35. www.usdairy.com/~/media/usd/public/dairypowercasestudy_renewableenergy.pdf.
36. www.uvka.net/static/Downloadable/Vasudara-Dairy-Profile.pdf.

CHAPTER 12

MINIMIZING POWER REQUIREMENT FOR PUMPS IN DAIRY INDUSTRY

ADARSH M. KALLA, BHAVESH B. CHAVHAN, P. BISEN, and C. SAHU

CONTENTS

12.1　INTRODUCTION

The dairy industry faces an increasingly competitive environment, and looks out opportunities to reduce production costs without negatively affecting the yield or the quality of the finished product. The challenge of maintaining high product quality, while simultaneously reducing production costs, can often be met through investments in energy efficiency, which can include the purchase of energy-efficient technologies and the implementation of plant-wide energy efficiency practices. Energy-efficient technologies can often offer additional benefits, such as quality improvement and can often lead to reductions in emissions of greenhouse gasses and other important air pollutants. Investments in energy efficiency are therefore a sound business strategy in today's manufacturing environment. The cost of electrical energy is increasing dramatically and awareness of energy consumption in the dairy industry is becoming an issue in the cost of milk production. Hence, measures have to be taken to reduce electricity consumption wherever it is possible.

According to IEA [7], electricity cost is one of the major factors which influence the firm's decisions and growth of the industries. In most developed countries, industrial users pay lower prices for electricity compared to other users because the cost of supplying electricity to industrial users is typically lower [7]. However, in India, industrial users pay higher price for electricity relative to domestic and agricultural users. Politicians desire to win favor with households and farmers who form crucial voting blocks. For instance in 2000, industrial users in India paid about 15 times the price paid by agricultural users for electricity [1]. In a 2006 World Bank survey, Indian manufacturing firms were asked to indicate which element posed the biggest constraint to their operations out of a list of 15 elements including electricity, access to finance, and corruption. Electricity was the most common major obstacle indicated, with more than 36 percent of firms listing electricity as the biggest constraint [6].

Pumping systems account for nearly 20% of the world's energy used by electric motors and 25% to 50% of the total electrical energy usage in certain industrial facilities [11]. Hence it makes us necessary to take prerequisite steps to reduce power consumption in pumps. In dairy industry different types of pumps are used based on their function and type of the product to be pumped. Basically pumps are classified into two types: dynamic pumps and positive displacement pumps.

This chapter introduces types of pumps that are used in the dairy industry.

12.2 POSITIVE DISPLACEMENT PUMPS

They are distinguished by the way they operate. The liquid is taken from one end and positively discharged at the other end for every revolution (Figure 12.1). Positive displacement pumps are widely used for pumping mostly viscous fluids, such as cream with high fat content, cultured milk products, curd, whey, etc.

1. LOBE PUMP

2.SLIDING VANE PUMP

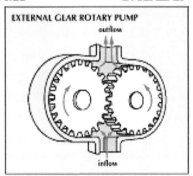

3. GEAR PUMP

FIGURE 12.1   Types of pumps.

- Reciprocating pumps: If the displacement is by reciprocation of a piston plunger, then it is called as reciprocating pump. Reciprocating pumps are used only for pumping viscous liquids and oil wells.
- Rotary pumps: If the displacement is by rotary action of a gear or vanes in a chamber of diaphragm in a fixed casing then it is called as rotary pump. In dairy industry, rotary pumps are used to pump cream and other viscous products.

12.3   DYNAMIC PUMPS

In the centrifugal pump, liquid enters the eye of the impeller and exits the impeller due to the centrifugal force (Figure 12.2). As water leaves the eye of the impeller, a low-pressure area is created, causing more water to flow into the eye (Atmospheric pressure and centrifugal force causes this to happen). Velocity is developed by the spinning impeller, and water velocity is collected by the diffuser and converted into the pressure by specially designed pathway [2].

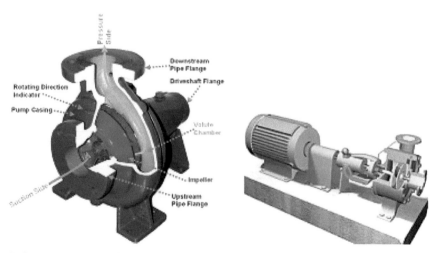

FIGURE 12.2   Centrifugal pump.

12.4 SANITARY PUMPS

One striking difference between Indian and the western dairy processing has been the type of pumps used in the plants. While most of the dairies in India use sanitary centrifugal pumps almost exclusively, dairies in the western countries use a wide range of positive displacement pumps mainly bilobe and trilobe, rotary cam and twin-screw designs. Many leading equipment companies offer a big selection of such pumps with different capacities and different rotor designs and different materials to suit different applications. Also, these rotary pumps offer options such as horizontal and vertical porting. Vertical porting has self-draining and venting capacity while horizontal porting facilities suction of high viscosity products. Use of these pumps allows metering of the fluid while pumping, which makes the automation possible.

Centrifugal pumps are also used where a large capacity is required like receiving and transferring. Sanitary centrifugal pumps designs include standard centrifugal pump, high pressure pump which can work at an inlet pressure of up to 40 kg/cm^2 and are suitable for reverse osmosis and ultra-filtration; and multistage pumps for low capacity and high pressure applications. The present trend is to use frequency converters on these motors to have these pumps operate at variable capacities as per processing needs. Usually the output of centrifugal pump is reduced by throttling the flow in the discharge side but this occurs at the increased pressure drops. Frequency converters on the other hand shift the operating point of the centrifugal pump by changing its speed, while keeping the pressure drop the same.

Many designs of such pumps are used for feeding the packaging machine lines, and balance tank feed pump while maintaining a predetermined level of pressure. Another design of centrifugal pump features a helical screw which helps feed liquids into the eye of the impeller which has specific application for pumping hot water under vacuum as they reduce the net positive suction head and reduce cavitation problems.

12.5   ENERGY EFFICIENCY FOR PUMP SYSTEMS

The basic components in a pump system are pumps, drive motors, piping networks, valves, and system controls. Some of the most significant energy efficiency measures applicable to these components and to pump systems as a whole are described below.

12.5.1   PUMP SYSTEM MAINTENANCE

In a typical life cycle cost, energy and maintenance costs will account for over 50–95% of pump ownership costs with initial costs less than 15% of pump life cycle costs. Hence maintenance of pump plays a major role in energy efficient pumping system [12].

The improper maintenance of pump lowers the system efficiency and causes pumps to wear out more quickly, and increases pumping energy costs. The implementation of a pump system maintenance program will help to avoid these problems by keeping pumps running optimally. Furthermore, improved pump system maintenance can lead to pump system energy savings of anywhere from 2 to 7% [13]. A solid pump system maintenance program will generally include the following tasks:

- Replacement of worn impellers, especially in caustic or semi-solid applications.
- Bearing inspection and repair.
- Bearing lubrication replacement, on an annual or semiannual basis.
- Inspection and replacement of packing seals. Allowable leakage from packing seals is usually between 2 to 60 drops per minute.
- Inspection and replacement of mechanical seals. Allowable leakage is typically 1 to 4 drops per minute.
- Wear ring and impeller replacement. Pump efficiency degrades by 1 to 6% for impellers less than the maximum diameter and with increased wear ring clearances.
- Checking of pump/motor alignment.
- Inspection of motor condition, including the motor winding insulation.

12.5.2 PUMP SYSTEM MONITORING

Monitoring can be used in combination with a proper maintenance program to detect pump system problems before they cause major performance issues or equipment repairs. Monitoring can be done manually on a periodic basis (e.g., performing regular bearing oil analyzes to detect bearing wear or using infrared scanning to detect excessive pump heat) or can be performed continuously using sensor networks and data analysis software (e.g., using accelerometers to detect abnormal system vibrations) [9]. Monitoring can help to keep pump systems running efficiently by detecting system blockages, impeller damage, inadequate suction, clogged or gas-filled pumps or pipes, pump wear, and if pump clearances need to be adjusted. In general, a good pump monitoring program should include the following aspects:

- Wear monitoring.
- Vibration analysis.
- Pressure and flow monitoring.
- Current or power monitoring.
- Monitoring of differential head and temperature rise across pumps (also known as thermodynamic monitoring).
- Distribution system inspection for scaling or contaminant build-up.

12.6 DIAGNOSTIC TOOLS

The pumps are monitored for their efficient working condition by using different diagnostic tools. The tools keep a check on the pump by using different analyzers. The analyzers are capable of detecting the changes in temperature and sound patterns, occurring in the pump which is considered to be harmful.

- **Thermography** – An infrared thermometer allows for an accurate, non-contact assessment of temperature. Its application for pumps includes assessments on bearing assemblies at the impeller housing and motor system connections.
- **Ultrasonic analyzer** – Fluid pumping systems emit very distinct sound patterns around bearings and impellers. In most cases, these sounds are not audible to the unaided ear, or are drown-out by other

equipment noises. Using an ultrasonic detector, the analyst is able to isolate the frequency of sound being emitted by the bearing or impeller. Changes in these ultrasonic wave emissions are indicative of changes in equipment condition-some of these changes can be a precursor to component degradation and failure.

- **Vibration analyzer** – Within a fluid pump, there are many moving parts; some in rotational motion and some in linear motion. In either case, these parts generate a distinct pattern and level of vibration. Using a vibration analyzer, the analyst can discern the vibration amplitude of the point on the equipment being monitored. This amplitude is then compared with trended readings. Changes in these readings are indicative of changes in equipment condition [10].

12.7 HIGH-EFFICIENCY PUMPS

Considering that a pump's efficiency may degrade by 10–25% over the course of its life, the replacement of aging pumps can lead to significant energy savings [12]. The installation of newer, high-efficiency pumps typically leads to pump system energy savings of 2–10% [4].

A number of high-efficiency pumps are available for specific pressure head and flow rate capacity requirements. Choosing the right pump often saves both operating costs and capital costs. For a given duty, selecting a pump that runs at the highest speed suitable for the application will generally result in a more efficient selection as well as the lowest initial cost.

12.8 CONTROL SYSTEMS

Control systems can increase the energy efficiency of a pump system by shutting off pumps automatically when demand is reduced, or, alternatively, by putting pumps on standby at reduced loads until demand increases.

In 2000, Cisco Systems upgraded the controls on its fountain pumps so that pumps would be turned off automatically during periods of peak electrical system demand. A wireless control system was able to control all pumps simultaneously from one location. The project saved $32,000 and 400,000 kWh annually, representing a savings of 61.5% in the total energy consumption of the fountain pumps [3]. With a total cost of $29,000, the simple payback period was 11 months. In addition to energy savings, the project reduced maintenance costs and increased the pump system's equipment life.

12.9   PROPERLY SIZED PIPES

Pipes that have a smaller diameter size for a required velocity will require higher amount of energy for pumping. In much the same way that drinking a beverage through a small straw requires a greater amount of suction. Hence where ever it is possible, the pipe diameters can be increased to reduce pumping energy requirements, but the energy savings due to increased pipe diameters must be balanced with increased costs for piping system components. Increasing pipe diameters will only be cost effective during greater pump system retrofit projects. It has been estimated that an energy savings of 5% to 20% can be obtained by proper pipe sizing [13].

12.10   PUMP SELECTION

The pump is selected based on how best the system curve supplied by the user and pump curve intersects, when graphically superimposed on each other. The point at which system curve and pump curve intersect is called as the pump operating point or best efficiency point. At this point the pump operates at its high speed and gives best output (Figure 12.3). However, it is impossible for one operating point to meet all desired operating conditions.

The right selection of pump depends on operating point and how accurate the system curve is calculated. If actual calculated system curve is different from that calculated, the pump will operate at a flow and head different to that expected. Generally, in industries, to have an additional safety margins to the calculated system curve, the facility manager will sufficiently select a large sized pump that results in installing an oversized pump, which will operate at an excessive flow rate, which increase energy usage and reduce pump life.

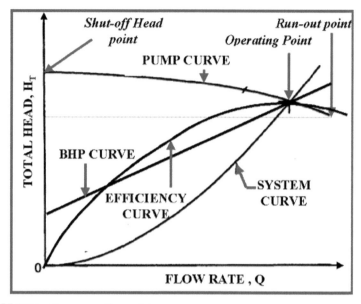

FIGURE 12.3   Pump operating point.

12.11 SELECTION OF PROPERLY SIZED PUMPS

Pumps that are oversized for a particular application consume more energy than is truly necessary. Replacing oversized pumps with pumps that are properly sized can often reduce the electricity use of a pumping system by 15–25% [13]. If a pump is dramatically oversized, often its speed can be reduced with gear or belt drives or a slower speed motor. The typical payback period for these strategies can be less than one year [5].

The efficiency of a pump is affected when the selected pump is over-sized. This is because flow of oversized pumps must be controlled with different methods, such as a throttle valve or a bypass line. These devices provide additional resistance by increasing the friction. As a result the sys-tem curve shifts and intersects the pump curve at a different point, a point of lower efficiency (Figure 12.3). In other words, the pump efficiency is reduced because the output flow is reduced but not the power consump-tion. The inefficiency of oversized pumps can be overcome by, for exam-ple, the installation of variable speed drives, operating the pump at a lower rpm, or installing a smaller impeller or trimmed impeller.

The energy usage in a pumping installation is determined by the flow required, the height lifted and the length and friction characteristics of the pipeline. The power required to drive a pump (P_i), is defined as follows:

$$P_i = [\rho\, g\, H\, Q]/E \tag{1}$$

where, P_i is the input power required (W); ρ is the fluid density (kg/m^3); g is the standard acceleration of gravity (= 9.81 m/s^2); H is the energy Head added to the flow (m); Q is the flow rate (m^3/s); and E is the efficiency of the pump plant in decimals.

The energy head added by the pump (H) is a sum of the static lift, the head loss due to friction and any losses due to valves or pipe bends all expressed in meters of fluid. Power is more commonly expressed in kilowatts (10^3 W, kW) or horsepower (1.00 kW = hp/0.746). The value for the pump efficiency (E) may be stated for the pump itself or as a combined efficiency of the pump and motor system. The *energy usage* is determined by multiplying the power requirement by the length of time the pump is operating.

12.12 AVOIDING THROTTLING VALVES AND BYPASS CONTROLS

As discussed earlier, inaccurate calculation of system curve will lead to selection of oversized pumps which has excessive flow rate and increased head. To overcome these problems industries started using throttling valve and bypass control loop. Throttling valves and bypass loops are indications of oversized pumps as well as the inability of the pump system design to accommodate load variations efficiently, and should always be avoided [8].

However throttling valve reduces the flow, it does not reduce the power consumed, as the total head (static head) increases. This method increases vibration and corrosion and thereby increases maintenance costs of pumps and potentially reduces their lifetimes.

The flow can also be reduced by installing a bypass control system, in which the discharge of the pump is divided into two flows going into two separate pipelines. One of the pipelines delivers the fluid to the delivery point, while the second pipeline returns the fluid to the source. In other words, part of the fluid is pumped around for no reason, and thus is energy inefficient. Because of this inefficiency, this option should therefore be avoided. The elimination of bypass loops and other unnecessary flows can also lead to energy savings of 10–20% [13]. But in some cases, small bypass line is required to prevent a pump running at zero flow required for safe operation of pump.

12.13 IMPELLER TRIMMING

Impeller trimming is one of the methods to reduce the pump flow rate. Impeller trimming refers to the process of reducing an impeller's diameter, so that it matches to the required flow rate and hence reducing the energy added by the pump to the system fluid. Changing the impeller diameter gives a proportional change in peripheral velocity, which in turn directly lowers the amount of energy imparted to the system.

According to affinity law, the equations relating pump performance parameters and impeller diameter are given as follows (Figure 12.4):

$$Q \alpha D$$

$$H \alpha D^2$$

$$P \alpha D^3 \tag{2}$$

Hence from Eq. (2), it is clear that power consumed is directly proportional to cube of diameter. The small change in diameter can reduce power consumption greatly.

According to the U.S. DOE [9], one should consider trimming an impeller when any of the following conditions occur:

- Many system bypass valves are open, indicating that excess flow is available to system equipment.
- Excessive throttling is needed to control flow through the system or process.
- High levels of noise or vibration indicate excessive flow.
- A pump is operating far from its design point.

Trimming an impeller is slightly less effective than buying a smaller impeller from the pump manufacturer, but it can be useful when an impeller at the next smaller available size would be too small for the given pump load. Changing the impeller diameter is an energy efficient way to control the pump flow rate. However, for this option, the followings should be considered:

- This option cannot be used where varying flow patterns exist.
- The impeller should not be trimmed more than 25% of the original impeller size, otherwise it will cause cavitation producing vibrations and resulting in decrease of the pump efficiency.
- The balance of the pump has to been maintained, for example, the impeller trimming should be the same on all sides.

In Figures 12.4 and 12.5, one can observe that after trimming the impeller the pump serves at lower flow rate and head, which once operated at higher flow rate.

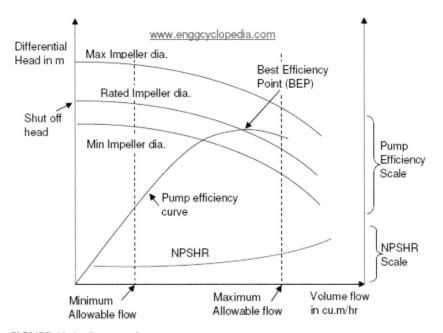

FIGURE 12.4   Pump performance parameters.

FIGURE 12.5   Trimming of impeller.

12.14   SPEED CONTROLLERS

12.14.1   ADJUSTABLE-SPEED DRIVES (ASDS)

Pumps that experience highly variable demand conditions are often good candidates for ASDs. As pump system demand changes, ASDs adjust the pump speed to meet this demand, thereby saving energy that would otherwise be lost to throttling or bypassing. The resulting energy and maintenance cost savings can often justify the investment costs for the ASD. However, ASDs are not practical for all pump system applications. For example, pump systems that operate at high static head and those that operate for extended periods under low-flow conditions [9].

The most generally used speed controllers are the variable frequency drives. VFDs adjust the electrical frequency of the power supplied to a motor to change the motor's rotational speed. VFDs are by far the most popular type of VSD.

12.14.1.1 Benefits of Speed Controllers

- **Energy Savings:** Energy savings of between 30% and 50% have been achieved in many installations by installing VSDs.
- **Improved Process Control:** By matching pump output flow or pressure directly to the process requirements, small variations can be corrected more rapidly by a VSD than by other control forms, which improves process performance.
- **Improved System Reliability:** Any reduction in speed achieved by using a VSD has major benefits in reducing pump wear, particularly in bearings and seals.

12.15 SUMMARY

The dairy processing industry engaged in the conversion of raw milk to consumable dairy products consumes billion worth of purchased fuels and electricity per year. Energy efficiency improvement is an important way to reduce these costs and to increase predictable earnings, especially in times of high-energy price volatility. The dairy manufacturing industry has radically improved its energy efficiency over the last 20 years through wide upgrading of equipment and the closure of smaller and less efficient factories.

The Pumps in dairies are often operated inefficiently. The reasons will vary from process to process, but the constant outcome is the cost to industry through wasted energy, which runs into millions of rupees per year. Pumping systems account for nearly 20% of the world's energy used by electric motors and 25–50% of the total electrical energy usage in certain industrial facilities. Significant opportunities exist to reduce pumping system energy consumption through smart design, retrofitting, and operating practices. In particular, the many pumping applications with variable-duty requirements offer great potential for savings. The savings often go well beyond energy, and may include improved performance, improved reliability, and reduced life cycle costs.

KEYWORDS

- best operating point
- bypass loops
- cavitation
- centrifugal pump
- cost
- discharge
- energy
- flow rate
- head
- impeller
- impeller trimming
- life cycle cost
- motor
- payback period
- positive displacement pump
- pump curve
- suction
- system curve
- throttling
- throttling valves
- valves
- velocity

REFERENCES

1. Abeberese, B. A. (2012). Electricity cost and firm performance: Evidence from India. *Job Market Paper,* p. 1.
2. Ahmad, Tufail, (1985). *Dairy Plant Engineering and Management.* Kitab Mahal. 5th edition, Allahabad, India.
3. California Energy Commission (CEC) (2002). *Case Study: Pump System Controls Upgrade Saves Energy.* Network Equipment Manufacturing Company's Corporate Campus, Sacramento, CA.

4. Elliot, R. N. (1994). *Electricity Consumption and the Potential for Electric Energy Savings in the Manufacturing Sector.* American Council for an Energy – Efficient Economy, Washington, D. C. Report No. IE-942.
5. Galitsky, C., Worrell, E., Radspieler, A., Healy, P., & Zechiel, S. (2005). *Best Winery Guidebook: Benchmarking and Energy and Water Savings Tool for the Wine Industry.* Lawrence Berkeley National Laboratory, Berkley, CA. Report No. LBNL-3184.
6. http://www.enterprisesurveys.org/Data/ExploreEconomies/2006/india.
7. IEA, 2012. *Energy Prices and Taxes.* Quarterly Statistics, Second Quarter 2012. Paris: IEA.
8. Tutterow, V., D. Casada, and McKane, A. (2000). Profiting from your Pumping System. In: *Proceedings of the 2000 Pump Users Expo*, Louisville, Kentucky.
9. United States Department of Energy (DOE) (2006). *Improving Pumping System Performance, A Sourcebook for Industry.* Office of Energy Efficiency and Renewable Energy, Industrial Technologies Program, Washington, D. C. Report DOE/GO-102006–2079.
10. US Department of Energy (US-DOE) (2010). *Federal Energy Management Program, Operation and Maintenance Best Practices: A Guide to Achieving Operational Efficiency.* US Department of energy, p. 225.
11. US Ministry of Energy, Department of Industrial Energy Analysis. 2004. *Variable Speed Pumping – A Guide to Successful Applications.* Executive Summary, US DOE, LBNL-55836, pp. 1–12.
12. US Ministry of Environmental energy, Department of Transport and Regions, (1998). *Energy Savings in Industrial Water Pumping Systems.* Good Practices Guide, New York, p. 24.
13. XEnergy, Inc. (1998). *United States Industrial Electric Motor Systems Market Opportunities Assessment.* Prepared for the United States Department of Energy's Office of Industrial Technology and Oak Ridge National Laboratory. Burlington, Massachusetts.

CHAPTER 13

WATER REQUIREMENTS FOR A DAIRY PLANT: QUALITY AND QUANTITY ISSUES

ARCHANA KHARE, ANIL K. KHARE, and A. K. AGRAWAL

CONTENTS

13.1 INTRODUCTION

Water is an important utility for dairy plants as it governs the hygiene of dairy plants [6]. Water is used in large quantities for cooling, washing and sanitation, production of steam and other operations in dairy plants [4].

Water is also required for cleaning of machines, utensils, floors, etc. The availability of water would certainly affect processing cost of milk and milk products whatever may be the level of production [2]. In the past, abundant and inexpensive sources of water were taken for granted in the dairy processing industry and not much attention was given to economize its use. But, in recent times we have witnessed acute water scarcity in arid and semi-arid regions.

With limited water resources, many dairy plants in such areas find it difficult to operate or otherwise expand their operations. Besides, indiscriminate use of water also results in excessive wastewater generation, which becomes a burden for the dairy in terms of treatment and disposal costs. Dairy processors, therefore, are aggressively challenged to conserve water not only for reducing water consumption but also for employing measures for recovery and recycling of process water, without compromising on the hygienic quality and safety of the products. Once water was easily available commodity at no cost, but now sometimes we need to buy it. The sudden increase in water procurement costs is a matter of great concern today to almost all industries including dairy industry, as it directly affects the production cost [5].

This chapter sheds light on the quality and quantity issues of water requirements in a typical dairy plant.

13.2   STANDARDS FOR QUALITY PARAMETERS OF WATER USED IN DAIRY INDUSTRY

The availability of adequate water of suitable quality is a primary consideration in the selection of site and in the establishment of a dairy processing plant. Water is needed in the processing plant for generating steam, for cleaning, as an ingredient in finished products, as a heat exchange medium in heating and cooling operations, for cleaning plant and equipment, and for protection against fire. There are many sources of water supplies and may be classified as surface water (from lakes, stream and reservoirs) and subsurface water (from shallow and deep wells). The characteristics of water from these sources vary with rainfall, the nature of materials with which the water comes in contact and the time of year [9], as shown in Table 13.1. Water hardness is defined in Table 13.2.

TABLE 13.1 Characteristics of Water from Different Sources

Type	Organic matter	Microbial count	Mineral content
Deep well	Usually low	Usually low	Usually high
Shallow well	Variable	Variable	Ordinarily low
Surface	May be high	May be contaminated	Ordinarily low

TABLE 13.2 Classification of Hardness

ppm of $CaCO_3$	Condition
Less than 50	Soft
50 to 100	Slightly hard
100 to 200	Hard
Above 200	Very hard

Most of the functions of processing plants call for water of a high degree of purity [3]. An overall quality of water cannot be prescribed as so many specialized requirements prevail, so that parameters, that are objectionable for one use, may not necessary prove detrimental for others. In general, only potable water should be used in the preparation of any food intended for human consumption. Portable water is the water which contains no bacteria capable of causing human internal diseases and is esthetically satisfactory for drinking purpose, for example, free from undesirable odors and flavors [17]. The suitability of water for use in dairy processing plant depends upon:

13.2.1 PHYSICAL PROPERTIES

Color – Turbidity should not exceed 10 ppm (silica scale); Odor – should not exceed 20 ppm (cobalt scale); Flavor – free from objectionable odor and taste.

13.2.2 CHEMICAL PROPERTIES

Dissolved solids and gases; pH; and hardness – Calcium and manganese salts cause the hardness in water. Generally, hardness is expressed as ppm of $CaCO_3$, which decides degree of hardness (Table 12.2).

13.2.3 MICROBIOLOGICAL CONTAMINATES

These include algae, and pathogenic and non-pathogenic organisms. The bacteriological quality of water for the plant should meet the standards for drinking water. The fitness of water for drinking purpose with respect to bacterial content is determined by the presence or absence of the coliform group of bacteria that include Escherichia and Aerobatic species which indicate the possibility of fecal contamination of water supply.

The three organisms associated with the fecal matter are: *E. coli communis*, *Streptococcus* sp., and *Cl. welchiio*. The last one usually occurs in the form of spores. Of these, *E. coli* is by far the most abundant. The number present in polluted water may vary widely.

Nonpathogenic organisms present in water may influence flavor and odor, and produce slim and bio-fouling of pipes. A high bacterial court in can cooling water may result in recontamination and spoilage. For instance, an iron bacterium found commonly in water contains iron, and extracts iron continually and accumulates it as iron hydroxide. This may lead to the eventual blocking of pipes or discoloration of products.

The total bacterial count in water considerably influences plant sanitation. The purpose of washing and cleaning equipment is to reduce contamination. Obviously, water used for this purpose should itself have a low bacterial count.

Pure waters, such as those drawn from deep wells or those purified by artificial sources, seldom show the presence of *E. coli* in 100 ml of water. This should therefore be the state of purity. When a sample shows a positive test for *E. coli* in 50 ml, it should be considered as the maximum limit, which can possibly be permitted and it should be free from organic pollution. Nitrate in excess of 40 to 80 ppm justifies careful testing due to possibility of sewage contamination.

13.3 CLASSIFICATION OF WATER IN A DAIRY PLANT [15]

13.3.1 PROCESS WATER

Water is used for direct preparation of products, cleaning purposes and various technical purposes such as: washing/cleaning of equipment, transport of product, dissolution of ingredients, etc. A characteristic of process water is that it comes into contact with product directly or indirectly. Therefore, process water should strictly meet requirement of drinking water standards.

13.3.2 COOLING WATER

Cooling water is the water used for removal of heat from process streams and products. The quality requirements for cooling water used in plate heat exchangers to cool milk is critical, since with this type of equipment there is a risk of failure and leakage of cooling water to product. In such situations, cooling water should be of drinking water quality.

13.3.3 BOILER FEED WATER

Boiler feed water is required for steam production. The main quality requirements are low hardness and low air and carbon dioxide content.

13.3.4 MISCELLANEOUS USE

Other uses of water in a dairy plant are for ancillary purposes such as amenities and gardens, and extraordinary incidents (e.g., fire protection). Depending upon product mix, dairy processing plants can use substantial volumes of water for cleaning of utensils, in cooling towers and other processes.

13.4 ESTIMATION OF QUANTITY OF WATER

Dairy plant use water in all three states: solid (ice), liquid (water) and vapor (steam) [1]. Since the requirements of water for a dairy plant is comparatively large [12], it may be difficult to meet the demand from a municipal supply. It is advisable to have a supplementary source such as bore well or open well with overhead storage facility at a convenient location for distribution within plant. The dairy industry involves processing of raw milk into products such as liquid milk, butter cheese, curd condensed milk, dried milk (milk powder), ice cream casein, etc. Various processes such as chilling, pasteurization, homogenization, evaporation and drying are being used in manufacture of these products. The water is not only needed in these processes but also in the production of milk [7]. The major water requirement during production is in the washing of the floor, drinking needs by the animal, bathing of animal, etc. It has been reported that temperature above and below the critical limit adversely affects physiological processes and decreases milk production by 3 to 10% [4]. With the water sprinkling over the lactating animals heat stress is reduced and thereby milk production can be enhanced. In India out of about 125 million tons of milk production/annum, about 15% of total milk produced is brought in the dairy plants for processing [8]. A small dairy plant may be engaged in processing of raw milk into products such as liquid milk, makkhan, dahi chhana, paneer, mishti dahi, khoa, sandesh, and various chhanna and khoa based sweats, etc.

The processing of milk requires huge amount of water. In a dairy processing plant, lot of water is used mainly to clean the equipment after processing as well as washing floors to maintain good hygiene. This results in generation of large volume of waste water. A proper balance should be observed between maintaining good hygiene and use of water for cleaning to conserve the water. Table 13.3 shows the use of water in different activities/products in a dairy plant. Milk house and parlor waste can be estimated from data found in Table 13.4.

TABLE 13.3 Approximate Water Consumption in a Dairy Plant

Operation	Water requirement ratio (liters of water/liters of milk)
Bottle washing per 100 bottles	75.0
Channa/paneer making	8.0
Floor washing	1.0
Ghee making	12.0
Khoa making	4.0
Milk pasteurization	2.0
Milk reception	1.8
Processing of milk for fluid milk production	6.0
Washing of utensils	2.0

TABLE 13.4 Volume of Milk House and Parlor Wastes

Washing Operation	Water Volume
Bulk Tank	
Automatic wash	190–225 liters per wash
Manual wash	110–150 liters per wash
Miscellaneous equipment	110 liters per day
Pail milkers	110–150 liters per wash
Pipeline, in parlor (Volume is higher for long stanchion barns)	280–470 liters per wash
Cow Wash	
Automatic	4–15 liters per wash per cow
Holding pen (sprinklers)	5 liters per head (depending on nozzle size and pressure)
Manual	1–2 liters per wash per cow
Milkhouse floor	35–70 liters per day
Parlor floor	150–180 liters per day

Source: Dean E. Falk, Extension Dairy Specialist, University of Idaho: http://www.oneplan.org/Stock/DairyWater.asp.

13.5 COMMON REASONS FOR WATER LOSS IN A DAIRY PLANT

13.5.1 RMRD

Manual cleaning of cans with running water hoses; water lubrication of conveyors.

13.5.2 PROCESS SECTION

Operating pasteurizers for short durations; Condensate from pasteurizers allowed to drain; Water being allowed to overflow the balance tank when pasteurizers and other equipment are on rinse; Manual cleaning of separator with running water; Cleaning of milk tanks by high pressure, etc.

13.5.3   MILK FILLING SECTION

Manual cleaning of crates with running water hoses; Machine cooling; water drained.

13.5.4   BUTTER AND GHEE SECTION

Draining of cooling water sprayed over butter churns; Draining of condensate from ghee boilers; Draining of cooling water from settling tanks.

13.5.5   EVAPORATING AND DRYING SECTIONS

Draining of condensate from evaporator; Operating evaporators for short durations and frequent cleaning.

13.5.6   CLEAN-IN-PLACE (CIP) SYSTEMS

CIP done without recirculation or used CIP solutions being drained frequently; CIP systems not recovering final rinse water for reuse as pre-rinse water.

13.6   WATER CONSERVATION IN A DAIRY PLANT

Water conservation in a dairy processing plant gives dual benefits. It lowers the water and energy bill of the plant [12]. It also helps to reduce the effluent treatment cost as all the water from the plant reaches the effluent treatment facility before its disposal. For successful implementation of the water conservation measures, the commitment of management is required. The people involved should consider the water as a raw material with a cost and the management should encourage people to innovate measures for conservation of water. All water for reuse should be screened to reduce solid buildup.

Substantial water wastage can be reduced by using the following technology innovations [13]:

- Use of automatic shut off devices on all water hoses of steam and water mixing batteries.
- Use of high pressure, low volume cleaning systems for cleaning crates, silos, etc.
- Use of recuperation tanks as a part of the CIP system recovering the hot water for reuse as pre- rinse in next CIP cycle.
- Use of automation for CIP system and other operations. It is reported that in an automated dairy plant a saving to the tune of 20–25% can be achieved in the expenditure on water supply.
- Use of continuous rather than batch processes to reduce the frequency of cleaning.
- Use of level controllers switching on/off the pumps to avoid overflow from tanks.
- Use of automatic shut off valves with photo sensors for wash basins/urinals.

13.6.1 WATER REUSE

Besides prevention of water loss, optimization of water use in a dairy plant can also be affected by water reuse for specific application, when possible.

This can reduce the demand on water supply and also reduce volumes of wastewater, the treatment and disposal. Water that is considered suitable for reuse is the one that has been recovered from a processing step, including from the food components, and that after subsequent reconditioning treatment(s), as necessary, is intended to be re(used) in the same, prior, or subsequent food processing operation. Reuse water includes:

- **Recirculate water:** Water reused in a closed loop for the same processing operation (e.g., chilled water, condenser cooling water in circulation, pasteurizer cooling water in circulation, etc.)
- **Reclaimed water:** Water that was originally a constituent of food, has been removed from the food by a process step, and is intended to be subsequently reused in food processing operation (e.g., Condensates from milk evaporators).
- **Recycled water:** Water, other than the first use or reclaimed water, that has been obtained from a food processing operation (e.g., permeate from reverse osmosis plant, CIP final rinse water, etc.) [14].

13.6.2   REQUIREMENTS FOR HYGIENIC REUSE OF PROCESSING WATER IN A DAIRY PLANT

Reuse water shall be safe for its intended use and shall not jeopardize the safety of the product through the introduction of chemical, microbiological or physical contaminants in amounts that represent a health risk to consumer. Reuse water should be introduced into a processing system so that it will not add to microbiological or chemical burden of the product. Such water shall at least meet the microbiological and, as deemed necessary, chemical specifications for potable water. In certain cases, physical specifications may be appropriate. Reuse water should not adversely affect the quality and suitability of the product. Reuse water shall be subjected to ongoing monitoring and testing to ensure safety and quality [14, 16].

The frequency of monitoring and testing are dictated by the source of water or its prior condition and the intended reuse of water; more critical applications normally require greater levels of reconditioning than less critical uses [15]. If reconditioned to potable water quality, distribution of reuse is permitted. Water should be in clearly marked (e.g., different colors) systems, including piping and outlets that are separate from the distribution lines for potable water. Cross-contamination by backflow, back-siphoning, or cross connections from reuse water should be prevented. Reuse water storage vessels, if used, should be properly constructed of materials that will not contaminate the water and should allow for periodic cleaning and sanitizing where appropriate.

Proper maintenance of water reconditioning system is critical to avoid having the systems become source of contamination. For example, filtration systems can become sources of bacteria and their metabolites if bacteria are allowed to grow on entrained organic materials removed from the incoming water; proper maintenance and testing is needed to ensure absence of this situation.

13.6.3   TREATMENT OF REUSE-WATER

To achieve sustainable water management in a dairy factory, both the quantity and quality of water need to be considered. Generally speaking, two scenarios can be distinguished for water reuse: (a) water not in contact with raw, intermediate or final product; (b) typical reuse applications are for cooling purposes and for generation of 'non-food steam' [16].

Water in contact with the products (e.g., water used for cleaning of equipment, reconstitution, washing of products such as butter, cheese and paneer, moisture adjustment in products such as butter, etc.): In some cases, water can be reused without pre-treatment (e.g., the use of condensates as washing water). However, in most cases, water that is recycled or reused will need to be treated to improve its quality particularly when it comes into contact with food or is used to clean surfaces that will come in contact with the products. Advanced water treatment technologies make it possible to treat water to very high degree, significantly reducing potential health risks associated with water reuse. It is even possible to treat wastewater to such a high degree that it can be safely used as a supplement to potable drinking water supplies. However, treating water to a high degree is expensive. The quality of water and the degree of treatment required, should correspond to the intended water use.

While deciding on the type of water treatment system for use, it is important to consider the following points: Reconditioning of water should be under taken with the knowledge of the types of contaminants the water that may have acquired from its previous use. For example, UV disinfection may have limited effectiveness for inactivating protozoan cysts, helminthes or viruses. Similarly, the use of chlorine or ozone on organically enriched water may result in the formation of hazardous organic compounds.

The water treatment system under consideration should be such that it will provide the level of reconditioning appropriate for the intended water reuse. For example, UV disinfection as a sole treatment is not appropriate for water that is turbid or contains particulates because the organisms in the shadow of particles or entrained within particles are protected from lethal effects of the irradiation.

Extremely large volumes of reuse water may justify the use of an advanced wastewater treatment system. Such systems, depending upon the prior state of the water, may require the use of one or more processes such as: filtration, de-nitrification, phosphorus removal, coagulation–sedimentation and disinfection. Overall, matching water quality requirements with the type of water use requires analysis of the critical control points and an evaluation of potential for contamination of food products.

Therefore, in addition to developing a framework for water refuse in food production/processing, where possible water reuse in the factory should be integrated into existing HACCP program.

13.6.4 SOME POSSIBLE USES OF CONDENSATE AND REVERSE OSMOSIS PERMEATE

Possible uses for recovered vapor condensate or reverse osmosis permeate could be divided into three broad classes:

Vapor condensate as a heat source: As boiler feed water, for melting butter in a jacketed tanks; For preheating (through heat exchanger) milk prior to entering evaporator; and for preheating (through heat exchanger) or drying air for spray dryers.

Vapor condensate and reverse osmosis permeate for product washing and product uses: As a cheese curd wash water, as casein wash water; Water for reconstitution of powdered products; Recirculation water for evaporator on failure of product feed; Infiltration water in ultrafiltration installations. However, utilization of vapor condensate and RO permeate for product use need strict collection and treatment practices for food safety and economic reasons. In particular, their bacteriological, chemical and organoleptic characteristics require close monitoring.

Vapor condensate and reverse osmosis permeate as water and heat source: For CIP, pre- and intermediate rinsing instead of potable water use, for preparation of CIP solutions; for cleaning of floors and walls of the building [11]; for external cleaning of milk transport vehicles; for use after cooling, as pump seal water.

To achieve sustainable water management, industry needs to focus on low cost solutions for conservation of water. The feasibility of exploiting potential of rainwater harvesting for meeting the water requirement of dairy plant is discussed in the following sections.

13.7 TECHNOLOGY OF RAINWATER HARVESTING

When natural water cycle is disrupted due to external and man-made factors, it becomes utmost important to trap the rain water before it contaminates beyond recognition or runs off beyond our reach. In India, average rainfall is in range of 600 to 1000 mm [10]. The portion of this water will substantiate the water that is needed for dairy processing. It is estimated that 10,000 to 30,000 liters of rain water per 100 m^2 of land area per year can be collected from roof tops/open areas, depending upon the rainfall in the area.

13.7.1 INSTALLATION OF RAINWATER HARVESTING SYSTEM

- To lay out piping system for collection of rainwater from rooftops and directing it through the filtration system followed by storage. A bypass arrangement is provided to segregate initial rainwater, which may contain surface impurities during first spell.
- Depending on the need, water collected from rooftops and storage, can be treated employing appropriate treatment processes for further conservation to meet specific requirement of potable water.
- In case of permeable strata is available, the rainwater can be directly charged to this strata by constructing percolation well.
- Depending on the site conditions, shallow percolation column can be constructed for charging the ground water.

13.7.2 DESIGN OF RAINWATER HARVESTING SYSTEM

The basic concept for rainwater harvesting involves design of storm water drainage system for effective collection of rainwater, first spell separator (for roof top rain water), grit chambers/arrestors (for land runoff), sand bed filters (arresting silt and other physical impurities), storage tanks/ ponds and percolation elements. Engineering design can be easily mastered with practice and based on requirement, depending on the site conditions. Sand filters are designed as conventional filter beds with sand layer followed by layers of pebbles/ uniform sized aggregate of size 25/40 mm, 50/65 mm. The layers are separated by nylon mesh so that they do not mix and filters can be easily serviced. The larger sized pebbles/stones in the lower layers provide holding of water before it is transferred to storage or percolation elements.

13.7.3 COMMON ELEMENTS OF RAINWATER HARVESTING SYSTEM

• Rainwater drainage piping and grid, first spell separate unit, sand filter, storage/ or to the inlet of percolation pit/element.
• Sand filters drum/pit type with media viz. sand, pebbles, stone aggregates in layers.
• Water storage tanks or water body in lieu of it.
• Recharge unit for charging existing dry or operating open/bore well.
• Ground water recharge pits of 1.2 m diameter × 1.8 m deep size filled with sand, pebbles and boulders in layers.
• Ground water recharge pits as above with auger bored percolation column. There can be single or multiple auger bored column of 6 m depth filled with boulders.
• Percolation bore well of depth up to permeable strata with collection tank.

13.8 CASE STUDY

A case study was done for a typical dairy plant of 200,000 liters/day capacity with a site area of around 20 acres, located in zone having 90 cm of rainfall spread over 50 to 60 rainy days. It has been found that with this roof top and land area, the dairy plant has a potential of harvesting rainwater to meet water requirement for 192 days. The maximum possible quantity of water can be harvested into storage tanks, reservoirs or by charging to ground water to meet the requirement. To start with, existing water storage tanks can be used for storage of water or by creating low cost water bodies and subsequent conservation, charge dry or operating wells/bore wells and if the strata permit charge the ground water table with rainwater. The fact remains that water situation cannot improve overnight. However, whatever may be achieved, it will add to the available resource to begin with. Engineering practices can be developed with the experience gained to further conserve maximum possible quantity of rain water.

13.9   CONCLUSIONS

The use of water should be restricted to absolute minimum consistent with maintenance of high degree of cleanliness and sanitation required in a plant. This will reduce direct cost of providing water for processing and consequent cost of treatment of the dairy effluent. Following suggestions are made with a view to conserve water and affect the economy in dairy plant:

a. Providing nozzle attachment to the hose to prevent excessive flow during floor washing.
b. Using water regulating valves where continuous flow is required and solenoid operated valve for intermittent operation.
c. Use of cooling tower for recycling of cooling water.
d. Utilizing water from heat exchangers for truck floor or can washing.
e. Fixing float control valve for maintaining water levels in storage tanks.
f. For effective implementation of water conservation techniques in the dairy plant, a "water master" should be appointed and a new water conservation cell should be created.
g. System of water audit should be implemented and water conservation law should be framed.
h. Standardization of water conservation process should also be listed as a criterion for licensing other assistance to dairy industry.
i. Strong commitment and support from the top management is essential for a successful conservation program.
j. National and state level awards should be instituted for water conservation in order to promote and encourage energy savings.

13.10   SUMMARY

Water is an important utility as it governs the hygiene of dairy plants. The suitability of water for use in dairy processing plant depends upon, physical properties: like color, odor, flavor and turbidity; chemical properties: like dissolved solids and gases, pH and hardness; Microbiological contaminants:

like algae, and pathogenic and non-pathogenic organisms. Most of the functions of processing plants call for water of a high degree of purity. An overall quality of water cannot be prescribed as so many specialized requirements prevail, so that characters that are objectionable for one use may not necessary prove detrimental for others.

In general, only potable water should be used in the preparation of any food intended for human consumption. The dairy processing plants require ample quantity of water for processing of milk and manufacturing of milk products. Water is one of the major services, which required in dairy plants for processing of milk and manufacture of milk products. Quality and quantity of water used in a dairy plant affect significantly cost of processing and energy requirements.

It is estimated that the water requirement of a dairy plant engaged in fluid milk processing is approximately 6 times the quantity of milk handled; in traditional dairy products like khoa 4 times; channa/paneer 8 times; and Ghee requires 12 times quantity of milk handled. Every plant should monitor the water usage and decide whether water used in one process can be utilized in another process. Emphasis should be made on lowest possible use of water in dairy plants to reduce the quantity of waste water coming out and thereby reduce the environment pollution. To achieve sustainable water management, industry needs to focus on low cost solutions for reduction in water consumption. Subsequently, the scope can expand, bringing around infrastructural improvements such as equipment upgrades, efficient water treatment systems.

The feasibility of exploiting potential of rainwater harvesting to meet water demand of a dairy plant is discussed in detail. Beside simple modifications in routine methods of water consumption, suggestions in this chapter include appointment of 'water master' (for control on water usage), 'water audit' to determine the means of loss of water, fixing of 'water standards' (for replicating in other dairy plants) and provision of some incentives (to encourage others to follow water conservation methods). In a nut-shell, there is a strong need to conserve substantial amount of water by adopting proper conservation techniques. The authors of this chapter advocate generating water consciousness and awareness among the producers and processors both.

KEYWORDS

- boiler feed water
- dairy plant
- milk
- potable water
- process water
- rainwater harvesting
- reuse
- reverse osmosis plant
- water demand
- water loss

REFERENCES

1. Agrawal, A. K., Sandey, K. K., Uprit, S., Goel, B. K. (2008). Water requirements of a small dairy plant. *Souvenir of National Seminar on "Dairy Engineering for the Cause of Rural India."* Unpublished Report by I. G. Agricultural University, Raipur, India.

2. Ananta Krishnan, C. P., & Simha, N. N. (1987). *Technology and Engineering of Dairy Plant Operations.* Laxmi Publications, Delhi, pp. 70–85.

3. Aneja, R. P., Mathur, B. N., Chandan, R. C., & Banerjee, A. K. (2002). *Technology of Indian Milk Products.* Dairy India Publication, Delhi, pp. 94–96.

4. Anjali, A., & Singh, M. (2006). Effect of water cooling on physiological responses, milk production and composition of murrah buffaloes during hot humid season. *Indian Journal of Dairy Science, 59*(6), 386–389.

5. Andersen, M., & Kristensen, G. M. (2004). A generic approach to water minimization and reuse in industry – implications in dairy operations. *The Australian Journal of Dairy Technology, 59*(2), 73.

6. COWI Consulting Engineers and Planners (2000). *Cleaner Production Assessment in Dairy Processing.* United Nations Environmental Program, Division of Technology, Industry and Economics.

7. Das, M. M., Singh, K. K., & Maity, S. B. (2008). Water for livestock production. *Indian Dairyman, 60*(6), 36–39.

8. Hale, N. (2003). *Sources of Wastage in Dairy Plants.* Bulletin of the International Dairy Federation No. 382/2003, IDF, Brussels, pp. 7–30.

9. IDF (1988). *The Quality, Treatment and Use of Condensate and Reverse Osmosis Permeates*. Bulletin of the International Dairy Federation No. 232/1988, IDF, Brussels.

10. Mandloi, R. S. (2005). Rain water harvesting and water conservation-need of the hour. *Indian Dairyman, 57*(1), 45–47.

11. Narsaiah, K. (2007). Environmental management in dairy processing plant. *Indian Dairyman, 59*(5), 43–46.

12. Parekh, J. V., & Naware, M. L. (2013). *Technology of Dairy Products*. IBDC Publishers, Lucknow.

13. Porwal, R. K. (1991). Energy conservation in dairy plants: A case study. *Indian Dairyman, 43*(11), 525–529.

14. *Proposed Draft Guidelines for the Hygienic Reuse of Processing Water in Food Plants*. CX/FH 01/9(2001), 34th Session of Codex Committee on Food Hygiene, Bangkok.

15. Ranganna, S. (1991). *Handbook of Analyzes and Quality Control for Fruit and Vegetable Products*. Ist Edition. Tata McGraw-Hill Publishing Company Limited.

16. Rausch, K. D., & Powell, G. M. (1997). *Dairy Processing Methods to Reduce Water Use and Liquid Waste Load*. Cooperative Extension Service, Kansas State University.

17. Sharma, K., & Saxena, R. (1991). *Water Management in Dairy Plants (Project Report)*. Institute of Rural Management, Anand.

CHAPTER 14

COMMON UTILITIES FOR THE DAIRY INDUSTRY

VANDANA CHOUBEY

CONTENTS

14.1   INTRODUCTION

Milk is highly perishable and so it requires immediate treatment as soon as it comes out of cow's udder to sustain for a longer period of time. In order to increase its shelf life milk is either chilled or heat treated which involves the use of hot water or steam or cooling or chilled water, etc., depending on the type of treatment. Not only for milk but for its conversion to various products, steam or water or refrigeration or electricity is employed in some or other way. Like for pasteurization, UHT, evaporation, drying, etc. hot water or steam or hot air is required. For cooling, chilling or freezing, refrigeration system is required. For cleaning or even for blending of some ingredients, water is required; and for most of the machinery to work electricity is required. Therefore, these common utilities play a key role for proper functioning of dairy industry.

This chapter will discuss about type and quality of water used in dairy for different purposes: Boilers, electricity and refrigeration system.

14.2 WATER

Water is used in many processes in dairy industry such as cleaning, cooling, pretreatment, and rinsing of equipment or for recombination purpose. The quality of the water depends on its final use. In dairy, water is used in many forms: RO water, demineralized water for direct contact of product and soft water or raw water for cleaning purpose. In order to provide various grades of water, softner, RO and demineralization unit is installed as per requirement and their design may vary according to the end use of water. These units work as follows.

14.2.1 DUAL MEDIA (DM) FILTERING UNIT

Raw water is stored in raw water storage tank, which is chlorinated using sodium hypochlorite for decreasing soluble iron and organic matter. The pH of the water is maintained by dosing alkali and is brought to 7.0 to 8.0 as raw water is generally acidic and pH of 7.5 to 8.0 is required for removal of iron as it gets converted to insoluble form and filtered through filters. Dual media filters are provided to remove insoluble iron and turbidity. Residual chlorine helps to restrict growth of microbiological matter in the filter media. Hence, small quantity of water at outlet of DMF is maintained.

The outlet water of DMF goes through "basket strainer" for fine particle removal and then goes to ultrafilter that removes colloidal silica, organics, coloring matter and left turbidity.

From UF, feed water goes to RO block where feed water is conditioned by different chemicals for de-chlorination and anti-scalant dosing for reducing salt deposition in RO membrane at high concentration level and then the feed passes through a cartridge filter for fine particle removal and then it finally enters the RO block.

14.2.2 RO BLOCK

RO Block consists of pressure tube arrangement with RO membrane. Here, flowrate and feed is forced at required pressure through a high pressure pump. Finally, feed is obtained through two outlets: one is Permeate, for example, the required product with lower TDS (total dissolved solids) and saline water, for example, reject. Then the permeate passes through "Degasser Tower" to wipe out dissolved carbon dioxide.

14.2.3 MIXED BED POLISHER

This unit consists of mixed foam of Cation and Anion resin which is fully charged. Cation resin exchanges Ca, Mg, Na ions for H ions and Anion resin exchanges SO_4, Cl, Silica ions for OH ions.

Hence, mixed bed polisher acts as a unit for removal of ionic impurities and works as polisher after RO block. However, the output obtained depends on the initial load of ions. Mixed bed resins are regenerated in the regeneration cycle by respective regeneration when Mixed bed is exhausted and after regeneration mixed bed unit can be taken back to use. Outlet water of mixed bed is slightly acidic, hence alkali dosing for pH correction is done.

14.3 RO UNIT

RO unit is a separation process that is able to separate dissolved solutes from solvent mostly through pressure driven membrane separation process (Figure 14.1). The solute may range in size from 1–10 angstrom or less and either organic or inorganic in nature. RO is capable of removing 90% of dissolved solute weather organic or inorganic. RO is mostly applicable for treatment of raw water. RO membranes are generally constructed from cellulose, acetate, polyamides or other polymers.

FIGURE 14.1 RO unit.

14.3.1 PROCESS

When salt solution is separated from demineralized water by semi permeable membrane, the salt solution of high osmotic pressure causes flow of demineralized water into salt solution section. The rise in solution water equals osmotic pressure continues to increase as water will flow from water to salt solution. By exerting pressure on the salt solution compartment water can be made to flow in reverse direction. This is reverse osmosis.

14.3.2 OPERATION

The osmosis pressure is based on the specific solute and its concentration in water, practically to produce first drop of pure water from the solution of solute at specific concentration the minimum pressure requirement is osmotic pressure. Very fine pores are used for membrane in reverse osmosis of order 5 angstrom units. The rejection rate of the membrane depends on ionic, for example, higher ionic charge of an ion, the better rejection therefore monovalent salt will pass the membrane at higher rate than multivalent.

For production of permeate and concentrate, pressure is applied to feed the stream by pump. The level of solids is high in concentrates and low in permeates.

14.4 TREATMENT

14.4.1 PRETREATMENT

It is very important to pretreat water before the RO process for removal of chlorine as well as other hardness elements. Hence, softener (removes hardness of water) or other suitable method for treatment must be used to delay hard water scale built up.

14.4.2 RO MEMBRANE

RO membrane stops the passage of the dissolved and suspended solids while letting or allowing the water to pass through. RO efficiently rejects turbidity, colloids and organic matter although this process may foul the membrane. The type of membrane and impurity decides the final product quality.

14.5 BOILERS

Steam is utilized in various heat treatments of dairy industries like pasteurization, sterilization, UHT, evaporation, CIP, etc. and is an excellent medium for conveying heat. In dairy applications, latent heat of steam and some sensible heat of hot water are utilized.

The steam is produced generally from centralized boiler houses. Boilers are pressure vessels designed to produce steam or produce hot water or an enclosed container that provide a means for heat from combustion to be transferred into working media (usually water) or a gas (steam) until it becomes heated.

In an industrial/technical context, the concept "steam boiler" (also referred to as "steam generator") includes the whole complex system for producing steam for use e. g. in a turbine or in industrial process. It includes all the different phases of heat transfer from flames to water/steam mixture (economizer, boiler, super heater, re-heater and air preheater). It also includes different auxiliary systems (e.g., fuel feeding, water treatment, flue gas channels including stack) [1].

When phase changes of the water is discussed, only the liquid-vapor and vapor-liquid phase changes are mentioned, since these are the phase changes that the entire boiler technology is based on [5].

14.5.1 BOILER EFFICIENCY

The boiler efficiency can be determined by the total fuel bunt and the total water evaporated into steam in given period of time or can be stated as the ratio of difference of fuel energy input and energy lost up by stack to the fuel energy input:

$$\eta = \frac{heat\,used\,in\,producing\,steam}{heat\,liberated\,from\,fuel} = \frac{W\,(H-h)}{C} \tag{1}$$

where, W = weight of water actually evaporated or steam produced in kg/hr or kg/kg of fuel burnt; H = total heat of steam in kJ/kg of steam corresponding to given working pressure (from steam tables); h = sensible heat of feed water in kJ/kg of steam corresponding to temperature of feed water (from steam tables); C = calorific value of fuel in kJ/kg of fuel.

Guidance for the construction, operation, and maintenance of boilers is provided primarily by the American Society of Mechanical Engineers (ASME), which produces the following resources:

- Rules for construction of heating boilers, *Boiler and Pressure Vessel Code*, Section IV-2007
- Recommended rules for the care and operation of heating boilers, *Boiler and Pressure Vessel Code*, Section VII-2007

Boilers are often one of the largest energy users in a building. For every year, a boiler system goes unattended, boiler costs can increase approximately 10% [4].

The boilers are generally made up of steel Indian Boiler Regulation (IBR) grade and are generally closed vessel type (Figure 14.2). Main function of the boiler is to produce steam through water by combustion of fuels. Fuel is used in food plants mostly for generating process steam and process drying. Natural gas and liquefied propane (LPG) are preferred fuels in food processing, because their combustion gases are not objectionable in

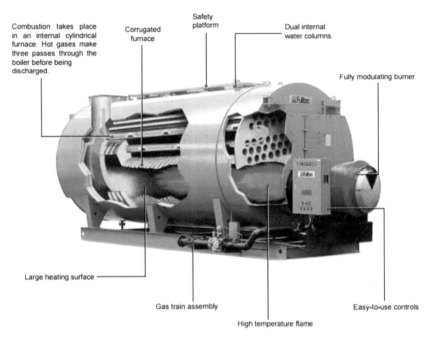

Combustion takes place in an internal cylindrical furnace. Hot gases make three passes through the boiler before being discharged.

Corrugated furnace

Safety platform

Dual internal water columns

Fully modulating burner

Large heating surface

Gas train assembly

High temperature flame

Easy-to-use controls

FIGURE 14.2 Cut section of fire tube boiler (Source: http://www.sanelijomiddle.us/industrial/industrial-steam-boiler-diagram)

direct contact with food products. Fuel oil and coal can be used for indirect heating, for example, through heat exchangers. The heating values of the common industrial fuels are [10, 11]:

Natural gas	37.2 MJ/m^3	LPG	50.4 MJ/kg
Fuel oil	41.7 MJ/kg	Anthracite coal	30.2 MJ/kg
Lignite coal	23.2 MJ/kg		

Two main types of steam boilers are used in dairy industries:

1. **Fire tube boilers**: Low-pressure boilers are limited to a maximum working pressure of 15 psig (pound-force per square inch gauge) for steam and 160 psig for hot water [3].
2. **Water tube boilers**: High-pressure boilers are constructed to operate above the limits set for low-pressure boilers, and are typically used for power generation. Operating water temperatures for hot water boilers are limited to 250°F [3].

14.5.2   MAIN COMPONENTS OF A BOILER

14.5.2.1   Boiler Shell

Made up of steel plate bend into cylindrical form and riveted or welded together. The ends of the shell are closed by means of end plate and should have enough space for water and steam.

14.5.2.2   Burner

There are two types of burners:

1.   Natural Draft burners or Atmospheric burners.
2.   Forced draft burner or power burners.

Now-a-days, low NO_x burners or premix burners are more commonly used as they ensure efficient mixing of air and fuel at the burners entrance and reduce NO_x emissions.

14.5.2.3   Furnace

This is also-called fire box, where fuel is actually burnt (Figure 14.3).

14.5.2.4   Combustion Chamber

Burners are placed here and the combustion process takes place. Temperature inside the combustion chamber increase very rapidly. Combustion chambers are generally made of steel or cast iron.

FIGURE 14.3   Furnace.

14.5.2.5   Heat Exchangers

Here, heat transfer takes place that converts water to steam. Heat exchangers (Figure 14.4) can be of steel tube bundle, cast iron. Also some smaller boilers may be made of copper clad steel.

14.5.2.6   Exhaust Stack

It is piping that conveys hot combustion gases away from the boilers to outside. These pipes are generally made of steel or stainless steel.

14.5.2.7   Boiler Mountings

Boiler mountings are mounted on boilers for its proper and safe functioning. All combustion equipment must be operated properly to prevent disasters from occurring, causing personal injury and property loss. The basic cause of boiler explosion is ignition of a combustible gas that has accumulated within the boiler. There is a tremendous amount of stored energy within a boiler. The state change of superheated water from a hot liquid to a vapor (steam) releases an enormous amount of energy. For example, 1 ft^3 of water will expand to 1600 ft^3 when it turns to steam. Therefore, *"if one can capture all the energy released when a 30 gallon home hot water tank flashes into explosive failure at 332°F, one would have enough force to send the average car to a height of nearly 125 feet. This is equivalent to more than the height of a 14 story apartment building, starting with a lift-off velocity of 85 miles per hour"* [7].

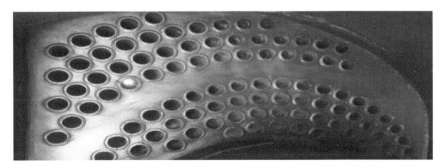

FIGURE 14.4   Heat exchanger.

14.5.2.8 Pressure Gage

Pressure gage measures the pressure of steam inside the boiler. These are generally Bourdon type. The bourdon pressure gage consists of an elastic bourdon tube whose one end is connected to steam space in the boiler and fixed. Due to pressure, the elliptical cross section of the tube tends to round out. Since the tube is encased in circular curve, therefore it tends to become circular. The pointer moves due to the elastic deformation with help of pinion and sector arrangements. The gage pressure is directly shown on the calibrated scale as the pointer moves.

14.5.2.9 Water Level Indicator

It is the safety device generally placed in front. These are two in number to indicate the level of water inside boiler. It consists of 3 cocks and glass tube.

1. **Steam cock** connects glass tube with steam space and kept open during working.
2. **Water cock** connects glass tube with water in boiler and kept open during working.
3. **Drain cock** assures the clarity of steam and water cock and kept closed during working.

14.5.2.10 Flame Detector

It consists of flame rod and ultraviolet or infrared scanner to monitor the flame condition and deactivate the burner in event of non-ignition and other unsafe condition. Flame safeguard controls are programmed to operate the burner and cycle through the stages of operations.

14.5.2.11 Safety Valves

Safety valves blow of the steam when pressure of steam inside the boiler exceeds the working pressure. Safety valves are fitted on the steam chest to prevent explosion, because of extreme internal pressure of steam.

14.5.2.12   Steam Stop Valve

It is usually attached at the highest part of the shell with the help of flange and is the largest valve of the steam boiler. The main function of the steam stop valve is to supply control flow of steam to the main steam pipe from the boiler and to completely shut down steam when steam is not required.

14.5.2.13   Blow Off Cock

This is attached to the bottom of the boiler and consists of body or casing fitted with conical plug. Blow off cock is generally used to discharge mud scale and sediment from the bottom of the boiler and to clear the boiler whenever necessary.

14.5.2.14   Feed Check Valve

This is fitted to the shell little below water level of the boiler and is a non-return valve fitted to screwed spindle to regulate the lift and control the supply of water by feed pump that is pumped into the boiler.

14.5.2.15   Fusible Plug

The main function of the fusible plug is to put off fire in the furnace when the water level in the boiler goes to a very low or unsafe limit and hence avoid explosion that may occur due to over -heating of furnace plate. This is generally placed on furnace crown plate.

14.5.3   *BOILER ACCESSORIES*

14.5.3.1   Feed Pump

Delivers water to the boiler as water is continuously converted into steam and water pressure is generally kept at 20% higher than that in boiler. For this purpose double acting reciprocating pumps are generally used.

14.5.3.2   Super heater

Placed in the route of flue gases from furnace and uses heat given up by flue gases to superheat steam and increase temperature of saturated steam without increasing its pressure.

14.5.3.3   Economizer

Placed between boiler and chimney; and is used to preheat water by utilizing heat of exhaust from flue gases before leaving the chimney.

14.5.3.4   Air Preheater

Extracts heat from the flue gases and transfer to air a portion of heat that would go wasted and is generally placed between economizer and chimney.

14.5.4   OPERATION OF A BOILER

Boiler uses controlled combustion of the fuel to heat water. The fuel and water are mixed together and with the assistance of ignition device provide platform for combustion. The heat generated through combustion chamber is transferred to water through heat exchangers. Ignition, burner, firing rate, fuel supply, air supply, steam pressure, boiler pressure, and water temperature and exhaust draft are regulated by controls.

Hot water produced is delivered to the equipment throughout the building as it is pumped through pipes. Steam boilers produce steam that flows through pipes from area of high pressure to area of low pressure unaided by any external energy source like pump. Steam can be utilized through heat exchangers by providing heat or directly utilized by steam using equipment.

14.6   ELECTRICITY

Electricity is a main energy source used in the dairy industry. In the food industry, about 25% of the electricity is used for cooling and refrigeration and 48% for machine drive [11]. Mechanical system of refrigerator – compressor is driven by motor. Pumps are one of the most important features of all dairy processing operations. Liquid and semi-liquid dairy products are moved through pumps and motors are required to drive pumps. Non-process uses for space heating, venting, air conditioning, lightening and onsite transport consume about 16% of electricity [9]. Pumps and refrigerator driving motors are major electricity consumers in dairy industry.

14.6.1 TYPES OF ELECTRICAL LOADS

14.6.1.1 Resistance

Alternating current and voltages are supplied by standard utility power systems. During the positive and negative cycle, the alternating current and voltages perform work in load with an electrical resistance. Power product of current and voltage is positive during both positive and negative cycle.

14.6.1.2 Inductance

In a circuit the property of the coil, that opposes a change of current, is inductance. An inductor is without electrical resistance and has a significant effect on power line. The power dissipation and consumption is zero as it takes power from the source and returns power to the source every half cycle.

14.6.1.3 Capacitance

In a circuit the property, that opposes the change of voltage, is capacitance. The power dissipation in capacitor is also zero like an inductor.

The power supplied to resistance loads is real power whereas apparent power can be defined as product of current and voltage ($P = I \times V$). There is real power in resistor but no real power in inductor or capacitor whereas there is a current and voltage through a resistor, indicator and capacitor. In a circuit with resistance load and inductance or capacitance, the apparent power should be bigger than the real power.

14.6.1.4   Electric Motor

Motors are designed for rated frequency, voltage and number of phases. Motor efficiency can be defined as shaft power divided by the electrical input power:

$$\eta = \frac{746 \, X \, HP \, output}{Watts \, input} \tag{2}$$

The motor efficiency is mentioned on the name plate of motor. Dairies normally purchase their electric power from local distributors. Generally it is supplied at high voltage, between 3,000 and 30,000 V, but dairies with a power demand of up to approximately 300 kW may also take low-voltage supplies of 200–440 V. The principal components of the electrical system are:

- High voltage switch gear.
- Power transformers.
- Low voltage switch gear.
- Generating set.
- Motor control centers (MCC).

There are several ways to improve the efficiency of an electrical distribution system and a drive system, such as [8]:

- Maintain voltage levels.
- Minimize phase imbalance.
- Maintain a high power factor.
- Identify and fix distribution system losses, particularly resistance loss.
- Select efficient transformers.
- Use adjustable speed motors.
- Select efficient motors.
- Match motor operating speeds.
- Size motor for high energy efficiency.

14.7 REFRIGERATION

The objective is to cool the low temperature source, and the device is named a 'refrigeration machine' [2, 6] or the process for removal of heat from a substance under controlled conditions is termed as refrigeration. Refrigeration means continuous extraction of heat from the body at a lower temperature than its surrounding. The substance that extracts heat from cold body and delivers to hot body is known as refrigerant.

14.7.1 UNITS OF REFRIGERATION

'Ton of refrigeration' is the practical unit of refrigeration and is defined as the amount if refrigeration effect produced by uniform melting of one ton of ice from and at 0°C in 24 hours.

14.7.2 COEFFICIENT OF PERFORMANCE OF A REFRIGERATOR

Coefficient of performance (C.O.P.) is defined as the ratio of heat extracted in refrigerator to the work done on refrigerant.

$$C.O.P. = \frac{Q}{W} \tag{3}$$

where, Q = amount of heat extracted by refrigerator; W= amount of work done.

The basic principle of the most common type of mechanical refrigeration is a cyclic thermodynamic process known as the *Rankine cycle* (William John Macquorn Rankine, 1820–1872, Scottish engineer) or a *vapor compression cycle*. The refrigeration phenomenon may be caused by [12]:

- Evaporation of low pressure liquid at low temperature.
- Expansion of high pressure gas.
- Flow of electric current through an interface of different semi-conductors.
- Chemical reaction.

The refrigeration system may be divided into two categories based on driving power of the system: Mechanical compression refrigeration, and thermal energy driven refrigeration system.

14.7.3 MECHANICAL VAPOR COMPRESSION CYCLE

It is the most commonly used refrigeration system and it consists of following four operations:

1. **Compressor:** The compressor (Figure 14.5) is used to compress the vapor refrigerant of low pressure and temperature from the evaporator drawn in it by suction valve where it is compressed to high pressure and temperature. The high pressure and temperature vapor refrigerant is discharged into condenser through discharge valve.

2. **Condenser:** Here the high pressure and temperature vapor refrigerant is cooled and condensed in coils of pipe (Figure 14.5). The surrounding condensing medium either water or air is used to condense vapor refrigerant. This process takes place as it gives up its latent heat to the condensing medium and converts to liquid refrigerant. The liquid refrigerant obtained here is at high temperature and high pressure.

3. **Receiver:** Before entering the expansion valve the liquid refrigerant is stored in receiver from where it is fed to expansion valve in required quantity.

4. **Expansion valve:** The liquid refrigerant at high pressure and temperature is fed to expansion valve where its pressure and temperature is reduced and is fed to evaporator at controlled rate some of the liquid refrigerant evaporates as it enters the expansion valve but most of the liquid changes to vapor in evaporator as it absorbs heat from the region to be cooled.

5. **Evaporator:** The liquid vapor refrigerant at low pressure and temperature in coils of the pipes of evaporator extracts the heat from the outer surface in contact with the coils and cools that area and the transferred heat causes the liquid refrigerant to convert to vapor refrigerant of low pressure and temperature (Figure 14.5).

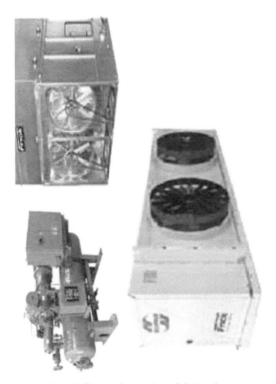

FIGURE 14.5   Compressor (top left), condenser (top right) and evaporator (bottom).

In evaporator the liquid vapor refrigerant absorbs its latent heat of vapor from the medium which is to be cooled. From here vapor refrigerant at low pressure and temperature is fed to the compressor and this way the cycle goes on.

The four processes of the cycle (Figure 14.6) are:

1–2 Isentropic compression process: The refrigerant is compressed isentropically, for example, no heat is absorbed or rejected whereas pressure increases, temperature increases and volume decreases.

2–3 Isothermal compression procession: Here the refrigerant is compressed isothermally which means that the temperature remains constant. Here the pressure increases and the volume decreases and heat is rejected by refrigerant.

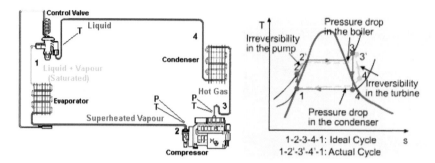

FIGURE 14.6   Left: Vapor – compression cycle; and Right: T – S diagram. (Source: **Left:** [http://pallrefrigerationrajpura.synthasite.com/; **Right:** https://ecourses.ou.edu/cgi-bin/ebook.cgi?doc=&topic=th&chap_sec=10.1&page=theory)

3–4 Isentropic expansion process: The refrigerant is now expanded isentropically, for example, no heat is absorbed or rejected whereas pressure decreases, temperature decreases and volume increases. Here no heat is absorbed or rejected by refrigerant.

4–1 Isothermal expansion process: The refrigerant is expanded isothermally. Here the volume increases and pressure decreases whereas the temperature remains constant and heat is absorbed by the refrigerant.

14.8   SUMMARY

The basic important utilities required for a dairy plant are as under: Water, steam, refrigeration, electricity, and air. When the product flow chart has been planned and checked, the approximate requirements for water, steam, refrigeration, electricity, air and other services can be calculated. With this information, much of the service equipment can be selected. Out of these, steam and electricity are the most important utilities considered for the dairy plant operations.

Steam systems are part of almost every major industrial process today. The steam, in turn, is used to heat processes, to concentrate and distil liquids, or is used directly as a feedstock. All of the major industrial energy users devote significant proportions of their fossil fuel consumption to steam production: food processing (57%), pulp and paper (81%), chemicals (42%), petroleum refining (23%), and primary metals (10%). Steam can be regenerated through waste heat recovery from processes, cogeneration, and boilers [2]. Steam has high capacity to store energy in the form of heat and move it in a controlled amount easily and efficiently throughout a manufacturing facility. Due to this, steam is a popular choice for a wide variety of industrial uses. So proper understanding for the efficient generation, transmission and utilization of the steam is necessary.

It is well known that no large-scale industry can exist without electricity. Modern dairy plants depend on electricity for lighting, heating and power drive for the equipment. With the increased use of electricity, the danger of accident and breakdown at any point has also increased. The need for trained laborers and technical personnel for efficient operation, supervision and maintenance has also increased to a great extent. Even the manager and shift superintendent of a milk plant are expected to have clear conception of basic facts regarding electric supply and tariff.

For procuring electricity, there are two sources: one of power generating plant at the site, and another of purchasing required amount of power from private supply company or State Electrical Board. The first one is only adopted where there is no second choice or alternative. The main reason is that the generation of power in small amount is very costly and there is additional cost involved for installation of generation plant. It is found that electric power can be purchased on permanent basis from nearby Supply Company at reasonable rates.

KEYWORDS

- boiler
- boiler accessories
- boiler efficiency
- boiler mountings
- C.O.P. of refrigeration
- DM
- electricity

- rankine cycle
- refrigeration
- refrigeration cycle
- RO
- types of electrical load
- types of refrigeration system
- water

REFERENCES

1. Ahonen, V. (1978). Höyrytekniikka II. *Otakustantamo*, Espoo.
2. ASHRAE (2006). *ASHRAE Refrigeration Handbook*. ASHRAE, New York.
3. ASHRAE (2008). *ASHRAE Handbook: HVAC Systems and Equipment*.
4. Capehart, B., Turner, W., & Kennedy, W., (2006). *Guide to Energy Management*.
5. Combustion Engineering. In: *Combustion: Fossil Power Systems*. 3rd ed. Windsor. (1981).
6. Dosset, R. J., & Horan, T. J. (2001). *Principles of Refrigeration*. 5th edn. Prentice Hall, New Jersey.
7. *FEMP O&M Best Practices: A Guide to Achieving Operational Efficiency*, U.S. Department of Energy, August 2010. http://www1.eere.energy.gov/femp/pdfs/omguide_complete.pdf.
8. Lobodovsky, K. K. (2006). Electric energy management. Chapter 11. In: *Energy Management Handbook* (6th ed.), edited by Turner, W. C., & Doty, S., Lilburn, GA: The Fairmont Press, Inc.
9. Okos, M., Rao, N., Drecher, S., Rode, M., & Kozak, J. (1998). *Energy Usage in the Food Industry*. American Council for an Energy-Efficient Economy (ACEEE). Available at: http://www.aceee.org/pubs/ie981.htm.
10. Robberts, T. C. (2002). *Food Plant Engineering Systems*. CRC Press.
11. Saravacos, G. D., & Kostaropoulos, A. E. (2002). *Handbook of Food Processing Equipment*. Kluwer Academic/Plenum.
12. Sun, D. W., & Wang, L. J. (2001). Novel refrigeration cycles. In: *Advances in Food Refrigeration*, edited by Sun, D. W., pp. 1–69. Leatherhead Publishing. UK: Leatherhead.

INDEX

T - #0811 - 101024 - C422 - 229/152/19 - PB - 9781774636633 - Gloss Lamination